AMCP 706-245

ORDNANCE CORPS PAMPHLET

ORDNANCE ENGINEERING DESIGN HANDBOOK

ARTILLERY AMMUNITION SERIES

SECTION 2, DESIGN FOR TERMINAL EFFECTS(U)

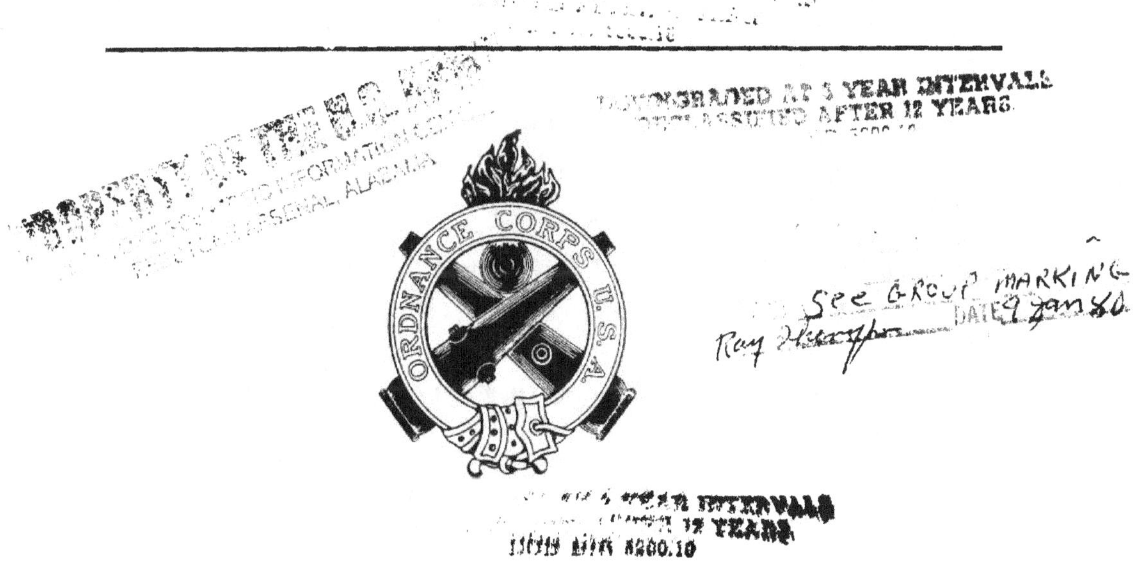

ORDNANCE CORPS

MAY, 1957

REGRADING DATA CANNOT BE PREDETERMINED

OFFICE OF THE CHIEF OF ORDNANCE
Washington 25, D. C., 31 May 1957

ORDP 20-245, Section 2, Design for Terminal Effects, forming part of the Artillery Ammunition Series of the Ordnance Engineering Design Handbook, is published for the information and guidance of all concerned. The complete series is listed below. The foreword, with acknowledgements; preface; table of contents; glossary and index for the series, are made part of the first pamphlet in the series, ORDP 20-244.

OFFICIAL:
R. E. PETERS
Colonel, Ord Corps
Executive Officer

E. L. CUMMINGS
Lieutenant General, USA
Chief of Ordnance

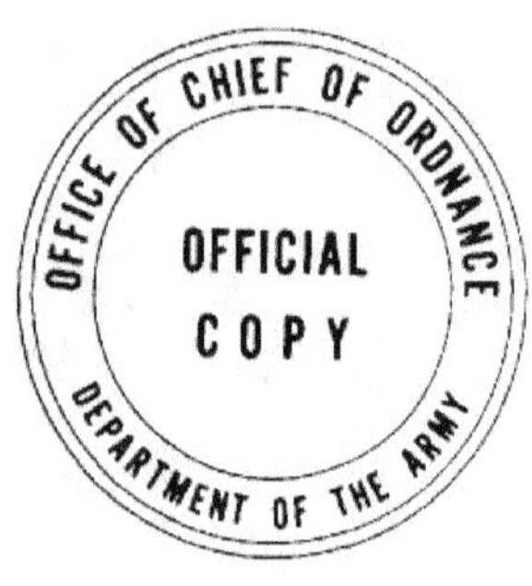

DISTRIBUTION: Special

Artillery Ammunition Series

ORDP 20-244	Section 1, Artillery Ammunition — General, with Table of Contents, Glossary and Index for Series
ORDP 20-245	Section 2, Design for Terminal Effects
ORDP 20-246	Section 3, Design for Control of Flight Characteristics
ORDP 20-247	Section 4, Design for Projection
ORDP 20-248	Section 5, Inspection Aspects of Artillery Ammunition Design
ORDP 20-249	Section 6, Manufacture of Metallic Components of Artillery Ammunition

ORDP 20-245

ORDNANCE CORPS PAMPHLET

ORDNANCE ENGINEERING DESIGN HANDBOOK

ARTILLERY AMMUNITION SERIES

SECTION 2, DESIGN FOR TERMINAL EFFECTS (U)

TABLE OF CONTENTS

Section 2 - Design for Terminal Effects

TABLE OF CONTENTS

Section 2 - Design for Terminal Effects (continued)

SECTION 2

DESIGN FOR TERMINAL EFFECTS

INTRODUCTION

The ultimate purpose of any round of service ammunition is the production of a desired effect at the target. This section attempts to give a broad picture of the major engineering and tactical requirements that govern the design of missiles intended to produce these desired terminal effects.

CLASSIFICATION OF MISSILES

2-1. Classification of Missiles by Type of Target. The purposes of terminal effects fall into two broad categories: the actual defeat of a target; and the production of an effect (signaling, illuminating, or screening) that will aid in the ultimate defeat of the target. Table 2-1 lists the terminal effects which may be produced and the purposes which each of these terminal effects may serve. In all, eleven effects are listed. Paragraph 2-2 gives a brief description of each of these effects.

2-2. Classification of Missiles by Effect.

1. Blast. The production of an explosion which will propagate a high-velocity, high-pressure wave in the surrounding air. Since a metal body must be used as She carrier for the blast-producing high explosive, the production of blast is always accompanied by fragmentation.

2. Fragmentation. The disruption of a metal shell body by a high explosive filler in order to produce the optimum distribution of a maximum number of high-velocity lethal fragments. Due to the use of the high-explosive filler, fragmentation is always accompanied by blast.

3. Penetration of Armor by a Solid Projectile (Kinetic Energy Shot). The projection of a solid projectile of steel or some other hard, dense material (tungsten carbide) at a velocity sufficient to supply the necessary kinetic energy to enable it to penetrate armor plate. Kinetic energy shot may contain a high-explosive charge sufficient to disrupt it after penetration of the armor plate; current design, however, tends to eliminate this feature.

4. Penetration of Armor by a High-Velocity Jet (Shaped Charges). The use of the Munroe effect to obtain an extremely high-velocity jet of metal particles capable of penetrating armor plate.

5. Spalling of Armor (HEP). This effect is used to defeat armor without actually effecting a penetration. By the use of a high-explosive plastic (HEP) filler in a deformable shell, an explosion on the outside of armor plate can produce sufficient shock to cause the formation of a spall on the inside surface of the plate. This spall, roughly circular in shape, may be separated from the surface of the plate and projected with sufficient velocity to cause serious damage inside the tank.

6. Perforation by Preformed Missiles Other Than Armor-Piercing (Canister). The loading of a non-explosive shell with a large number of small preformed missiles in order to obtain a short-range lethal effect on personnel. This type of shell is roughly analogous to a common shotgun shell.

7. Incendiary. The use of a shell filler which will produce high enough temperatures to ignite any flammable material in the target, or to incapacitate personnel.

8. Release of Poison Gases. The use of a poison gas to cause injury to personnel or to contaminate an area and thereby deny its use to the enemy.

9. Production of Light. The production of light for signaling, or for visual or photographic observation.

10. Production of Smoke. The production of white or colored smokes for. signaling or screening purposes.

11. Dissemination of Leaflets. The broadcast of propaganda leaflets for the purpose of undermining enemy morale.

Table 2-1

	Blast	Fragmentation	Penetration of Armor by a Solid Projectile	Penetration of Armor by a High-Velocity Jet	Spalling of Armor	Perforation by Preformed Missiles other than Armor-Piercing	Incendiary	Release of Poison Gases	Production of Light	Production of Smoke	Dissemination of Leaflets
Antipersonnel & Materiel	X	X				X	X	X			
Antiaircraft	X	X					X				
Armor Defeating			X	X	X						
Demolition	X										
Signaling, Illuminating, or Screening									X	X	
Psychological Warfare											X

PROJECTILE DESIGN

2-3. General. The projectile designer is called upon to supply the best possible projectile for a given purpose (effect). A projectile is usually designed for a particular weapon and the characteristics of the weapon limit the design. Other limitations may be placed on the designer by considerations of handling by the gun crew, which may limit overall size and weight of the round, and by requirements for a large range of operating temperature.

2-4. Requirements for Gun Projectiles. Present-day projectiles must meet the following general requirements.

1. Safety in handling, in gun bore, and in flight.
2. Safety when fired in gun, i.e., no prematures resulting from pressure or shock of discharge in gun or from hot propellant gases entering the base.
3. Stability in flight throughout the trajectory.
4. Ballistic efficiency, for maximum range or minimum time of flight with minimum dispersion.
5. Tactical effectiveness at target, including:
   a. Effective fragmentation, or
   b. Maximum blast effect, or
   c. Required armor-defeating ability.
6. Capable of being manufactured by production methods, mainly forging and machining.
7. Capable of being loaded with explosive filler.
8. Minimum amount of wear on the gun bore.

2-5. Ordnance Committee Minutes. The specifications for the design of a projectile are usually given by the Ordnance Committee Minutes (OCM). They usually include the following information.

1. Rated maximum pressure of the gun.

2. True maximum pressure on the base of the shell.
3. Twist of rifling (for an existing gun only).
4. Caliber of the gun.
5. Required range.
6. Approximate weight of the projectile.
7. Approximate weight of the propelling charge (if it is separate from the projectile).
8. Approximate weight of the shell-charge combination (for fixed ammunition).

In addition, the following information may be included.

1. For HE, HEP, or chemical shell, the shape of the shell and the yield strength of the steel.
2. For armor-defeating ammunition, the thickness and obliquity of the armor to be defeated.
3. For antitank and antiaircraft ammunition, the first-round probability of hit and the confidence level of this probability.

2-6. Design Procedure.

a. Rough Design. From the tactical requirements, a projectile may be roughly outlined to meet prescribed conditions. The weight of this first design must be calculated and adjustments made to bring the design to the proper weight.

b. Stress Analysis. The next step in the design procedure is to determine the stresses acting at the critical elements of the shell. The maximum combined stresses on the projectile walls of any section should not, in most cases, exceed the yield point of the metal from which the shell is made. Stress analysis procedures are given in Section 4.

c. Determine Stability. The final step in the design of a projectile is to calculate its stability and to estimate the retardation caused by air resistance. Section 3 describes the procedure to be followed.

d. Optimize Terminal Ballistic Effect. The first step of the design procedure, rough design, takes into account the effect which the shell is to produce; however, at this point it is difficult to determine whether or not the design is near optimum. The usual method for obtaining optimum terminal ballistic effect is to design several projectiles which are satisfactory and then, by means of analytical methods described in succeeding sections, or by means of actual firings of test shell, to determine which of these designs is best.

2-7. High-Explosive Ammunition. High-explosive ammunition may be designed to perform any one of several functions and in most cases may be expected to perform more than one of them. These functions are:

Defeat of personnel
Defeat of aircraft
Defeat of fortifications.

Considerations pertinent to each of these functions are discussed in the succeeding paragraphs.

2-8. Defeat of Personnel. Defeat of personnel by high-explosive ammunition requires that the projectile be designed to produce the maximum lethal area. Recent wound ballistic studies indicate that for fragments traveling at the velocities commonly obtained from high-explosive shell, extremely small fragments are required to optimize the lethal area. Since the size of fragments is a function of the thickness of the shell wall, calculations reveal that design for optimum fragmentation results in a shell which is not strong enough to resist the setback forces. Accordingly, the practice in designing shell of this type is to design them with the thinnest walls capable of sustaining the stresses in the gun. Where economic manufacturing methods may be used, consideration should be given to obtaining controlled fragmentation by the use of multiple walls. If setback forces are not excessive, consideration may be given to fragmentation control by means of grooved rings or notched wire.

2-9. Defeat of Aircraft. For antiaircraft projectiles, consideration must be given to (1) the vulnerability of the target, (2) speed of the target, (3) accuracy, (4) time of flight, (5) rate of fire, and (6) lethality of the projectile. Depending upon the size and point of burst of the projectile, it may be desirable to maximize either (1) blast damage or (2) fragmentation damage. For smaller projectiles, the decision must also be made as to whether the projectile is to (1) detonate outside the aircraft, (2) in contact with the aircraft, or (3) inside the aircraft. If detonation internally is desired, it is necessary that the projectile be sufficiently strong to penetrate without deforming to a point where its effectiveness is impaired. Here, as with antitank ammunition, the lethality criterion should be first-round probability of kill. However, the design problem is rarely presented in these terms. Consideration should be given to multiple wall and liner techniques of fragmentation control; however, it should be borne in mind that use of the liner method will result in a loss in blast effectiveness of the round.

Occasionally the designer is called upon to produce a shell that, in addition to being effective against aircraft, may also be expected to be effective against personnel. Here the compromise must be between optimum fragment size for antipersonnel effect and optimum fragment size for defeat of aircraft. Optimum size for antiaircraft use is considerably in excess of that for antipersonnel.

2-10. Defeat of Fortifications. HE projectiles, which have as their primary purpose the defeat of personnel, may also be required to defeat concrete or log-and-earth fortifications. Hence it may be necessary to arrive at some compromise between maximum fragmentation effectiveness and the ability to penetrate without undue breakup of the shell. The fuze designer has cooperated in this direction by providing concrete piercing fuzes, which will help to attain this objective. The use of a special fuze, with conventional shell, still leaves much to be desired.

2-11. Kineiie Energy Ammunition.

a. General. Kinetic energy ammunition is intended primarily for the purpose of defeating armor, although it may also be called upon to defeat concrete fortifications. There are three types of KE ammunition in current use. They are (1) steel armor-piercing shot (AP), (2) capped steel armor-piercing shell with an explosive filled cavity, (3) carbide-cored, hypervelocity, discarding sabot shot (HVAPDS). In addition to these, subcaliber composite-rigid shot, and skirted or squeeze-bore projectiles have been made; however these are not currently being designed, the former because of its too high ballistic coefficient, and the latter because of the difficulty of interchanging ammunition in a tapered-bore gun. Still another type of kinetic energy shot, the hypervelocity, discarding sabot, fin-stabilized (HVAPDSFS) is currently being investigated. If this type lives up to its promise, it may become an important member of the family of kinetic energy ammunition.

b. Design for Defeat of Armor. For kinetic energy antitank projectiles, accuracy, along with the ability to penetrate the specified target, is the prime consideration. First consideration should be given to AP or APC shot. The advisability of the use of an armor-piercing cap depends upon the type (face-hardened or homogeneous), obliquity, and thickness of the armor. If the target cannot be defeated by this type of ammunition, consideration should be given to the more expensive types such as the carbide-cored discarding sabot types and perhaps the HVAPDSFS. Consideration is first given to determining the optimum subprojectile for maximum penetration at the specified range. Calculations should also be checked at shorter ranges to ensure against the presence of a "shatter gap." In design of subcaliber projectiles for existing guns, great attention must be paid to the stability of the projectile. This consideration quite often governs its dimensions.

c. The Sabot. In the design of discarding sabot projectiles, in addition to the primary problems of imparting spin and discarding promptly, it is important that it be recognized that the sabot itself forms a secondary missile. This projectile must be discarded in such a manner that it will not endanger friendly troops.

d. Lethality. The actual criterion governing the lethality of these rounds is first-round probability of kill, which takes into account armor penetration, size of target, accuracy of the projectile, and time of flight; however, the design problem is usually presented in terms of first-round probability of hit and ballistic limit.

2-12. High-Explosive Antitank (HEAT) Ammunition. The following problems are peculiar to the design of HEAT ammunition:

a. Time of Flight Versus Standoff. In order to obtain a high first round probability of hit, time of flight should be as short as possible. However, the requirement of standoff demands that the charge initiation take place before excessive crush-up of the nose has taken place. The resolution of this problem lies in the province of the fuze designer who is required to provide extremely quick-acting fuzes for high-velocity HEAT rounds.

b. Stability Versus Standoff. The requirement of long standoff distances, particularly on the slow-speed fin-stabilized rounds, results in an extremely light nose section. This type of configuration is extremely hard to stabilize. One approach to the problem has been the use of the drag-stabilized spike-nosed design.

c. Spin Versus Optimum Penetration. One of the major difficulties in design of HEAT rounds is the minimization of the degradation in performance caused by spin of the projectile. Several methods by which this problem may be attacked are:

1. HEAT projectiles fired from low-velocity recoilless rifles are given a very slow rate of spin.
2. Spin has been eliminated in some cases by reverting to fins to stabilize the projectile. This method is limited to low-velocity guns.
3. For high-velocity rounds, the approach to the problem of spin degradation has been the design of specially shaped liners which compensate directly for the spin of the projectile.

2-13. High-Explosive Plastic (HEP) Ammunition. HEP ammunition is intended for the defeat of armor; however, because of the use of extremely thin walls, it also has a very valuable secondary fragmentation effect. In design, the following peculiarities of HEP ammunition should be given careful consideration.

a. Crush-up of Nose. The effect of HEP ammunition is obtained by having the explosive charge explode in intimate contact with a large area of the armor plate. The nose of the projectile must be soft enough and thin enough to perform this function effectively. At the same time, the projectile must withstand successfully the strains of firing.

b. Velocity. The velocity of impact of HEP shell has been found to be directly related to the performance of the projectile. There exists a rather small range of velocities at which performance is satisfactory. Velocities either above or below this range result in ineffective rounds. It is thought that a partial solution to this problem may be found in the fuzing of the projectile, and work on new fuzes is now under way.

c. Banding. Because of the very thin walls of HEP shell, pressed-on rotating bands are not satisfactory; the high pressures used to apply them distort the shell wall. This problem has been overcome by the use of welded overlay rotating bands. At present, this type of band is used only for pre-engraved rotating bands used on recoilless rifle ammunition.

d. Stability. Because of the low rotational moment of inertia of the thin shell walls, it is difficult to stabilize the flight of HEP shell. This problem has been solved by the use of a blunt-nosed ogive, which drag-stabilizes the projectile.

2-14 Canister Ammunition. The major problems of canister design are:

1. Opening of the canister
2. Minimization of damage to the gun tube
3. Securing lethality at great enough ranges
4. Securing adequate dispersion.

2-15. Base Ejection Ammunition. Base ejection shell may be used for any of the following purposes:

1. Illumination
2. Dissemination of smoke
3. Dissemination of propaganda leaflets
4. Dissemination of poison gases.

The particular problems associated with design of this type of shell are:

a. Expelling Charge. The black powder expelling charge should, ideally,, eject the contents of the shell with a rearward velocity just equal to the forward velocity of the shell. If this were done the contents would have zero forward velocity and would just drop straight down. Because of limitations imposed by charge size and strength of the shell, this ideal cannot be attained in practice and the black powder charge must be considerably smaller than this ideal charge.

b. Shear Pins or Threads. In order to assure proper burning of the black powder charge, the shear pins or threads must be designed to permit some minimum pressure to be built up before shearing takes place. This problem is analogous to that of obtaining proper bullet pull for a cartridge.

c. Setback. The contents of a base ejection shell must be designed so that they will not be damaged by the setback forces created when the gun is fired and those that result when the contents are expelled from the shell. These two forces act in opposite directions. In the case of propaganda disseminating shell, this problem may be solved by packing the leaflets into split steel tubes, which are strong enough to resist these forces and which will discard completely after ejection.

2-16. Ammunition With Burster Charges.

a. General. This type of ammunition is usually similar in appearance to the high-explosive round with the exception of the replacement of the explosive filler by either a burster charge or a filler. The burster charge may be contained in a metal tube located axially in the shell or cast in position and separated from its surrounding medium by acid-proof black paint. A filler may be used to produce:

1. Heat sufficient to damage materiel and produce casualties among personnel. White phosphorous (WP) is usually used for this purpose.

2. Smoke intended for signaling or screening.

3. Poison gases, either persistent or nonpersistent.

b. Design. There are several factors, peculiar to the design of this type of ammunition, which may have to be considered:

1. The burster charge should be sufficient to completely break up the shell body without causing excessive dispersion of the contents. No portion of the fragmented shell body should form a cup that might retain some of the filler.

2. When a liquid filler is used it is extremely important that the shell be perfectly sealed to prevent leakage of the contents. Further information on sealing is contained in Section 2, "Special Purpose Shell."

3. When a liquid filler is used the rotational inertia of the shell is greatly reduced, due to the tendency of the filler to remain stationary relative to the rotation of the shell body. At present, this problem is dealt with by empirical method. Work now in progress should, however, yield a theoretically sound method of approach in the near future.

4. The use of a heavy steel burster tube causes the exterior ballistics of the shell to differ significantly from those of the HE shell designed for the same weapon. In order to secure ballistic matching it is desirable that, where it is compatible with the filler, an aluminum burster tube be used.

BLAST EFFECT

THE EXPLOSIVE WAVE

2-17. Explosive Nave Propagation – History.' The rapid expansion of the mass of hot gases resulting from detonation of an explosive charge gives rise to a wave of compression called a shock wave which is propagated through the air. The front of the shock wave can be considered infinitely steep, for all practical purposes. That is, the time required for compression of the undisturbed air ahead of the wave, to the full pressure just behind the wave, is practically zero.

If the explosive source is spherical, the resulting shock wave will be spherical, and, since its surface is continually increasing, the energy per unit area continually decreases. As a result, as the shock wave travels outward from the charge, the pressure in the front of the wave, called the peak pressure, steadily decreases. At great distances from the charge, the peak pressure is infinitesimal, and the wave, therefore, may be treated as a sound wave.

Behind the shock-wave front, the pressure in the wave decreases from its initial peak value. Near the charge, the pressure in the tail of the wave is greater than that of the atmosphere. However, as the wave propagates outward from the charge, a rarefaction wave is formed which follows tile shock wave. At some distance from the charge, the pressure behind the shock-wave front falls to a value below that of the atmosphere, and then rises again to a steady value equal to that of the atmosphere. The part of the shock wave in which the pressure is greater than that of the atmosphere is called the positive phase, and, immediately following it, the part in which the pressure is less than that of the atmosphere is called the negative or suction phase.

The velocity at which the shock wave is propagated is uniquely determined by the pressure in the shock-wave front and the pressure, temperature, and composition of the undisturbed medium. The greater the excess of peak pressure over that of the atmosphere, the greater the shock velocity. Since the pressure at the shock front is greater than that at any point behind it, the wave tends to lengthen as it travels away from the charge; that is, the distance between the shock front and the part at which the pressure in the wave has decreased to atmospheric continually increases.

2-18. Positive Impulse.' A gage that is capable of indicating the pressure instantaneously applied, and that is fixed with respect to the charge, will record the pressure in the wave as a function of time. The resulting pressure-time curve bears a close resemblance to the pressure-distance curve described above: there is an initial abrupt rise in pressure followed by a relatively slow decrease in pressure to a value below that of the atmosphere. The time elapsing between the arrival of the shock front and the arrival of the part in which the pressure is exactly atmospheric is called the positive duration, and this, like the length of the wave, increases as the wave travels away from the charge. A quantity of interest in the application of blast measurements is the positive impulse, which is the average pressure during the positive phase multiplied by the positive duration, $\int_0^a P\,dt$, where a is the positive duration. For most shock waves, the trace of the positive phase of the pressure-time curve is roughly triangular. Hence the positive impulse may be approximated by one-half the peak pressure multiplied by the positive duration.

2-19. Conditions Associated With Shock Front.[1] Associated with the propagation of the shock front is a forward motion of the matter behind the shock front, and the conditions that determine the shock velocity also determine the particle velocity. In gases, such as air, the particle velocity for high-shock pressures is very high. For example, at about 3 atmospheres excess pressure in the shock front, the particle velocity immediately behind it is about 1,000 mph.

The temperature behind the shock front is also greater than that ahead of it because of the compression of the medium. Since this compression is irreversible, the temperature of the air through which the shock wave has passed, and which has returned to atmospheric pressure, is somewhat greater than that of the undisturbed air prior to the arrival of the shock wave. The smaller the excess pressure in the shock wave, the less the irreversible heating of the air.

At a very great distance from the charge, the wave becomes acoustic, that is, the pressure rise, temperature rise, and particle velocity are all infinitesimal, and the velocity of the wave is that of sound.

2-20. Reflection of Weak Shock Waves.[1] Very weak shock waves, that is, those of nearly acoustic strength, are reflected from plane surfaces in such a way that a geometrical construction of the wave system can be made in a very simple way. Consider a point source of the shock C (fig. 2-1) and, at some distance from it, a plane reflecting surface S. The incident wave I, striking the surface, will be reflected from it in such a way that the reflected wave R may be considered to arise from a second image source C', on the opposite side of the reflecting surface, perpendicularly below the true source and equally distant from the surface.

Figure 2-1 shows two successive stages of this reflection process. In the first, I_1, the incident wave, is just tangent to the surface. Theexcess pressure over that of the atmosphere at the reflecting surface is just double (for very weak shock waves) that of the incident wave where it is not in contact with the surface. At a later stage, the incident wave is represented at I_2, and the reflected wave at R_2 imagined to arise from the image source C'. Again the pressure at the line of contact of I_2, R_2, and the surface S is just double that of I_2. The angles at which the shocks I_2, R_2 meet the surface S are equal.

2-21. Reflection of Strong Shock Waves – Mach Waves.[1] When the pressure in the shock wave is appreciably above that of the atmosphere, the phenomena are different. One reason for this is that the pressure, density, and velocity of the air into which the reflected shock advances are not those of the undisturbed atmosphere. In figure 2-2 there are represented three successive stages in the reflection of strong shocks. In the terminology used above, the incident wave I_1 is first shown just as it touches the reflecting surface S. The excess pressure above that of the atmosphere at this point is more than twice that of I_1 elsewhere, and the magnitude of the increase of pressure over that of I_1 is determined by the strength of I_1. For example, if the peak (excess) pressure of I_1 is 100 psi, the reflected shock pressure is about 500 psi, a five-fold increase of pressure.

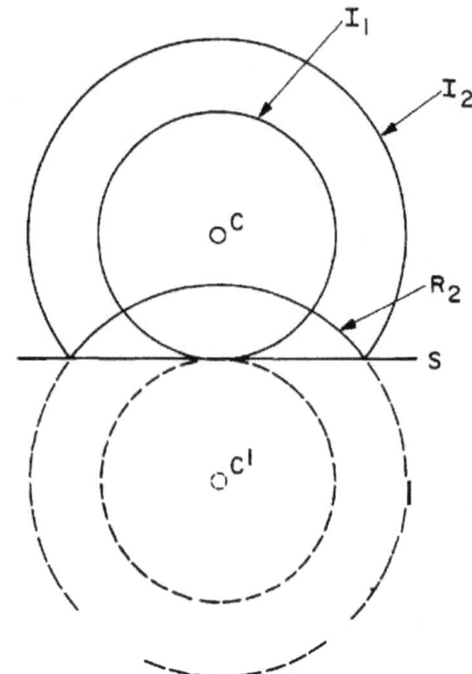

Figure 2-1. Reflection of weak shock waues

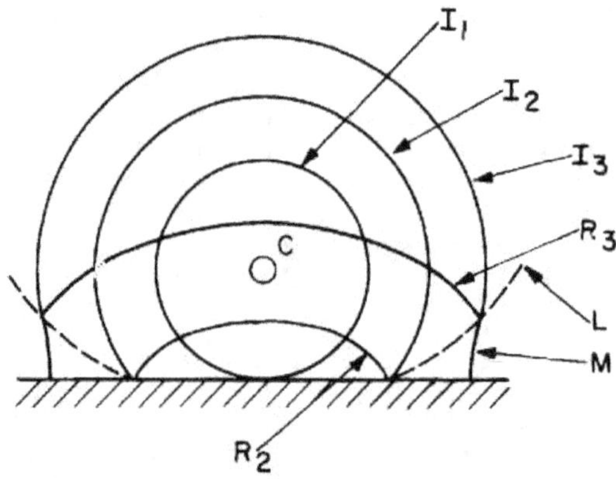

Figure 2-2. Reflection of strong shock waues

As the incident wave expands to some greater size I_2, the reflected wave R_2 also expands, but the reflected wave is not spherical and cannot be constructed by the device used in figure 2-1. The angles at which I_2 and R_2 meet the surface S are not equal, in general, and the angle of the reflected shock R_2 depends upon the strength and angle of incidence of the incident shock.

At some distance from the charge C, determined by the distance of C from S, and by the strength of the incident shock, a new phenomenon occurs. The intersection of R and I no longer lies on S, but lies above it and follows some path, 1. A new shock M, the Mach stem, connects the intersection of R and I to the surface. The intersection of R, I, and M is called the triple point. As the shock system expands further, the Mach stem grows rapidly, tending to swallow up the two-shock system above it. If C is very close to the surface, but not on it, the Mach stem is formed almost directly under C and, in a short time, has grown so that most of the shock system is a Mach stem, and only in a small region directly over the charge are R and I distinct. If the charge C is on the surface S, no separate reflection R is formed, and it can be considered that the entire shock wave is a Mach wave.

A very practical property of the reflection of shocks is that the pressure (and positive impulse) in the neighborhood of the triple point and in the Mach stem are considerably greater than those in I_3, or in the shock emitted when C is in contact with S. That is, if C is a bomb bursting above the ground represented by S, the intensity of the blast in the region M and just above it is greater, at a given horizontal distance from the bomb, than is the case if the bomb is burst in contact with the ground.

2-22. Effect of Shock Wave.[1] When a shock wave strikes a nonrigid obstacle, such as a building, the wave is reflected by the surfaces of the building in the various ways described above. The reflection from a nonrigid surface will not, however, conform quantitatively to that from a rigid surface such as that discussed above. At the instant the wave strikes the wall, the wall is accelerated, and continues to accelerate as long as there is an excess of pressure on its outer surface. At first, the deformation of the wall is elastic, so that for insufficient excess pressure or insufficient positive duration there may be no permanent displacement of the wall. If the blast intensity is sufficient, the wall eventually deforms inelastically and suffers permanent displacement. If, for the wall in question, the displacement is greater than some critical amount, the wall will collapse.

A simplified picture of the processes of damage consists of a wall of indefinite extent which has a certain natural period of vibration. If a shock wave of very long duration strikes it, the wall can be considered to be subjected suddenly to a blast of constant pressure equal to the pressure in the shock wave enhanced by reflection. For sufficiently small pressures, the wall will deform elastically (the amount of the displacement being about twice that from a static pressure equal to the pressure in the reflected blast) and will not rupture. Some pressure must exist, however, such that the wall will collapse. For shock waves of finite duration, the wall may not collapse even though the pressure is equal to the critical pressure. Instead, the wall will acquire momentum from the shock wave and will vibrate, without reaching the amplitude corresponding to collapse. If the duration of the wave is very short compared with the time required for collapse, the momentum imparted to the wall must be sufficient to deform it beyond the critical limit. On the basis of reasoning such as this, the peak pressure is usually considered to be the determining factor in the damage produced in the blast from very large bombs, such as atomic bombs. For small bombs it is generally assumed that the positive impulse is the important quantity, since the duration of the blast is quite short. Unfortunately, neither operational experience nor experiment is adequate to test these criteria properly.

2-23. Theories on the Dependence of Blast on Ambient Pressure and Temperature. To infer from the information obtained on the ground information concerning the blast at high altitudes, it is necessary to determine the effect of the change in pressure and temperature on the blast. There are two theories which formulate the scaling laws. These are Sachs' Theory and Kirkwood-Brinkley's Theory. These theories differ in their initial assumptions and in their choice of parameters. A complete discussion of these theories can be obtained from references 3, 9, 10, 11, 12, and 29.

2-24. Blast Information To Be Obtained from Later Experimentation. General design information is lacking at present on the effect on blast of length of column, and of diameter of explosive column, for a given weight of explosive. Also, little information of a general nature concerning the minimum booster requirements for various sizes and configurations of explosive charges is available. Since very little experimentation has been done with cased charges, it is not known how applicable the information obtained from bare charges would be to cased charges. The great need at the moment is for positive information on cased charges.

MEASUREMENT OF BLAST[1]

2-25. Piezoelectric Gages. The most common method of measuring air blast pressures employs piezoelectric gages. Piezoelectrically active crystalline substances that have been used in gages are tourmaline, barium titanate, quartz, Rochelle salt, and ammonium dihydrogen phosphate (ADP).

2-26. Condenser Microphone Gage. A condenser microphone consists of two parallel metal plates mounted so as to be insulatedfrom each other, and separated by a dielectric (air, mica, etc.). The two plates, which are the plates of a condenser, are connected to the associated electronic apparatus by means of an electric cable. Under the application of pressure, the dielectric between the condenser plates is reduced, and the capacity of the condenser therefore increases.

2-27. Resistance Gages. A third device for measuring transient pressures depends on the change of electric resistance of an element under stress. In one form, the gage consists of a resistance element that is hydrostatically compressed. In another, a resistance wire is formed in a spiral and cemented to the back of a diaphragm constrained at its periphery. When pressure is applied, the diaphragm is deformed, the wire is stretched, and the resistance of the wire changes. Associated with the gage is a simple potentiometer circuit by means of which changes in resistance give rise to proportional changes in voltage. These voltage changes are amplified and recorded.

2-28. Mechanical Gages. A gage for measuring peak pressure has been designed that operates by recording the maximum extension of a spring acted upon by a moving piston which is accelerated by the action of a pressure pulse. If the natural period of the piston-and-spring is short, compared with the duration of a transient pressure pulse, the maximum extension of the spring is proportional to the peak pressure of the pulse. For the measurement of positive impulse, gages that employ a freely sliding piston have been used.

2-29. Peak-Pressure Gages have been devised to operate on the principle that a thin diaphragm, stretched over a hole in a rigid plate, will rupture at a certain pressure when the diaphragm is subjected to a blast wave. If several such diaphragms are provided, covering holes of various sizes, the pressure required to rupture the diaphragm over a given hole will depend on the hole size. Hence, given a calibration of the device, the peak pressure of a blast wave is established as less than that required to break the diaphragm of the largest hole unbroken, and greater than, or equal to, the pressurerequired to break the diaphragm over the smallest hole broken. The pressure is thus bracketed as closely as is desired, simply by having a sufficient number of holes of graduated size.

One such device, the paper blast meter, has been used for many years in the approximate measurement of blast pressures. It consists of two boards clamped together, with a sheet of paper held tightly between them. Holes of about ten different sizes are bored through both boards, in register. The gage is mounted with the plane of the diaphragm perpendicular to the direction of propagation of the wave, that is, head-on to the wave. By virtue of the multiplication of pressure on reflection, the pressure exerted on the diaphragm is greater than that of the incident wave; proper account of this must be taken.

A more recent modification of this gage is the foilmeter, which consists of a wooden or metal box with one open end, over which is clamped an assembly similar to the paper blast meter, but with aluminum foil instead of paper. Foil is used because it is much less sensitive than paper to changes in atmospheric conditions such as temperature and humidity. The box gage can be oriented either face-on or side-on to the direction of propagation of the blast, since the box prevents the blast from acting on the reverse side of the diaphragm.

The great advantage of this type of peak-pressure gage is its simplicity. The operation and the interpretation of results are simple, and no elaborate machine work is involved. Its greatest limitation is that the precision of results is usually not high, and the limits within which the pressure can be bracketed with a reasonable number of holes are rather wide.

2-30. Shock Velocity Method. The shock-wave velocity is uniquely determined by the characteristics of the medium and the excess pressure in the shock wave. That is, under specified conditions, the pressure may be expressed explicitly in terms of the shock-wave velocity. Advantage is taken of this relation to make very accurate determinations of peak pressures.

2-31. The Blast Cube is used to measure the blast from 20-mm to 40-mm AA shell. The blast cube consists of an angle iron frame, with aluminum sheets of different thicknesses bolted on the frame. The sheets have diagonal slits. The exploding of the projectiles generally marks some plates, bends others, and rips some off. The blast is measured by a previously established system for evaluating relative damage to the aluminum sheets. This method of blast evaluation is qualitative, but scores can be given from the established system. Also, this method is good for comparative purposes.

2-32. Empty Varnish Cans. Another method of blast evaluation similar to the blast cube is the use of old varnish cans. In this method, varnish cans, with their covers on, are exposed to the blast. The relative decrease in volume of the cans at various distances from the center of blast is used for qualitative comparative purposes.

2-33. The Blast Tube. The blast tube is a useful apparatus for the study of shock waves in air and for the calibration of air-blast gages. It consists of a long tube divided into two sections, a compression chamber and an expansion chamber, by an airtight diaphragm. Compressed air is admitted to the compression chamber to build up the required pressure. When the diaphragm is punctured by a knife, the diaphragm shatters, and a shock wave is formed which is propagated along the expansion chamber. Gages can be mounted in the expansion chamber, and their characteristics, under conditions similar to those under which they are to be used, can be studied.

2-34. Experimental Methods for Determination of Relative Air-Blast Intensities.[1] The methods of comparing explosives on the basis of their air-blast intensities are essentially the same at all establishments where such work is done. The charges, consisting of identical containers filled with the explosives to be compared, are detonated while being supported in a fixed position on the testing field. Air-blast gages, usually piezoelectric, are set up at several distances from the charge, and blast pressure-time records obtained. From these records, the peak pressures and positive impulses are computed. The conditions of the test are held the same for each trial so that direct comparisons among the different explosives can be obtained. The results are usually reported as relative peak pressures and relative positive impulses, referring all results to those from one type of filling chosen arbitrarily as a standard. Several identical charges of each type of explosive are usually fired in each series of tests in order to establish the statistical validity of the results.

It is found that with bare charges and pressures below 50 psi the relative pressures and impulses are essentially independent of the charge-to-gage distance, so that results obtained at a number of such distances can be considered as averages. Moreover, on the average, the results from various groups of experimenters are in agreement. The average relative peak pressures and positive impulses for all explosives considered are summarized in table 2-3. These averages include results from trials in the United States by the Underwater Explosives Research Laboratory and Stanolind Oil and Gas Company, Tulsa, Oklahoma, both of Division 2, National Defense Research Council, and by Ballistic Research Laboratories, Aberdeen Proving Ground, as well as in Great Britain, by Road Research Laboratory and Armament Research Department. All results are reduced to the basis of the average loading densities listed in table 2-2. The adjustment to relative peak pressures and relative positive impulses for differences in weights was made according to the empirical formulas

$$\frac{P_1}{P_2} = \left(\frac{W_1}{W_2}\right)^{0.6}$$

and

$$\frac{I_1}{I_2} = \left(\frac{W_1}{W_2}\right)^{0.67}$$

where P_1, and P_2 are peak pressures from weights W_1, W_2 respectively, and I_1, I_2 are the corresponding positive impulses. For the usual variations in loading density, such corrections as a rule are on the order of 1 or 2 percent.

EXPLOSIVES FOR BLAST

2-35. Comparison of Explosives for Blast. Table **2-2** is based upon the results of experiments performed before **1946** at the Underwater

Table 2-2

Average densities and compositions of explosives

Explosive	Average loading density (grams/cm)^3	Composition,* percent by weight of								
		Ammonium nitrate	Barium nitrate	Ammonium picrate	Haleite	PETN	RDX	TNT	Aluminum	Wax
Torpex (30% Al)	1.74	...	...	...	...	...	35	35	30	...
Torpex-2†	1.72	...	...	...	...	...	42	40	18	0.71
Minol-3	1.71	29	...	...	...	...	...	43	28	...
DBX	1.64	21	...	...	...	...	21	40	18	...
HBX†	1.63	...	...	...	...	...	40	38	17	5§
Tritonal 75/25¶	1.72	...	...	...	...	...	...	75	25	...
Minol-2	1.65	40	...	...	...	...	...	40	20	...
Tritonal 80/20	1.70	...	...	...	...	...	...	80	20	...
Trialen	1.64	...	...	...	...	...	15	70	15	...
Baronal	2.14	...	50	....	...	...	...	35	15	...
Comp B	1.61	...	...	...	...	...	60	40	...	1‡
Pentolite	1.60	...	...	...	...	50	...	50	...	...
Ednatol	1.59	...	...	...	57	...	...	43	...	...
TNT	1.56	...	...	...	...	...	...	100	...	...
Picratol	1.57	...	...	52	...	...	...	48	...	...
Amatex	1.55	44	...	...	...	...	6**	50	...	...
Amatol 60/40	1.55	60	...	...	...	...	...	40	...	...
Amatol 50/50	1.55	50	...	...	...	...	...	50	...	...

*Under actual loading conditions, compositions vary by a few percent from the average values given here.

†When 0.5% calcium chloride is added to torpex-2, it is called torpex-3; HBX contains 0.5% calcium chloride in addition to its other ingredients.

‡Not taken into account in percentages of other ingredients.

§D-2; desensitizing wax of the following composition: 6.9 parts Victory wax; 1.0 part nitrocellulose; 0.1 part lecithin.

¶Also may include 2% carbon black.

**Varies between 5% and 9%, at the expense of ammonium nitrate.

Explosives Research Laboratory of the Office of Scientific Research and Development. Later results, including those for HBX-1, HBX-3, and HBX-6 are given in paragraph 2-36. Table 2-3 gives comparisons on the basis of equivalent volumes of composition B.

Table 2-3

Comparison of peak pressure and impulse, with Comp B as base

Explosive	Peak Pressure	Positive Impulse
Torpex-2	1.13	1.15
HBX	1.06	1.11
Tritonal 80/20	1.04	1.08
TNT	0.92	0.94
Comp B	1.00	1.00

2-36. Explosives of the RDX/TNT/Aluminum System. A comparison of twenty-seven different mixtures indicated that optimum aluminum content is approximately 22 percent for best peak pressure or 26 percent for best positive impulse. Table 2-4 compares the optimum castable mixture, HBX-6, with several other military explosives. Comparisons are on a TNT basis.

2-37. Damage Test Ranking.[4] The following ranking has been assigned by the Ballistic Research Laboratories to explosives for internal blast against aircraft structures. The results were arrived at by firings of light-cased 40-mm shell against B-26 aircraft and by evaluation of damage to clamped circular aluminum plates.

1. Torpex-2
2. HBX-6
3. Tritonal
4. Comp B
5. Pentolite
6. TNT

It should be noted that MOX-2B, for this series of tests, was shown to be of the same order of effectiveness as Torpex-2 and HBX-6 on an equal volume basis.

Table 2-4

Comparison of peak pressure and positive impulse, with TNT as base (bare charges,)

Explosive	Peak pressure		Positive impulse	
	EW*	EV*	EW*	EV*
TNT	1.00	1.00	1.00	1.00
Comp B	1.13	1.21	1.06	1.13
HBX-1	1.21	1.36	1.21	1.36
HBX-3	1.16	1.39	1.25	1.49
Tritonal	1.07	1.17	1.11	1.25
HBX-6	1.27	1.44	1.38	1.57
MOX-2B	...	...	...	...

*EW - Equivalent weight basis
EV - Equivalent volume basis

A recent series of extensive tests carried out by the Ballistic Research Laboratories using cased charges against B-29 aircraft as targets resulted in the following quantitative comparison of efficiency (table 2-5). The numbers represent the relative weight or volume of a test explosive needed to cause the same blast damage as a given weight or volume of Pentolite.

2-38. Comparison of Explosives of the Ammonium Perchlorate/RDX/Aluminum System.[6] Table 2-6, the result of tests performed at the Naval Ordnance Laboratory, lists the comparative effect of several mixtures on an HBX-3 basis. The best mixture (40/40/20) was found to be equivalent to HBX-6.

2-39. Medina/TNT/Aluminum Explosives.[5] Comparison of the explosive mixtures in table 2-7 are on the basis of equivalent weights of HBX-6. Although the results are somewhat superior to HBX-6, the use of Medina is not yet practical due to poor stability and difficulty in obtaining high loading density.

Table 2-5

	Comp B	H-6	MOX-2B
Weight basis	0.84	0.68	0.89
Volume basis	0.79	0.65	0.70

AP	RDX	Al	Peak pressure		Positive impulse	
			EW*	EV*	EW*	EV*
40	40	20	1.04	1.08	1.05	1.09
40	30	30	0.95	1.00	1.11	1.14
	78	22	1.12	1.09	1.00	0.98
47	31	+5% WAX 22	1.09	1.04	1.10	1.05
HBX-3			1.00	1.00	1.00	1.00

2-40. Effect of Loading Density.1 The most common military high explosives that have been used or considered for use as fillings for aerial bombs are listed in Table 2-3. Chemical compositions and densities are shown in Table 2-2. The compositions of actual fillings vary by a few percent from those given. Similarly, the loading density given for each explosive is an average over a number of actual filling densities in various batches. The importance of loading density is twofold. Explosives are usually compared on the basis of equal volumes, so that the greater the density, the more favorable the comparative blast effectiveness. Second, the loading density is a measure of the quality of the particular filling; a poor pour will have air cavities and the components of the mixture will segregate. Both of these faults lead to low overall densities.

EFFECT OF BLAST ON AIRCRAFT

2-41. Aircraft Damage by Internal Blast. Test firings to determine the vulnerability of specific aircraft to internal blast have been performed at the Ballistic Research Laboratories.[16, 18] Such information provides the weapons designer with data which enables him to decide on an optimum warhead size.

2-42. Effect of Case on Internal Blast. Firings of bare and cased charges have indicated that when the detonation takes place in a well enclosed space, such as the inside of a wing, the internal blast effect of a cased charge is well in excess of the predicted effect. The blast effect of the projectile charge is approximately the same as that for an uncased charge of the same weight. In less enclosed spaces, such as the interior of the fuselage, the effect is considerably less. It is thought that this effect may be explained by two factors:

1. In the enclosed space, the massed effect of the fragments is sufficient to help produce structural damage.
2. Expansion of the explosive gases adds to the effect of the shock wave.

2-43. Surface Charges Versus Internal Charges.8 Firings of surface charges (in direct contact with the external surface of the skin of the aircraft) have indicated that the

Medina 98 P.A.2	TNT	Al	Peak pressure	Positive impulse
60.65	17	22.35	1.12	0.99
50	15	35	1.05	0.99

weight of surface charge necessary to produce damage equivalent to that of an internal charge is about three times that of the internal charge. Although it is true that shell designed for explosion on contact may be of thinner wall construction, particularly at the nose, than those which must have sufficient strength to penetrate the skin, it is doubtful that the lighter construction can enable sufficient additional explosive to be added to offset the loss in effectiveness due to the lack of penetration.

2-44. The Effect of Altitude on Internal Blast.[17] The detonations of high-explosive charges within World War II-type aircraft under sea level and under high-altitude atmospheric conditions show that compared to the amount of explosive needed to cause a given amount of damage at sea level, approximately 5 to 10 percent more is needed at an altitude of 30,000 feet, and 60 to 70 percent more at an altitude of 55,000 feet.

2-45. Aircraft Damage by External Blast. Firings have been conducted at the Ballistics Research Laboratory to determine the effect of external blast on aircraft of various types.[13, 14] Bare charges of various weights have been exploded in a sufficient number of orientations, with respect to the aircraft, to enable damage contours to be plotted. These contours, obtained for several vertical and horizontal planes through the aircraft, depict the maximum distance from the aircraft at which 100A structural damage could be expected to be inflicted by the given weight of charge. Although the firings have been conducted against aircraft of American manufacture, it is expected that the information obtained can be extended to foreign craft having similar sizes and structures. Figure 2-3 illustrates a typical set of damage contours for the B-17 bomber.

It must be emphasized that the damage con-

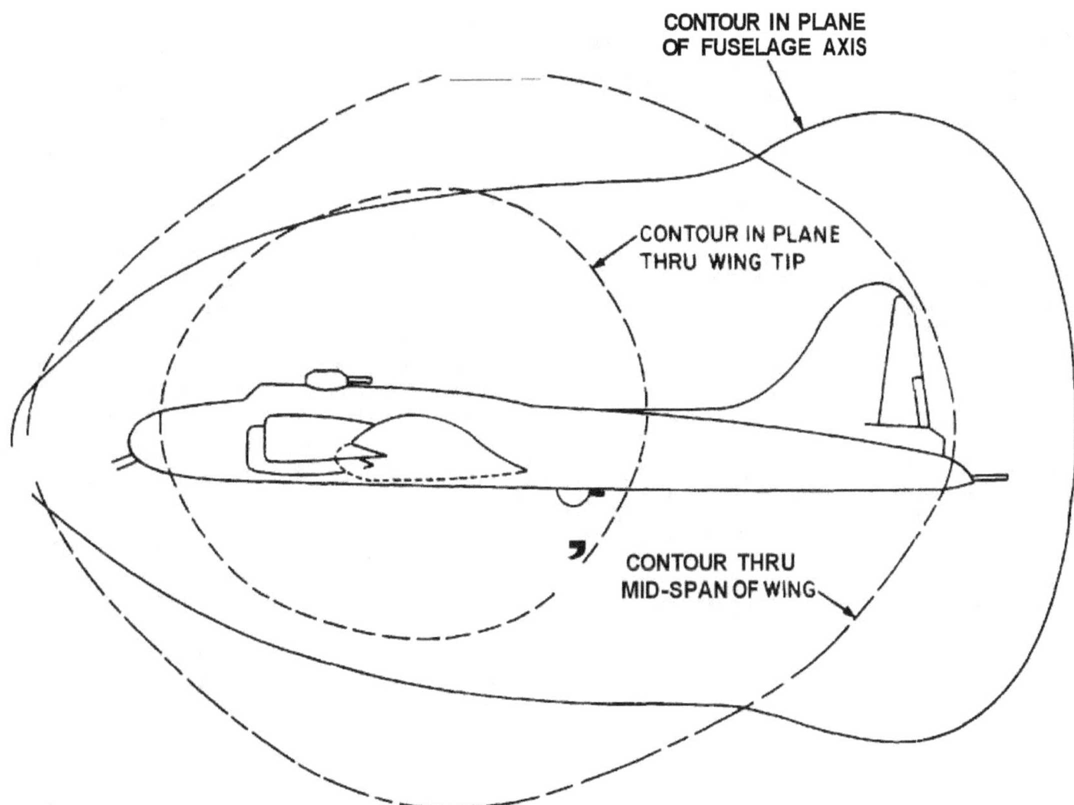

Figure 2-3. Damage contours for B-17 bomber

tours are the results of static tests, in which both aircraft and explosive charge are at rest at time of detonation. Experiments against aircraft structures with moving internal-blast type weapons indicate that there is a considerable directional modification of the peak pressure and impulse about a charge detonated while moving rapidly, as compared to the same charge statically detonated. The Ballistic Research Laboratories are conducting air blast tests to determine this modification. In addition to the effect of the moving charge, it should be noted that the tests were performed on aircraft which were static. It is to be expected that the dynamic loads imposed on aircraft in flight may add considerably to the effect of the explosive. Hence they may tend to expand the radii of effectiveness for the various explosive charges. It must be stressed that this information is for bare charges, and it is not known how applicable it might be for cased charges.

2-46. External Blast Damage Criteria.[13] One would expect that a plot of the peak pressure versus impulse just necessary to cause 100A damage by external blast to a given aircraft for a given orientation of the charge, with respect to the aircraft, would be of the form shown in figure 2-4.

The form of this curve is based on the belief that there is some low value of impulse below which it is impossible to achieve 100A damage regardless of how high the peak pressure and, conversely, there is some low value of peak pressure below which it is impossible to achieve 100A damage regardless of how high the impulse. Thus, in Region I, the impulse is the sole criterion of damage; in Region II, neither the peak pressure nor the impulse can be considered the sole damage criterion; in Region III, the peak pressure in itself is sufficient to define the damage. The general form of this curve has been borne out by limited results of the firings against A-25 aircraft by a wide range of charge weights. From these curves it should be possible to predict external blast vulnerability of an aircraft to weights of explosive charge other than those for which blast tests have been conducted and, also, it should be possible to modify external blast damage contours for high altitude conditions.

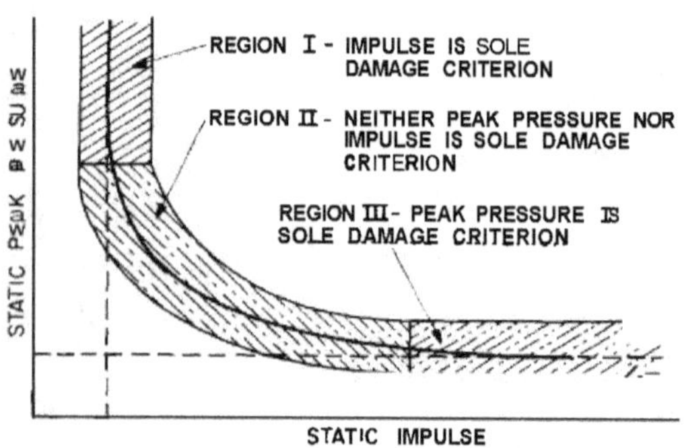

Figure 2-4. Static peak pressure versus static impulse

Curves of this type (peak pressure vs. impulse necessary to cause crippling damage) have been computed for A-25 aircraft.[14] These curves are illustrated in figures 2-5 and 2-6. The pressures used are side-on peak pressures. Impulse was obtained by assuming a triangular form for the positive phase of the pressure-time curve. The total positive impulse ($\int P\,dt$) is therefore equal to the area of the triangle, that is, peak pressure x 1/2 positive duration. Using these curves, it is possible to plot damage contours for any size charge at any altitude, provided that curves of peak pressure versus scaled distance, and scaled impulse versus scaled distance, are available for the explosive used and the required altitudes. If only sea level curves are available, those for greater altitudes may be computed by means of the dimensional scaling laws. Typical contours obtained by this semiempirical method are shown in figure 2-7. As before these curves are for bare charges, and their utility for cased charges is not completely known.

2-47. Modification of Blast Contours for Effect of Altitude. Since experimentally determined blast contours have been obtained only under sea level conditions, estimates of the effect of altitude on the contours must be made to apply the data to warhead design. A method for making such estimates has been presented in Ballistic Research Laboratories Memorandum Report No. 575. This method assumes that "damage threshold" curves of side-on peak pressure versus side-on impulse required to produce 100A damage to the aircraft remain invariant with

CONFIDENTIAL

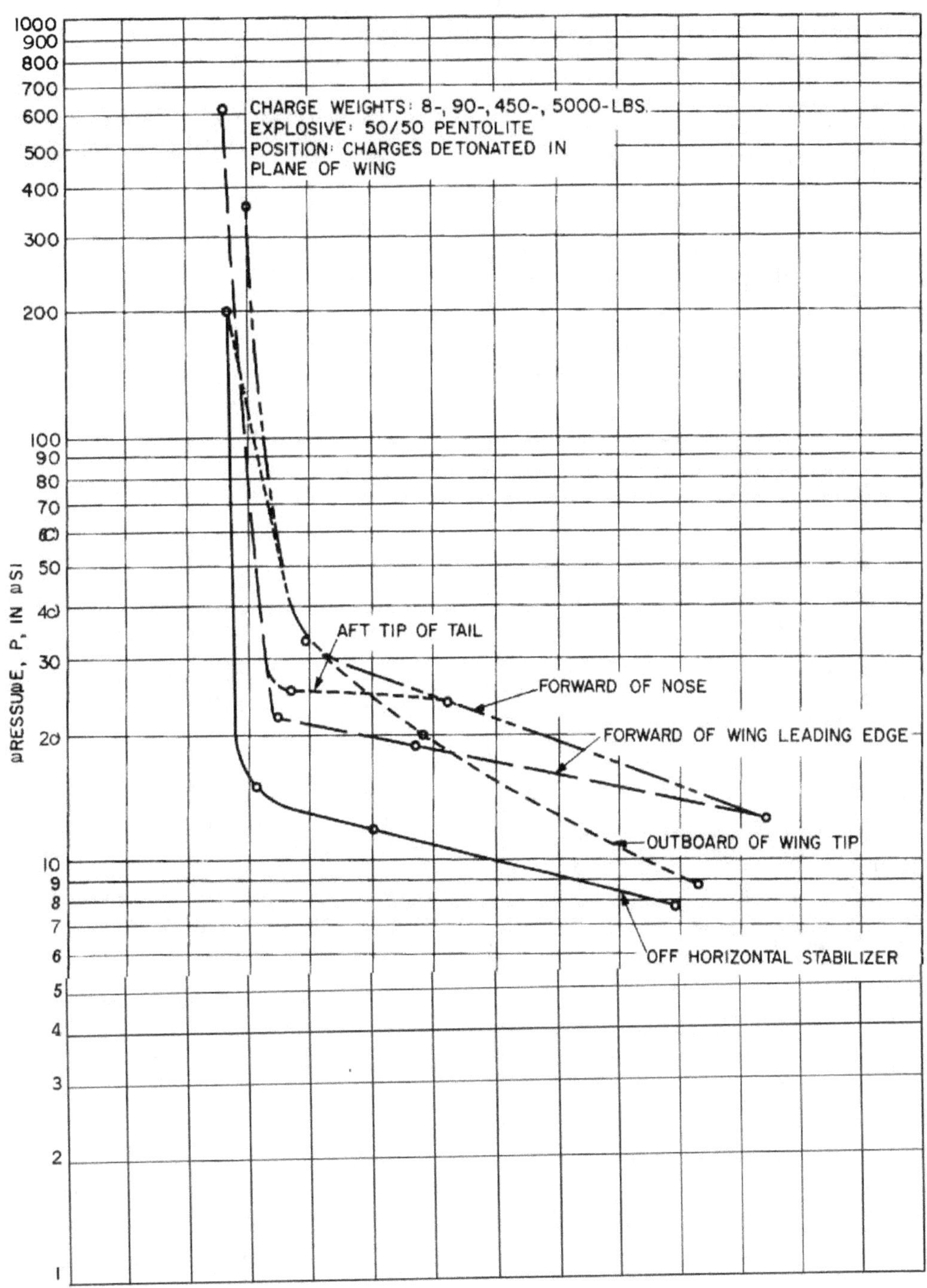

CONFIDENTIAL

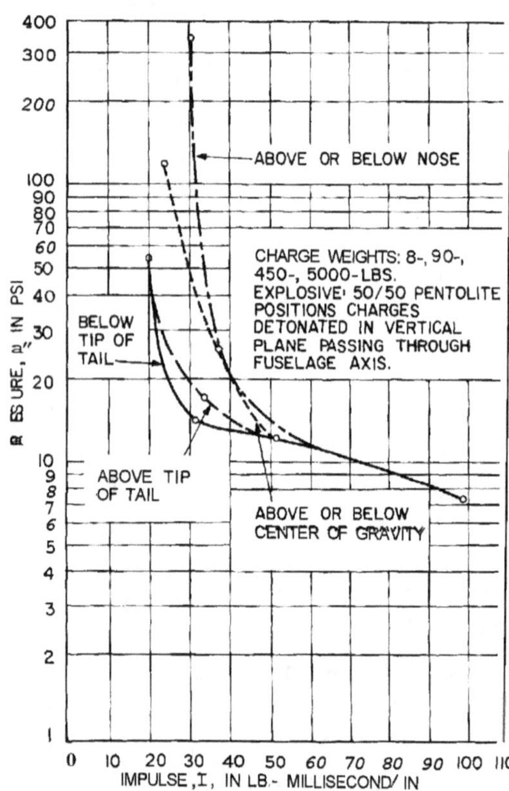

Figure 2-6. Crippling-damage air blast to A-25 aircraft, detonation in plane of wing

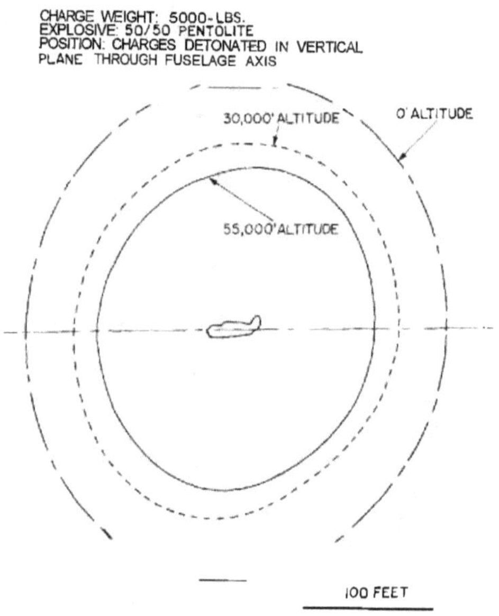

CHARGE WEIGHT: 5000-LBS.
EXPLOSIVE: 50/50 PENTOLITE
POSITION: CHARGES DETONATED IN VERTICAL PLANE THROUGH FUSELAGE AXIS
30,000' ALTITUDE
0' ALTITUDE
55,000' ALTITUDE
100 FEET

Figure 2-7. Blast contours around A-25 aircraft

altitude. Application of altitude scaling to the blast parameters then yields estimates of new contours for given altitude conditions. It is believed that this technique yields a very conservative estimate of contours of altitude. Another estimate of the altitude effect can be made by establishing damage threshold curves based on face-on blast parameters and applying altitude scaling to these curves. Since these parameters are degraded less than the side-on parameters with increase in altitude, the resulting contours are not reduced as much as when side-on parameters are used. It is believed that these two techniques yield both an upper and lower limit to the true contours at altitude.

2-48. Effect of Motion of Charge on Blast Damage to Aircraft. Firings of rockets and high explosive shell against aircraft targets by the Naval Ordnance Test Station, Ballistic Research Laboratories, and other agencies have indicated that the terminal velocity of antiaircraft missiles can greatly affect the damage they inflict on the targets. Damage is generally enhanced ahead of the missile and is reduced behind it. If the component being struck has relatively small internal volume, such as a wing panel, the overall damage is usually increased with increase in terminal velocity. If the component has large internal volume, such as the fuselage, the damage may not increase as much with increase in terminal velocity.

Some measurements of the free air blast around moving charges have been done by Ballistic Research Laboratories (Memorandum Report No. 767) but the data are as yet insufficient for correlation with damage studies.

2-49. Effect of Shape of Explosive Charge. When nonspherical explosive charges are detonated, separate blast waves are propagated from each of the faces. If the orientation of the faces with respect to each other is such that the blast waves intersect at an angle of 80° or greater, the waves will interact to form Mach waves or bridge waves. Bridge waves are defined as Mach waves caused by the interaction of two shock waves, resulting in the formation of a third shock wave, which bridges the volume between the two original waves. Double shocks may be produced at the intersection of the bridge wave and the original wave.

It has been determined experimentally[22] that the peak pressures and positive impulses arising from the detonation of fifty-pound spherical, cubical, cylindrical, conical, and laminar charges of RDX Composition C-3 ,plastic explosive do not differ significantly at large distances when averaged over all directions. However, the individual peak pressures and impulses in some directions from some of the nonspherical charges have been found to be as much as 50 percent higher than those in some other directions from the same charges at the same distances.

REFERENCES AND BIBLIOGRAPHY

1. NDRC, OSRD, Effects of Impact and Explosion, Summary Technical Report of Division 2, vol. 1, ch. 2, 1946.

2. Department of Ordnance, Terminal Ballistics, ch. 3, October 1951.

3. Goldstein and Hoffman, Preliminary Face-On Air Blast Measurements, BRLN 788, April 1953.

4. Baker and Johnson, Relative Air Blast Damage Effectiveness of Various Explosives, BRLM 689, June 1953.

5. Fisher, Air Blast Performance of Two Mixtures of Medina TNT and Aluminum, NAVORD 2959, November 1953.

6. Filler, Air Blast Small Charge Evaluations of Mixtures of Ammonium Perchlorate RDX or TNT and Aluminum, NAVORD 2738, May 1953.

7. Fisher, The Determination of the Optimum Air Blast Mixture of Explosives in the RDX/TNT/ Aluminum System, NAVORD 2348, March 1952.

8. Sarmousakis, Report on Tests of the Effect of Blast From Bare and Cased Charges, BRLM 436, July 1946.

9. Sachs, The Dependence of Blast on Ambient Pressure and Temperature, BRLR 466, May 1944.

10. Makino, Ray, The Kirkwood-Brinkley Theory of the Propagation of Spherical Shock Waves, and its Comparison with Experiment, BRLR 750, April 1951.

11. Dewey and Sperrazza, The Effect of Atmospheric Temperature and Pressure on Air Shock, BRLR 721, May 1950.

12. Hoffman, The Effect of Altitude on the Peak Pressure in Normally Reflected Blast Waves, BRLN 787, March 1953.

13. Baker and Johnson, Damage to B-17 and B-29 Aircraft by External Blast, BRLM 561, September 1951.

14. Sperrazza, Dependence of External Blast Damage to A-25 Aircraft on Peak Pressure and Impulse, BRLM 575, September 1951.

15. Smith, Effectiveness of Warheads for Guided Missiles Used Against Aircraft, BRLM 507, March 1950.

16. Cooney, Vulnerability of Aircraft to Internal Blast, BRLM 542, May 1951.

17. Sperrazza, Internal Blast Damage to Aircraft at High Altitude, BRLM 605, April 1952.

18. Sperrazza, Vulnerability of B-29 Aircraft to Internal Blast, BRLM 490, June 1949.

19. Hill, The Effects of Blast on Aircraft Fuel Tanks, BRLM 509, April 1950.

20. Fisher and Bengston, A New Air Blast Peak Pressure Nomograph Accounting for the Effects of the Steel Case and Height of Burst, NAVORD 2858, May 1953.

REFERENCES AND BIBLIOGRAPHY (cont)

21. Johnson and Baker, Internal Blast Damage to Aircraft Caused by Slender Cylindrical Charges Simulating Loki Warheads as Compared to Spherical Explosive Charges, BRLM 673, April 1953.

22. Adams, Sarmousakis, and Sperrazza, Comparison of the Blast from Explosive Charges of Different Shapes, BRLR 681, January 1949.

23. Fisher, E. M., The Effect of the Steel Case on the Air Blast from High Explosives, U. S. Naval Ordnance Laboratory, NAVORD Report 2753.

24. Kirkwood, J. G., and S. R. Brinkley, A New Theory of Shock-Wave Propagation with an Application to the Shock Wave Produced in Air by the Explosion of Cast Pentolite, NDRC, AES-3, Monthly Report, October 1944.

25. Kirkwood, J. G., and S. R. Brinkley, The Time Constant and the Positive Impulse According to the Theory of Shock-Wave Propagation of Blast Waves: Results for Cast Pentolite, NDRC, AES-4, Monthly Report, November 1944.

26. Kirkwood, J. G., and S. R. Brinkley, The Calculation of the Time Constant and the Positive Impulse from the Peak Pressure-Distance Curves of Blast Waves in Air with an Application to Cast TNT, NDRC, AES-4, Monthly Report, November 1944.

27. Kirkwood, J. G., and S. R. Brinkley, The Effect of Altitude on the Peak Pressure and Positive Impulse from Blast Waves in Air, NDRC, AES-11, Monthly Report, June 1945.

28. Kirkwood, J. G., and S. R. Brinkley, The Theory of the Propagation of Shock Waves from Explosive Sources in Air and Water, NDRC, A-318, March 1945.

29. Kirkwood, J. G., and S. R. Brinkley, Tables and Graphs of the Theoretical Peak Pressures, Energies, and Positive Impulses of Blast Waves in Air, NDRC, A-327, May 1945.

30. Kirkwood, J. G., and S. R. Brinkley, Theoretical Blast Wave Curves for Cast TNT, NDRC, A-341, August 1945.

31. Kirkwood, J. G., and H. A. Bethe, The Pressure Wave Produced by an Underwater Explosion I, OSRD, No. 588, May 1042.

32. Courant, R., and K. O. Friedricks,"Supersonic Flow and Shock Waves," Interscience Publishers, New York, 1948.

33. Lamb, H., "Hydrodynamics," Dover Publications, New York, 1945.

34. Hartmann, G. K., and P. Z. Kalawski, "The Optimum Height of Burst for High Explosives," NAVORD Report 2451, July 1952 (CONFIDENTIAL).

35. Fisher, E. M., Experimental Shock Wave Reflection Studies with Several Different Reflecting Surfaces, NAVORD Report 2123, September 1951 (CONFIDENTIAL).

36. Kirkwood, J. G., and S. R. Brinkley, Theoretical Blast Wave Curves for Cast TNT, OSRD Report 5481.

CHARACTERISTICS OF HIGH EXPLOSIVES

2-50. Introduction. The more useful chemical and physical characteristics of high explosives are presented in tables 2-8 and 2-9. Table 2-8 is essentially a digest of Picatinny Arsenal Technical Report No. 1740 (CONFIDENTIAL), and is a tabulation of the results of the tests described below. Table 2-9 shows the compatibility of given explosives with such other materials as metals, metal coatings, and plastics. The data for this table was abstracted from Picatinny Arsenal Technical Report No. 1783 (CONFIDENTIAL).

2-51. Description of Test Methods. Paragraphs 2-52 through 2-68 give brief descriptions of the test methods used in compiling Table 2-8. They are included to give the reader an appreciation of the test conditions and to permit him to compare and evaluate the tabulated data.

2-52. 75°C International Heat Test. A 10-gram sample is heated for 48 hours at 75°C, and is then observed for signs of decomposition or volatility, other than moisture.

2-53. 100°C Heat Test. A 0.6-gram sample of the explosive is heated for two 48-hour periods at 100°C. At the end of each period the sample is observed for volatiles other than moisture. It is also noted whether exposure at 100°C for 100 hours results in explosion.

2-54. Vacuum Stability Test. A 5.0-gram sample (1.0 gram in the case of initiators), after having been carefully dried, is heated in vacuo for 40 hours at either 100°C or at 120°C. The evolution of gas at each temperature is recorded.

Note

The capacity of the apparatus is 11 ml, therefore, any gas evolution in excess of 11 ml is reported as 11+.

2-55. Hygroscopicity. A 5- to 10-gramsample is exposed to an atmosphere of 30°C and 90 per-cent relative humidity (unless otherwise stated) until equilibrium is attained. In cases where either the absorption rate is extremely low, or very large amounts of water are absorbed, a specific time is stated. If the sample is solid, it is prepared by sieving it through a 50-on-100 mesh, U. S. Standard screen.

2-56. Impact Sensitivity Test. In both the Picatinny test and the Bureau of Mines test, a sample (approximately 0.02 gram) of explosive is subjected to the action of a falling weight of 2 kg. The impact test value is the minimum height at which at least one of 10 trials results in explosion. For the Bureau of Mines apparatus, the unit of height is the centimeter. In this test the explosive sample is held between two flat, parallel, hardened (Rockwell C63±2) steel surfaces, and the impact impulse is transmitted to the sample by the upper flat surface. Since the height of this apparatus is 100 cm, if, at 100 cm, no explosions result among ten trials, the value would have to be recorded as 100+. In the Picatinny apparatus the unit of height is the inch. In this test the sample is placed within a depression in a small steel die-cup, capyedby a thin brass cover. A slotted-and-vented cylindrical steel plug is placed, slotted side down, in the center of this cover. The principal differences between the two tests are that the Picatinny test involves (1) greater confinement, (2) a frictional component against the inclined sides, and (3) distributes the translational impulse over a smaller area, because of the inclined sides.

The test value obtained with the Picatinny apparatus depends to a marked degree on the sample's density. This value indicates the hazard to be expected when subjecting the particular sample to an impact blow; it is of value in assessing a material's inherent sensitivity only if the apparent "bulk" density is recorded along with the impact test value. The values tabulated were obtained on material screened between 50-on-100 mesh, U. S. Standard screens, where single component explosives were involved, and through 50-mesh in the case of mixtures.

2-57. Friction Sensitivity. The friction sensitivity of an explosive is determined by subjecting a 7.0-gram sample of the explosive (through 50-on-100 mesh) to the "sweeping" action of a "shoe" (of steel or fiber) attached to the end of a pendulum. The degree of sensitivity is reported qualitatively as (in decreasing order of sensitivity): explosion, snaps, crackles, or unaffected; indicated respectively by E, S, C, or U.

2-58. Rifle Bullet Impact Sensitivity. The sensitivity of an explosive to rifle bullet impact is determined by loading 1/2-pound samples of the explosive (in the same manner as they are loaded for actual use, or as indicated) in a 3-inch (2-inch I. D., 1/16-inch wall) pipe nipple, closed at each end by a cap. The loaded nipple may contain a small air space, which can be filled, if desired, by a wax plug. The loaded nipple is subjected to the impact of .30-caliber bullet (standard ball ammunition) fired from a distance of 90 feet perpendicular to the nipple's long axis.

2-59. Explosion Temperature Test. A 0.02-gram sample (0.01 gram in the case of initiator materials) of explosive, loose-loaded in a No. 8 blasting cap, is immersed for a short period in a Wood's metal bath. The temperature determined by this test is the temperature that produces explosion, ignition, or decomposition of the sample within 5 seconds. The behavior of the sample is indicated by an e, i, or d placed after the numerical value.

2-60. Booster Sensitivity Test. This test procedure is a version of the Bruceton "staircase" method for an unconfined charge. The source of shock consists of about 100 grams of tetryl, in two pellets, each 1.57 inches in diameter by 1.60 inches high. The initial shock is degraded through wax spacers of cast Acrowax B, 1 5/8 inches in diameter. The test charges are 1 5/8 inches in diameter by 5 inches long. The value reported is the thickness (in inches) of wax at the 50 percent detonation point.

2-61. Initiator Test. This test is run, using increasing quantities of initiator in each trial, until the amount of sand crushed no longer increases with an increase in the amount of initiator used, that is, until the rising curve (showing amount of sand crushed versus quantity of initiator used) levels off. The actual value reported is the difference between the total amount of sand crushed by the whole sample and the amount of sand crushed by the initiator alone.

2-62. Sand Test for Solids. A 0.4-gram sample of explosive pressed (at 3,000 psi) into a No. 6 cap is initiated by mercury fulminate or lead azide or, if necessary, by lead azide and tetryl, in a sand test bomb containing 200 grams of "on 30-mesh" Ottawa sand. The amount of fulminate or azide, or of azide and tetryl, that must be used to ensure that the sample crushes the maximum net weight of sand is designated as its sensitivity to initiation; and the net weight of sand crushed to finer than 30-mesh is termed the sand test value. The net weight of sand crushed is determined by subtracting from the total amount crushed the weight of sand crushed by the initiator material when fired alone.

2-63. Sand Test for Liquids. The sand test for liquids is made in accordance with the procedure given for solids, except that the special procedure described in "Methods of Inspection, Sampling, and Testing," MIL-P-11960 (24 April 1952) should be followed.

2-64. Fragmentation Test. The standard shell used in this test is either a 3-inch HE (M42A1, lot KC-5) or a 90-mm HE (M71, lot WC-91) shell. The values reported in this table are for the 3-inch shell. Either shell size is initiated by M20 booster pellets of height range 0.480 to 0.485 inch and weight range 22.50 ±0.10 grams. The shell is assembled with a fuze, actuated by a Blasting Cap (Special, Type II, Specification PA-PD-577) placed directly on a lead of comparable diameter. It is then placed in a box made of 1/2-inch pine. The box for the 3-inch shell is 15 x 9 x 9 inches in outside dimensions. This box containing the shell is placed on about 4 feet of sand in a steel fragmentation tub, the blasting cap wires are connected, and the box is covered by another 4 feet of sand. The shell is fired, then the surrounding sand is run onto a gyrating 4-mesh screen on which the fragments are recovered. The tabulated values represent the ratio of the number of fragments produced by the subject explosive to the number of fragments produced by an equal amount of TNT.

2-65. Ballistic Mortar Test. The amount of the sample explosive necessary to raise the ballistic mortar to the same height as it would be raised by 10 grams of TNT is determined by this test. The sample is then rated, on a proportional basis, as having a certain TNT value, that is, as being a certain percent as effective as TNT in this respect. The formula

$$\text{TNT value} = \frac{10}{\text{sample wgt.}} \times 100 \ (\% \text{ TNT})$$

gives the TNT value of the sample.

A ballistic mortar is a heavy, short-nosed mortar supported at the end of a long compound rod. The mortar contains a chamber about 6 inches in diameter by 1 foot long. About 7 inches of the chamber is occupied by a projectile, and the sample to be tested fills a small portion of the remainder. Upon detonation, the projectile is driven into a sand bank, while the mortar swings through an arc which is automatically recorded by a pencil attached to the mortar. The angle indicates the height to which the pendulum is raised by the explosion, and represents the energy measured by this test.

2-66. Trauzl Test. This test is used principally to obtain a qualitative concept of the power of a new explosive compared to an explosive (usually TNT) whose effects are known. It may be run either to determine the relative expansion of the test block, compared with the expansion produced by an equivalent weight of TNT, or to determine the weight of explosive required to produce the same order of expansion as a TNT reference standard. The results reported in table 2-8 are for the former method.

In the Trauzl test 10 grams of explosive are placed in a borehole, 25 mm in diameter by 125 mm deep, centrally located in the upper face of a lead block 200 mm in diameter by 200 mm in height. The block, cast in a mold, is made of desilverized lead of the best quality. The relative strength of the explosive is expressed as the ratio of the volume of the cavity after explosion to the initial volume; it is reported as the percent of expansion caused by an equivalent weight of TNT.

2-67. Detonation Propagation. The purpose of this test is to determine the minimum diameter below which a detonation wave will not propagate through a column of explosive. Columns of unconfined explosive 38 inches long by 3/4-, 1-, 1 1/4-, or 1 1/2-inch diameter were initiated at one end by a tetryl pellet. An impression made in a mild-steel plate at the other end of the column was accepted as evidence of complete detonation. The figure reported in the table is the minimum diameter for complete detonation.

2-68. Detonation Rate. Detonation rate varies with density; the values reported in the table, therefore, are only for representative densities, that is, 1.6 and "normal" (most common). The rates reported were determined by use of a rotating drum camera. The charges were 1 inch in diameter by 20 inches long, and were wrapped in cellulose acetate sheet. They were initiated by a system designed to produce a stable high-order detonation at maximum rate for the given conditions. A typical system consisted of four tetryl pellets 0.995 inch in diameter by 0.75 inch long, pressed to 1.50 grams per cc, with a Corps of Engineer's special blasting cap placed in a central hole in the end pellet

The remaining columns of table 2-8 give the normal and crystal (maximum) densities, and show the variation of loading density with pressure (for pressed explosives only). The usual, or recommended loading, method is also indicated.

2-69. Quantitative Definition of Compatibility. A compatibility problem may exist when foreign materials are in contact with, or in close proximity to, explosives or propellants. In this special sense, compatibility includes both the effect of the material on the explosive and the effect of the explosive on the material. The effect of the material on the explosive is determined by the $100^{\circ}C$ vacuum stability test. This is an accelerated test to determine whether the reactivity of the explosive is increased by contact with the material. In the standard test, if the net increase in gas evolution is 5 cc or greater, the reactivity is considered excessive and the material is deemed to be incompatible. The net increase in gas evolution is measured by comparing the volatility of 2.5 grams of the explosive and that of 2.5 grams of the material to the volume of gas released by the sum of 2.5 grams of each (5 grams total) mixed together.

2-70. Description of Table 2-9. The high explosives compatibility table shows the relative reactivity of military explosives with the metals they are most likely to be in contact with in service use. The compatibility of explosives with metals, unless otherwise noted, represents the effect of the explosive in contact with metal (in a wet or dry state, designated W or D) at ambient temperature for two years. The effect of explosives on metal is designated **F**, VS, S, **H**, **VH**, or C and the effects of explosive in contact with materials other than metal is designated **F**, N, M, U, or P. These designations are explained in the table. To indicate the direction of reactivity, a subscript m is used to indicate the effect of the explosive on the metal, and subscript x to show the effect of metal on explosive.

The table was compiled from data available in Picatinny Arsenal Technical Report No. 1783 (November 1950, CONFIDENTIAL). Extensive data on the compatibility of explosives and plastics are given in Picatinny Arsenal Technical Report No. 1838, "Completion of Data on the Compatibility of Explosives and Polymers," (1 October 1951, CONFIDENTIAL). Additional data on the compatibility of plastics may be obtained from the Ordnance Corps Plastic Laboratory at Picatinny.

Table 2-8

Characteristics of military high explosives

Explosive	Chemical name	Use	Governing specification	Composition, percent (mixtures only)	Stability / Sensitivity / Power / Rate of detonation / Density / Detonation pressure data	Method of loading	Melting point °C	Reasons for obsolescence	Explosive
OBSOLETE AND OBSOLESCENT									
Amatol 50/50		Demolition	none	50 TNT, 50 ammonium nitrate	[illegible]	Cast	[illegible]	Hygroscopic	Amatol 50/50
Ammonal			none	[illegible] TNT, 22 am. nitrate, 11 aluminum	[illegible]	Cast		Hygroscopic	Ammonal
Composition C-3			JAN-C-413	4 TNT, 77 RDX, 3 tetryl, 10 DNT, 5 MNT, 1 nitrocellulose	[illegible]	Hand tamp		Replaced by C-4	Composition C-3
Cyclotol 60/40		HE shell	none	40 TNT, 60 RDX	[illegible]	Cast		Replaced by Comp B	Cyclotol 60/40
Explosive D (Dunnite)	Ammonium picrate	AP shell, bursters	JAN-A-166		[illegible]	Pressed	265 Decomp	Hygroscopic; expensive to load	Explosive D (Dunnite)
Mercury fulminate		Primer, detonator	JAN-M-219		[illegible]	Pressed	Explodes	Poor stability	Mercury fulminate
Nitroglycerine		Double-base propellant, Demolition	JAN-N-246		[illegible]		13.2	High sensitivity	Nitroglycerine
Pentolite 50/50		Burster, shaped charge, demolition	JAN-P-408	50 TNT, 50 PETN	[illegible]	Pressed / Cast	85 / 76	Poor stability	Pentolite 50/50
Picric acid		Boosters	JAN-A-167		[illegible]	Pressed / Cast	122	Not compatible with metals (acidity)	Picric acid
Tetrytol 65/35		Burster, demolition		35 TNT, 65 tetryl	[illegible]	Cast	67	Replaced by 70/30	Tetrytol 65/35
Tetrytol 75/25		HE shell, demolition; Bursters, grenade, etc.		25 TNT, 75 tetryl	[illegible]	Cast	67	Replaced by 70/30	Tetrytol 75/25
CURRENT USE									
Amatol 80/20		Bursters	none	20 TNT, 80 ammonium nitrate	[illegible]	Cast or extruded	[illegible]		Amatol 80/20
Baratol 67/33			none	33 TNT, 67 barium nitrate	[illegible]	cast	80-81		Baratol 67/33
Black powder		Colored marker shell; igniters, delay elements	JAN-P-223	75 KNO_3, 10 S, 15.6 C	[illegible]	Loose or pressed	Decomp		Black powder
Composition A-3		HEP shell, shaped charge	JAN-C-440	91 RDX, 9 wax	[illegible]	Pressed	81		Composition A-3
Composition B		HE shell, shaped charge, fragmentation bombs	PA-PD-24	39 TNT, 60 RDX, 1 wax	[illegible]				Composition B
Composition C-4		Demolition	PA-PD-5	91 RDX, 9 Polyisobutylene binder[b]	[illegible]	Tamp or Pressed			Composition C-4
Cyclotol 75/25		HE shell, bursters		25 TNT, 75 RDX	[illegible]		157		Cyclotol 75/25
DDNP	Diazodinitrophenol	Primers, detonators	JAN-D-552		[illegible]	Pressed	Explodes		DDNP
Lead Azide (PVA)[a]		Primers, detonators	none	95 Lead azide, 5 impurities	[illegible]				Lead Azide (PVA)
Lead azide (dextrinated)		Primers, detonators	MIL-L-3055	91.5 Lead azide, 8.5 impurities	[illegible]		Explodes		Lead azide (dextrinated)
Lead styphnate (normal)		Primers, detonators	PA-PD-160		[illegible]	Pressed	Explodes		Lead styphnate (normal)
Lead styphnate (basic)			MIL-L-16355		[illegible]				Lead styphnate (basic)
MOX-2B		High-blast shell fillers (20- & 30-mm shell)	none	48 NH_4ClO_4/TNT(90/10), 6 RDX Wax (97/3), 50 Al (dichromated), 2 Ca stearate, 1 graph (added)	[illegible]				MOX-2B
PETN	Pentaerythritol tetranitrate	Booster mat., detonators	JAN-P-387		[illegible]	Pressed	138-141		PETN
Picratol 52/48		AP, SAP bombs	none	48 TNT, 52 ammonium picrate	[illegible]	Cast	80		Picratol 52/48
RDX (Cyclonite)	Cyclotrimethylene trinitramine	Booster, detonator	PA-PD-416		[illegible]	Pressed	204		RDX (Cyclonite)
RDX/wax 97/3		Booster mixtures	MIL-R-13728	97 RDX, 3 wax	[illegible]	Pressed			RDX/wax 97/3
Tetracene	1-Guanyl-4-nitrosoamino-guanyl tetrazene		none		[illegible]	Pressed	Explodes		Tetracene
Tetryl	2,4,6-Trinitrophenylmethylnitramine	Booster, small HE shell, detonator	JAN-T-339		[illegible]	Pressed			Tetryl
Tetrytol 70/30				30 TNT, 70 Tetryl	[illegible]	Cast	[illegible]		Tetrytol 70/30
TNT	Trinitrotoluene	Shell filler, bombs	JAN-T-248		[illegible]	Pressed			TNT
Tritonal 80/20		GP bombs	none	80 TNT, 20 aluminum	[illegible]	Cast	80.81		Tritonal 80/20
EXPLOSIVES IN DEVELOPMENT									
Cyclotol 70/30		HE shell		30 TNT, 70 RDX	[illegible]				Cyclotol 70/30
HBX-1		Blast, bursters		17 TNT, 67 Comp B, 17 aluminum, 5 D-2 desensitizer, 0.5 $CaCl_2$ (added)	[illegible]				HBX-1
HBX-3		War heads		6 TNT, 38 Comp B, 35 aluminum, 5 D-2 Desensitizer, 0.5 $CaCl_2$ (added)	[illegible]	Cast			HBX-3
HBX-6		Explosive bombs		74 Comp B, 21 Aluminum, 5 D-2 desensitizer, 0.5 $CaCl_2$ (added)	[illegible]	Cast			HBX-6
HMX (β-Form)	Cyclotetramethylenetetranitramine	Detonators, replacement for RDX in mixtures	none		[illegible]	Pressed	273-6.280 decomp		HMX (β-Form)
Nitroguanidine	Picrite	AP shell, bomb filler, propellant ingredient	PA-PD-302		[illegible]	Pressed	[illegible]		Nitroguanidine
Octol 75/25		HE & HEAT shell, bombs	none	25 TNT, 75 HMX	[illegible]	Cast	[illegible]		Octol 75/25

[a] Polyvinyl alcohol adulterated

[b] Polyisobutylene binder: 25 polyisobutylene, 59 di(2-ethylhexyl) sebacate, 16 SAE 10 motor oil

[c] D-2 desensitizer (Navy designation, Composition D-2): 84.16.? wax, nitrocellulose/lecithin

[d] Expl. D at 100% RH

[e] Bl. powder at 20°C, 75% RH

[f] Pb azide at 2?°C, 90% RH

[g] RDX at 25°C, 100% RH

[h] 8 ounce weight

[i] 1 kilogram weight

[j] Not usually tested, would probably detonate

[k] Nitroglycerine by liquid method

[l] Tetracene: 6.0 gms sand crushed when initiated by black powder fuse

[m] Tamped, crystalline TNT

[n] Pure lead azide

~~CONFIDENTIAL~~

Table 2-9

Compatibility of high explosives with metals and miscellaneous materials

Material / Explosive			Brass								Steel																			
	Aluminum		Brass		NRC coated		Shellac coated		Magnesium		Steel		Cadmium plated		Copper plated		Parkerized		Zinc plated		Stainless		Steel, acidproof black paint		Titanium		Black paint, acidproof		Cement Pettman	
	D	W	D	W	D	W	D	W	D	W	D	W	D	W	D	W	D	W	D	W	D	W	D	W	D	W	D	W	D	W
Amatol 50/50			S_m	S_m	VS_m	VS_m	VS_m	F_m	S_m‡	H_m‡	C_m	VH_m	S_m‡	C_m‡	VS_m	H_m	F_m	C_m	H_m	H_m	F_m	F_m	VS_m	F_m						
Ammonal																														
Composition c-3	F_m F_x		VS_m VS_x								F_m F_x		VS_m F_x						F_m F_x				S_m F_x							
Cyclotol 60/40																														
Mercury fulminate	F_m	F_m	F_m	F_m	F_m	F_m	F_m	F_m			H_m	F_m	F_m‡	F_m‡	F_m	F_m	F_m	F_m			F_m	F_m	F_m	F_m						
Nitroglycerine																														
Nitrostarch																														
pentolite	F_m	F_m	F_m	VS_m					F_m¶	S_m¶	F_m	F_m	F_m F_x	F_m	F_m	VS_m			F_m F_x	F_m	F_m	F_m	F_m	F_m			M			
Picric acid	F_m	F_m	F_m	F_m	F_m	S_m	F_m	F_m			F_m	F_m	C_m‡	F_m‡	F_m	S_m	F_m	F_m	F_m	VS_m	F_m	F_m	F_m	F_m						
Tetrytol 65/35									F_m	F_m																			H_m	
Tetrytol 75/25	F_m	F_m	F_m	VS_m					F_m‡	S_m¶	F_m	VS_m	F_m	VS_m	F_m	VS_m			F_m	S_m	F_m	F_m	F_m	F_m					H_m	
Amatol 80/20																														
Baratol 67/33																														
Black powder	H_m F_x	H_m	VH_m	H_m	F_m	F_m	F_m	F_m	F_m¶	F_m¶	H_m	H_m	F_m‡ F_x	F_m‡	VH_m	VH_m	F_m	S_m	F_m F_x	F_m	S_m	F_m	F_m	F_m						
Composition A-3	F_m	F_m	VS_m	S_m					VS_m¶	S_m¶	VS_m	S_m	VS_m	F_m	F_m	S_m			F_m	S_m	F_m	F_m	F_m	F_m						
Composition B		F_m	VS_m	S_m					F_m [4]	S_m [4]	F_m	VS_m	F_m F_x	VS_m	F_m	S_m			VS_m F_x	S_m	F_m	F_m	F_m	F_m			M•100 U•120 U$_x$•120 M$_x$•90		F_x U_x	F_m
Composition C-4																														
Cyclotol 75/25																														
DDNP																														
Explosive D	F_m	F_m	F_m	F_m	F_m	F_m	F_m	F_m			F_m	C_m	F_m‡	F_m‡	S_m	VS_m	F_m	F_m	F_m	S_m	F_m	F_m	F_m	F_m						
Lead azide	F_m† F_x	F_m†	F_m	P_m	F_m	P_m	F_m	P_m	F_m¶	F_m¶	F_m				F_m		F_m		F'_m	VS'_m	F_m		F_m		F_x					
Lead styphnate																														
MOX-2B																														
PETN	F_m	VS_m	F_m	VS_m					F_m¶	S_m¶	F_m	VS_m	F_m	VS_m	F_m	F_m			F_m	VS_m	F_m	F_m	F_m	F_m	F_x					
Picratol 52/48													F_x						F_x											
RDX	F_m	F_m	F_m	F_m	F_m	F_m	F_m	F_m			F_m	F_m	F_m‡	F_m‡	S_m	VS_m	F_m	F_m	F_m	F_m	F_m	F_m	F_m	F_m	F_x				F	
Tetracene	F_m																													
Tetryl	F_m F_x	F_m	VS_m	F_m	F_m	F_m	F_m	F_m			H_m	C_m	F_m‡ F_x	F_m‡	F_m	F_m	F_m	F_m	F_m F_x	F_m	F_m	F_m	F_m	F_m	F_x		U			
Tetrytol 70/30									F_m	F_m																			H_m	
TNT	F_m	F_m	F_m	F_m	F_m	F_m	F_m	F_m	F_m¶	H_m¶	F_m	F_m	F_m‡ F_x	VS_m‡	VS_m	VS_m	F_m	F_m	F_m F_x	F_m	F_m	F_m	F_m	F_m	N_x		N to M		F_m	
Tritonal 80/20													F_x						F_x								N to M			

* 2 months † 10 months ‡ 12 months ¶ 18 months

Legend

F = favorable, no visible evidence of reaction

VS = very slight corrosion, indicated by light tarnishing

S = slight corrosion, indicated by heavy tarnishing

H = heavy corrosion

VH = very heavy corrosion

C = considerable corrision, indicated by pitting or rusting

P = prohibited

N = negligible reaction

M = moderate reaction

U = undesirable reaction

D = dry sample

W = wet sample

Subscript m = explosive reaction on metal

Subscript x = metal reaction on explosive

~~CONFIDENTIAL~~

SHAPED CHARGE AMMUNITION

LIST OF SYMBOLS

a = flute depth
c = charge mass (cross section)
d = charge diameter
E = heat of explosion
I_n = impulse delivered to liner in normal impact
I_θ = impulse delivered to unit area of a liner whose surface normal forms the angle θ with the direction of propagation of the detonation wave
J = collapse (stagnation) point
L = jet length
M = liner mass (cross section)
m_j = jet mass
m_s = slug mass
n = calibers per turn of rifling. Also the number of flutes on a fluted liner
P = penetration
p_s = collapse (stagnation) pressure
p_1 = explosion pressure
R = pitch radius of fluted liner element
S = standoff
T = wall thickness of blank before fluting
T_1 = explosion temperature
U = velocity of jet impacting target.
V_D = velocity of detonation
V_i = impact velocity
V_j = velocity of the jet
V_o = velocity of liner collapse
V_s = velocity of slug
α = angle between liner axis and liner
β = angle between liner axis and collapsing liner
δ = angle of indexing (when matching fluted tools are used)
θ = angle between normal to liner surface and direction of propagation of detonation wave
λ = T/R
μ = the constant a/R
ν = spin frequency
ν_o = optimum frequency of rotation
ρ = target density
ρ_j = jet density
ψ = angle between flute offset and radius through its root
ω = angular velocity

STATUS OF THEORY

2-71. Detonation Wave. The detonation process is most easily pictured in terms of the passage of a "detonation front" (see figure 2-8) through the explosive, with the velocity V_D in the range 5 to 9 mm per μ sec (km/sec). Behind this detonation front, pressures p_1 of the order of 250,000 atmospheres and temperatures T_1 in the range of 2,500° to 4,000°C are commonly observed. The total chemical energies feeding the detonation are of the order of E = 1,000 cal. per gram. The detonation front is regarded as a shock surface followed by a "reaction zone" in which chemical reaction takes place; the thickness of the zone is estimated to be on the order of 1 mm for most solid explosives, corresponding to 0.1 μsec reaction time.

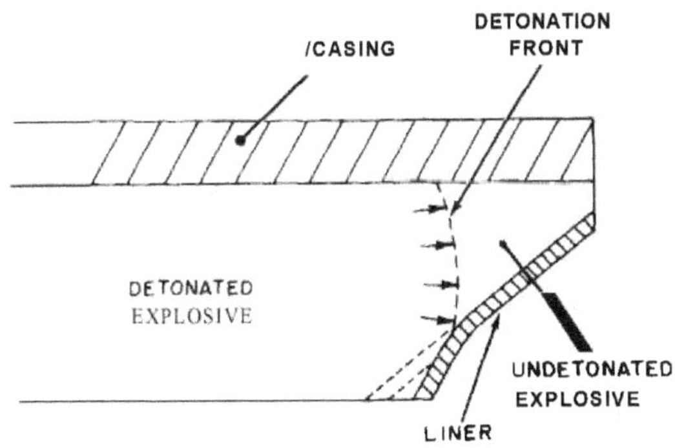

Figure 2-8. Detonation wave in an explosive

From V_D, E, and an assumed equation of state, one can estimate p_1, T_1, and the particle velocity behind the detonation front by the conservation of mass, momentum, and energy. The so-called Chapman-Jouguet condition gives a fourth equation, from which V_D itself can be predicted. However, the equations of state of solids under high temperatures and shock pressures are not accurately known, and so the preceding method is of limited practical value.

If one assumes that V_D is a constant for a given explosive, so that the detonation front propagates by Huygens' principle (just as in geometrical optics), one has a rational basis for "shaping" explosive waves by peripheral initiation, or by using composite charges having different detonation velocities in different regions (for example, having inert cores). Actually, V_D may be affected by the curvature of the detonation front, and composite charges are especially liable to imperfections.

In lined cavity charges, the primary effect of the explosive is through the collapse velocity (V), which it transmits to the liner, in the high-pressure zone behind the detonation front. This velocity is transmitted, by a complicated process of multiple shock reflection, in 5 to 50 μsecs. The net effect of these multiple reflections has been shown to be nearly the same as if the liner were rigid. The effect of finite charge dimensions and confinement is not easy to determine.

2-72. Jet Formation: "Zero Order" Theory. In the case of conical liners with cone angle 2α, the simplest picture is to assume that the liner collapses with a constant velocity V_0, and in a constant direction. Applying Bernoulli's equation to a moving reference frame, this direction bisects the angle 2θ between the normal to the uncollapsed liner, and the normal to the collapsing liner (see figure 2-9a). If shear stresses and shocks (increase of entropy) are neglected, the steady state hydrodynamical theory of jet formation is obtained. According to this theory, the collapsing cone divides into a high speed jet and a slower slug, whose mass-ratio is (in terms of the angle β between the collapsing liner and the axis)

$$\frac{m_j}{m_s} = \frac{1 + \cos \beta}{1 - \cos \beta} = \cot^2 \frac{\beta}{2} \qquad (1)$$

and whose velocities are respectively

$$V_j = V_0 \frac{\cos \alpha/2}{\sin \beta/2} \text{ and } V_s = V_0 \frac{\sin \alpha/2}{\cos \beta/2} \qquad (2)$$

The details of this theory have been published, and will not be repeated here.

The assumptions needed to derive equations (1) and (2) are the following.

a. Steady state flow, in a moving reference frame. Strictly speaking, this requires a plane detonation wave, and a liner whose thickness is inversely proportional to the distance from the cone apex.

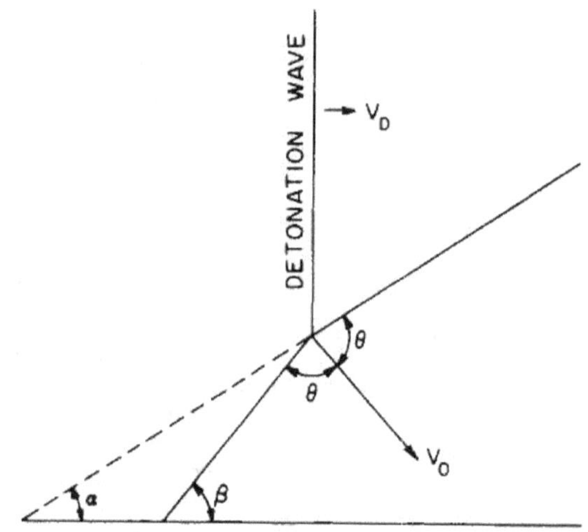

Figure 2-9a. Jet formation forces

b. Shear forces are negligible, since the yield stress of mild steel is only 8,000 atmospheres.

c. Isentropic, shock-free flow.

d. Constant pressure on the liner near the stagnation point J, the same inside and outside the liner.

e. Asymptotically uniform flow in the liner, jet, and slug, away from the stagnation point (J).

From the preceding assumptions, it follows that, relative to axes moving with the collapse (stagnation) point (J), we have the Bernoulli equation (by a to c), and hence (by d and e) the same relative velocity in the jet, slug, and collapsing liner (see figure 2-9b).

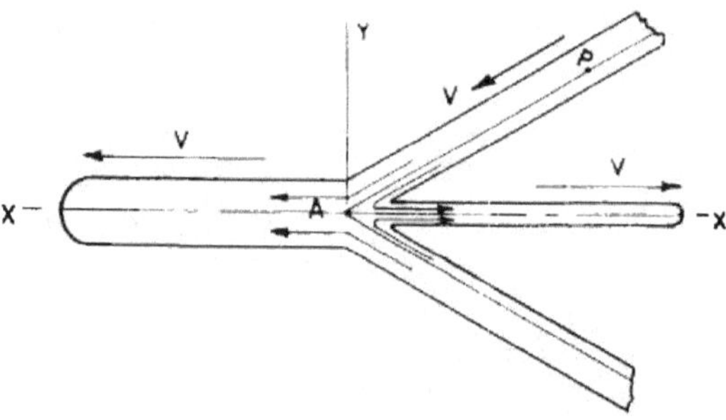

Figure 2-9b. Jet formation forces

Equations (1) and (2) have been confirmed experimentally near the apex of the cone. Especially is this true of the predictions that the jet length should equal the slug length, and of the initial V_j. However, near the base of the cone the collapse angle increases rapidly, the observed $\frac{m_j}{m_s}$ is considerably larger than that predicted by equation (1), and the jet becomes several times longer than the slug. We shall now attempt to rationalize these facts, following ideas first explicitly formulated by Pugh.

2-73. Jet Formation: "First Order" Theory. In HEAT shell with conical liners (figure 2-8), it is obvious that C/M (the ratio of mass of charge to mass of liner in a cross section) decreases, from infinity to a small quantity, as one moves along the liner axis (X-axis) from apex to base. Hence (see paragraph 2-71) the collapse velocity, $V_0(x)$, may be expected to decrease correspondingly, in a way which can be predicted roughly. It would be desirable to have accurate direct experimental measurements of $V_0(x)$. However, this is difficult to obtain.

If one assumes, in addition to assumptions a through e of paragraph 2-72, that there is negligible momentum transfer after the initial phases of the collapse process, one concludes that each liner element from the ring-shaped zone with initial position x moves with constant velocity $V_0(x)$ in a straight line until it reaches the liner axis. Since the collapse direction bisects the initial angle between the initial normal to the collapsing liner and the normal to the original cone, one can predict at all times, from $V_0(x)$ and the initial collapse angle, the shape of the collapsing liner. The predicted collapse profile agrees with observation, at least qualitatively.

As emphasized by Pugh, Eichelberger, and Rostoker, who originated the preceding "first order" theory, the inferred local collapse angle, $\beta(x)$, and local relative velocity, $V_1(x)$, on the axis will increase and decrease markedly as we move from apex to base. By (2), the increase in $\beta(x)$ accentuates the jet velocity gradient, so that $V_1(x)$ decreases to a fraction of its initial value.

With thin cones, the relative change in $\beta(x)$ and $V_1(x)$ per liner thickness is small; hence it seems reasonable to assume that the theory of paragraph 2-72 is locally applicable to these quantities. At least, this assumption gives a simple basis for calculating, as functions of x, the velocity $V_j(x)$ of jet formation and mass-ratio $\frac{m_j(x)}{m(x)}$.

2-74. Jet Breakup. Because of the jet velocity gradient already mentioned, the jet may be expected to lengthen continuously, while moving ahead in a straight line. In the case of well-formed, unrotated charges and liners, a straight, steadily lengthening jet is in fact observed. However, real jets always break up into streams of particles sooner or later (see figure 2-10). The time of breakup has an important effect on penetration (see paragraph 2-76).

With steel liners, breakup ordinarily occurs within a few cone diameters of travel, while for copper liners, as first predicted by Pugh and later confirmed experimentally, considerable ductile drawing occurs, and breakup occurs much later.

2-75. Similarity. If the diameter is taken as the unit of length, the "Law of Cranz" asserts that geometrically similar shaped charge rounds of widely varying diameter d behave approximately similarly.

The best theoretical basis for this fact consists in the principle that the inertial and explosive stresses involved depend mainly on the strain and much less on the time rate of strain. Although this principle is not exact, and is presumably not applicable to the reaction zone, to viscous effects, or to jet breakup, it has sufficient validity to be very useful in analyzing existing data.

Applied to rotating shaped charges,, it predicts that the relative deterioration in shaped charge performance due to spin, with similar rounds of different diameter d spinning at ω rps, should be determined by the spin parameter ωd measuring the peripheral velocity, rather than by ω itself. This peripheral velocity ωd is clearly $\frac{V_i}{n}$, where V_i is the impact velocity, and n the twist of rifling (in calibers per turn).

2-76. Penetration: "Zero Order" Theory. A continuous, perfectly formed fluid jet of density ρ_j, moving with constant velocity Vj, should penetrate a target of density ρ with a constant velocity U, which can be predicted roughly from

the continuity of pressure at the "stagnation point" (figure **2-11**), where the tip of the jet is boring into the target. In a reference frame moving with velocity U, neglecting target strength and compressibility, Bernoulli's Theorem assumes the simple form

Figure 2-10. Radiographs of stages in jet formation

$$\frac{1}{2}\rho_j (V_j - U)^2 = \frac{1}{2}\rho U^2 \qquad (3)$$

Hence, the rate of penetration U satisfies the equation

$$U = \left(\frac{\rho_j}{\rho}\right)^{1/2} (V_j - U)$$

where $V_j - U$ is the rate at which the jet is being used up. Solving, we get the Hill-Mott-Pack equation

$$P = \left(\frac{\rho_j}{\rho}\right)^{1/2} L \qquad (4)$$

connecting the total depth of penetration P with the total jet length L, for uniform, incompressible fluid jets.

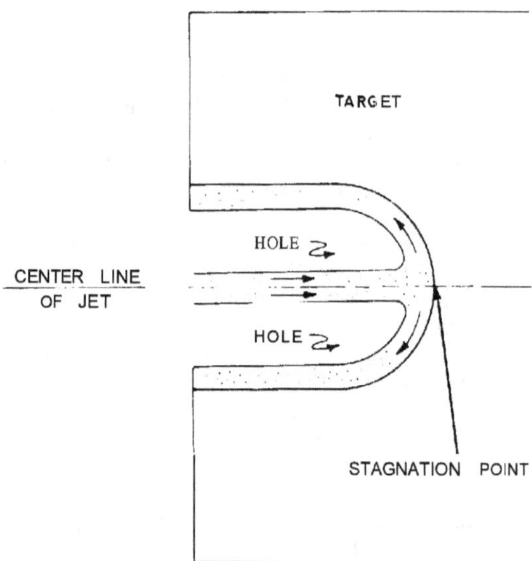

Figure 2-11. Target penetration

Because of the assumption V_j=constant, which corresponds to the model of paragraph **2-72**, this may be called a "zero order" theory. Combining with paragraph **2-72**, we see that $L=\frac{a}{2}\csc\beta$, is nearly the cone slant height. Actually, total penetrations several times this depth are obtained at large standoff, for reasons explained in paragraph **2-77**. However, equation *(3)* can be used to infer the useful equations

$$U = \frac{V_j}{1 + \sqrt{\frac{\rho}{\rho_j}}} \tag{5}$$

$$p_s + \frac{1}{2}\rho U^2 = \frac{\frac{1}{2}\rho_j V_j^2}{1 + 2\sqrt{\frac{\rho_j}{\rho}} + \frac{\rho_j}{\rho}} \tag{5a}$$

from which the instantaneous penetration velocity (U) and stagnation pressure (p_s) can be inferred approximately.

Thus, copper jets penetrating water at 4 mm per psec have been observed by Kerr cell photography. Again, a 10 mm per μsec steel jet will penetrate a steel target at $U = \frac{V_j}{2} = 5$ mm per psec, according to (5), giving a stagnation pressure $p_s \simeq 500{,}000$ atmospheres, roughly. This greatly exceeds the yield strength of steel, justifying the hydrodynamical model.

2-77. Penetration: "First Order" Theory. The considerations of paragraph 2-73 lead to an important modification of formula (4), by Pugh and Fireman, which explains the observed variation in penetration with standoff. In this modification, one assumes a gradual variation in the jet velocity and density along its length, so that Bernoulli's Theorem is locally applicable. This gives

$$P = \frac{1}{\sqrt{\rho}} \int \frac{dl}{\sqrt{\rho_j(x)}} \tag{6}$$

where $\rho_j(x)$ is the "effective" density of the jet when it reaches the target.

Looking only at the first factor in (6), we see that, for different target materials, $P \propto \frac{1}{\sqrt{\rho}}$. Thus, weight for weight, low-density materials provide the best defense against shaped charges, so long as $V_j(x)$ is so large that the target yield-strength is negligible. For mild steel, with a yield strength of 8,000 atmospheres, this corresponds to $V_j \gg 450$ meters per sec, which is not verified near the tail end of the jet. This explains qualitatively why penetrations into mild steel are 10 to 15 percent deeper than into armor, which has greater yield strength. However, the proportionality $P \propto \frac{1}{\sqrt{\rho}}$ has been confirmed approximately for many materials. The most notable exceptions to this are quartz-like materials.

Looking directly at (4), or at its refinement (6), it is clear that the improvement in penetration P with standoff S may be explained qualitatively by the tendency of the jet to lengthen as it progresses, and hence, indirectly, by the velocity gradient along the jet. This factor, rather than any overall increase in velocity, is considered responsible for the improvement in penetration with peripheral initiation.

The quantitative application of (6) requires a successful prediction of ρ_j, which is variable because of jet breakup, rotation, and other factors. In trying to describe the dependence of penetration P on standoff S, it is convenient to distinguish several cases.

a. In the case of well-formed copper jets, it is believed that ductile drawing makes ρ_j constant, out to a large standoff. Hence $P = P_0(1 + \alpha S)$. Other fluid jets are less effective; this may be due to lower density, wavering, or other factors.

b. In the case of perfectly alined particle jets, ρ_j decreases in inverse proportion to dl, so that a formula of the type $P = P_0\sqrt{1 + \alpha S}$ is inferred.

c. In the case of unalined jets, whether due to imperfections or rotation, ρ_j decreases also in proportion to S^2 due to "spreading," so that a formula of the type

$$P = \frac{P_0\sqrt{1 + \alpha S}}{S}$$

is inferred.

Curves of the preceding type can be roughly fitted to observed data; the large experimental scatter prevents drawing more exact conclusions. Ideally, especially in the case a above, it might be possible to infer an optimum $V_0(x)$ from theoretical considerations. But a large amount of empirical work at Bruceton, during World War II, failed to improve substantially on conical liners.

2-78. Effect of Rotation. It was observed as early as 1943 that rotation caused a large decrease in penetration (P), and that this effect was especially noticeable at large standoff. Typical records of $\frac{P}{d}$ as a function of ωd (refer to paragraph 2-75) are plotted in figure 2-12. The reduction in penetration by spin may be attributed to lateral dispersion of the jet, which decreases its effective mean density. This lateral dispersion is also evident in X-radiographs. Thus, the jet velocity and momentum are about the same as for unrotated liners.

Assuming that jet particles move in straight lines, we may correlate with the penetration theory of paragraph 2-77, since the mean density (ρ) will be proportional to the inverse square $\frac{1}{S^2}$ of the standoff (S). Thus, at large standoff (S), the penetration (P) should be proportional to $\frac{1}{S}$, and one may expect a decrease in penetration as standoff lengthens.

X-ray pictures sometimes show a bifurcation of the jet. A theory of the instability of a lengthening rotating jet is in process of construction. The effect of rotation is discussed in detail paragraphs 5-121 through 5-128.

2-79. Rotation Compensation. Attempts have been made to improve the performance of spin-stabilized shell by using various nonconical, axially symmetric liners (refer to paragraph 2-128 following). However, such attempts have not been promising at high rates of spin, and so the major emphasis has been on the design of fluted liners not having axial symmetry.

The idea underlying the use of fluted liners is that of "spin compensation" —that is, of annihilating the angular momentum of the liner so as to inhibit the jet spreading already discussed in paragraph 2-78. This is not quite the same as

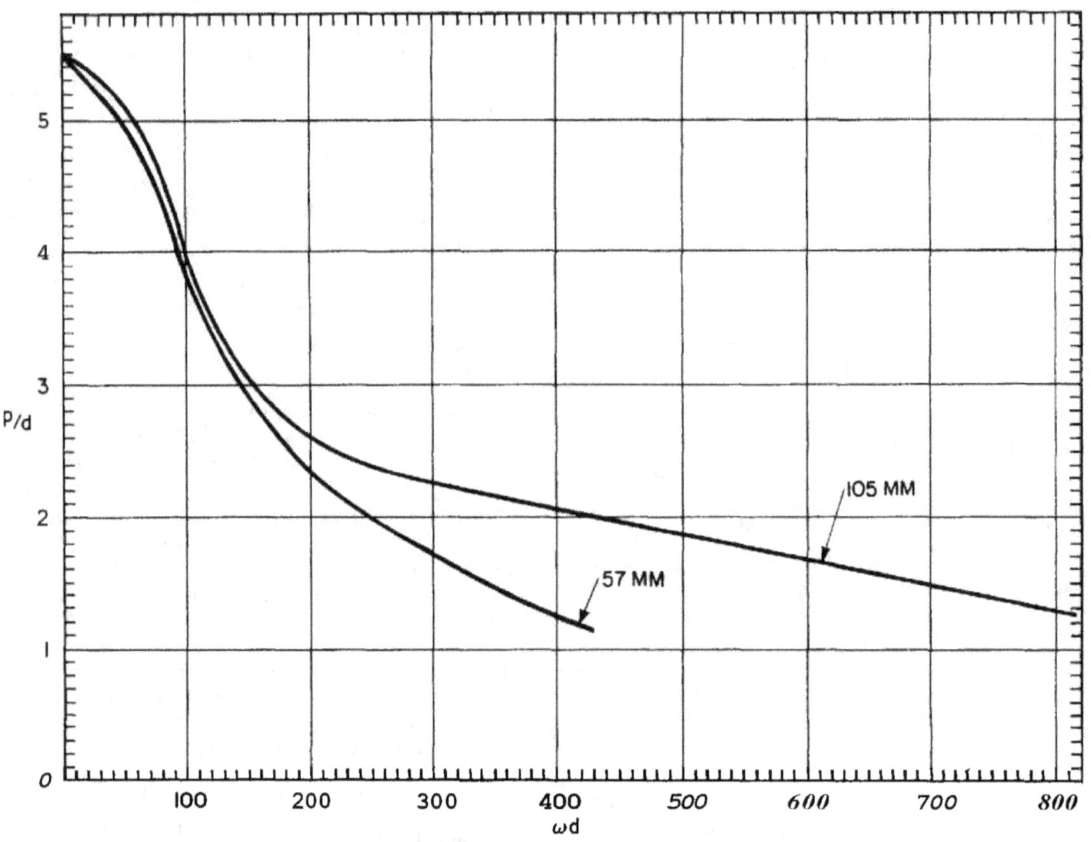

Figure 2-12. Relation of penetration to rotational velocity

the original idea of using "offsets" (see figure 2-13a) to make most of the liner collapse on the axis.

Generally speaking, one may hope to achieve some spin compensation by using wavy flutings, in which the wall is thicker on the forward side, as in figure 2-13b. This is because the momentum transferred is nearly normal to the liner surface, and proportional to its thickness.

Near the base of the liner, where rotation effects are most serious, one may also hope to convert some of the axial momentum of the explosive behind the shock wave into rotational momentum, by using spiral flutings, whose angle with a plane through the cone axis becomes progressively steeper toward the base of the cone.

However, the mechanism of spin compensation is not yet clearly understood. Thus, with several designs, even the direction of spin compensation is reversed by changing the number of flutings, and this seems hard to explain. Again, the consideration that uniform pressure of the explosive on the surface, however fluted, would produce exactly zero compensation, shows that a fairly sophisticated theory is required, one that must probably include a detailed discussion of the explosive-liner interaction, perhaps by numerical methods.

For the present, we must rely mainly on empirical data. Any successful theory must explain not only the direction of spin compensation but also its magnitude for the shapes discussed in detail in paragraphs 2-129 through 2-139.

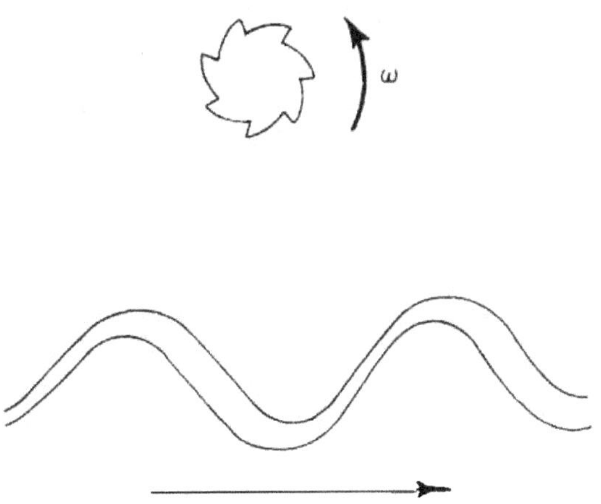

Figure 2-13. Liner fluting

LINER PERFORMANCE

2-80. Measures of Liner Performance. Liner performance is measured in terms of penetration into some homogeneous reproducible material, usually mild steel. Both mild steel and homogeneous armor are used, but the two are not equivalent. Different grades or types of mild steel all give about the same average penetration for a given shaped charge design, but this is not true for homogeneous armor. It is reported that the penetration of a given jet into steel at a fixed standoff varies essentially linearly with the Brinell hardness of the steel. Recent work at Ballistic Research Laboratories and Firestone indicate that the relative penetration into mild steel and homogeneous armor is also affected by standoff. The data show that the homogeneous armor is more effective at the longer standoffs. For convenience in measuring depth of penetration, targets are often made of stacks of plates 1/2 to 3 inches thick. There does not seem to be any objection to this practice if the plates lie flat on each other.

For some purposes, a better measure of liner performance is given by the volume of the hole or its smallest diameter. For most purposes, particularly as a measure of lethality, the best measure would probably be some factor which indicates the amount of damage done behind a given target plate by the residual jet and spalled material from the back face of the plate. It has so far been difficult to define such a measure, and even more difficult to determine it from the test. In this discussion total depth of penetration into mild steel will be used as the measure of liner performance, except where stated otherwise.

2-81. Factors Affecting Liner Performance. Shaped charge liners have been made in a variety of shapes, including hemispheres, spherical caps, cones, trumpets, and combinations. Cones have become almost standard, with hemispheres and trumpets occasionally used for special purposes. The results given in paragraphs 2-80 through 2-95 will be confined to simple cones, except for some brief remarks in paragraph 2-90 on double-angle cones and other unusual shapes. The gross factors affecting liner performance are the explosive charge (discussed in paragraphs 2-112 through 2-118), the standoff, and the diameter, angle, thickness, and material of the cone. It is generally assumed that a linear scaling relation exists for shaped charge performance (paragraphs 2-75, 2-123, and 2-132) and there is considerable evidence that a linear relation is valid. For this reason, liner dimensions in this chapter will be given in terms of cone diameters – the inside diameter of the base of the cone – and the diameter of the cone can be eliminated as a factor affecting the liner performance.* Details of liner design which affect liner performance include tapered walls, the base flange, and the presence of a spit-back tube. When a spit-back tube is not used, the configuration of the apex — whether sharp or rounded — seems to make little difference.

The effect of accuracy of the liner is discussed in paragraphs 2-84 and 2-85. The effect of accuracy of the complete round assembly will be discussed in paragraphs 2-108 and 2-109.

2-82. References. Where numerical data are given, references to the source of the data are usually provided. The references most commonly used are coded in the figures as follows:

NDRC Division 8 references are to interim reports for the period given.

E. I. du Pont de Nemours and Co. (Du Pont) references give the date of the report.

Carnegie Institute of Technology reports labeled CIT-ORD-No., are the "Fundamentals of Shaped Charges" series.

Firestone Tire and Rubber Co. reports, labeled FTRC No. are monthly progress reports.

2-83. Methods of Manufacture. Cones may be made by any of a number of processes. The most common methods used in the past are spinning, drawing, casting, machining from bar stock, and electroforming followed by machining.

a. Spinning. In the early days of shaped charge work, when the demands for cones were small, they were made by cutting a sector from sheet metal and rolling it to the desired shape, or by spinning. Such methods did not produce very good cones and were soon abandoned. However, it is reported that spun cones that compare very favorably with drawn cones can now be obtained. The poor performance of some of the early spun cones can now be better understood in the light of recent work which shows that spinning produces a small amount of built-in spin compensation.

b. Drawing. When the demand for cones became sufficient to justify the cost of dies, cones were made by drawing, and this is the method usually used today for production quantities. Its advantage is low cost. Its disadvantage is relatively lower accuracy, in those cases where extreme accuracy is required in the finished cone, than can be obtained by machining. Since the accuracy required is relative, it is sufficient for large cones, but may not be sufficient for small ones. This method is not usually suitable for small quantities of a given design on account of the cost of the dies.

c. Casting. Various methods of casting have been used. For a metal that shrinks when it freezes, casting by itself usually gives poor accuracy. If a suitable metal, probably an alloy, which yields accurate and homogeneous castings is found, casting may become an important method of manufacture for cones. It has been reported by Mr. G. C. Throner, formerly of the Naval Ordnance Test Station, that cones cast with Zamac 5, a zinc alloy, gave 6.2 cone diameters penetration in mild steel targets, which compares very favorably with copper cones. Work on this type of alloy is continuing at Firestone and at Ballistics Research Laboratories.

d. Machining. For a few cones of very high accuracy, or where the cost per cone is not of primary importance, machining from bar stock is preferable to the methods mentioned above. Annealing the bar before machining may be desirable. There may be some difficulty in the machining near the apex, especially on the inside. Because a cone is machined, it does not necessarily follow that it is accurately made; what is meant is that accurate cones can be made by this method if the required care is exercised.

e. Electroforming. Electroformed cones are deposited on an accurately made mandrel, and so have an accurate inner surface. If it is machined on the outside, means must be provided for chucking accurately and checking the mandrel when it is put back in the lathe to insure that the outside runs true with the inside.

*Carnegie Tech suggests the use of charge diameters or, better still, calibers as a fairer measure of performance. Caliber is an overriding limitation on penetration. One can get better results with a given cone diameter by increasing case thickness and hence confinement, or by increasing the charge diameter (see paragraph 2-93), which also gives the effect of more confinement. Either method, however, necessitates a larger caliber. (See also a. in paragraph 2-110.)

Carnegie Institute of Technology has reported favorable results with cones electroformed and peened, without machining.

2-84. Desirable Properties of a Liner. Aside from the fact that a givenmethodof manufacture may be suitable for use with one material and not suitable for another, the method of manufacture affects the quality of the cone in two ways: the accuracy of the cone, and its metallurgical properties.

a. Geometrical Accuracy. The formation of a shaped charge jet from the collapsing cone is a critical process. Ideally, the walls of the cone collapse and meet exactly on the axis of the cone. If, for any reason, one side of the cone collapses at a faster rate than the opposite side, they will not meet on the axis. This results, generally, in a crooked jet and the point of contact of the jet wanders on the surface of the target, giving impaired penetration. Thus, it is very important that sections of the cone perpendicular to its axis be true circles, with centers on the axis, and that the walls be of uniform thickness around a circumference. Uniform density of the metal is also required. Monotone variations in wall thickness along a slant height do not seem to be so important. Waviness along a slant height appears to be an undesirable characteristic. Axial symmetry in the explosive charge and the assembly are discussed in paragraphs 2-108, 2-109, 2-116, and 2-117.

b. Metallurgical Properties. The metallurgical properties of the liner depend strongly on the method of manufacture as well as on the material and heat treatment. The metallurgical problem is difficult to analyze on account of the extremely high pressures and rates of strain and the excessive amount of plastic strain. For these reasons it cannot be said that the properties of the jet are the same as the properties of the cone. Also, it must be remembered that the important properties are those under the high pressures and rates of strain mentioned above. That these may be very different from the properties under ordinary conditions is emphasized by the fact that glass cones give penetrations in concrete targets greater than might be expected from the "metallurgical" properties of glass. Nevertheless, it has been found possible to make some very interesting and important correlations between properties of the liner, principally crystal structure and melting point, and behavior of the jet. One of the most interesting features is a built-in spin compensation factor in certain cases, apparently resulting from an unusual crystal structure that gave poor penetration in static firings.

Theory indicates that for the fast moving portion of the jet penetration obtained is proportional to the length of jet and the square root of the jet density. The assumption that the jet density is the same as that of the cone is about as good a guess as any, if the jet is a continuous one. On account of the velocity gradient, the jet lengthens as it travels. The stretching of the jet eventually causes it to break up into a series of particles. Thus, if the jet did not break up into particles, its length and penetration would increase linearly with time and, consequently, with standoff.

Penetration standoff data show that penetration increases with standoff up to a maximum value of penetration, the corresponding standoff being called the "optimum" standoff (figure 2-14). Beyond the optimum standoff the average penetration decreases with standoff, while the best values of penetration approach an asymptotic value. The decrease in penetration from the linear value to the asymptotic value may be ascribed to breakup of the jet, while the decrease from the asymptotic value to the average value is due to increasing spread of the jet. Thus, for good penetration the jet should be capable

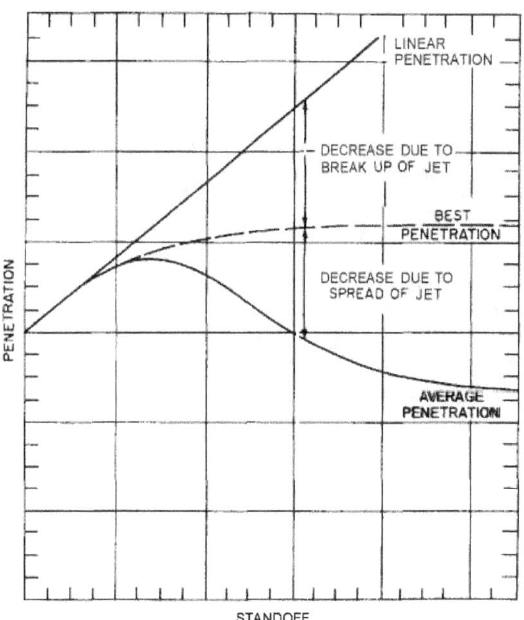

Figure 2-14. Degradation of penetration

of attaining a great length before breaking up. The ability to do this will be affected by the metallurgical properties of the jet. I' experimental penetrations are adjusted for the effect of density of the jet (cone), the following comparison is obtained:

Copper	100%
Aluminum	110%
Steel	75%
Zinc	65%
Lead	50%
Glass	40%

One may conclude that copper and aluminum have metallurgical properties superior for shaped charge cones, while lead and glass have inferior properties. A desirable material would have properties similar to copper and aluminum and a high density.

2-85. Experimental Results of Inaccuracies in the Liner. Tests performed by BRL and the Budd Co., on intentionally malformed cones, indicate that for maximum penetration from simple cones the axial symmetry of the liner should be as nearly perfect as can be obtained, especially near the base of the cone. Deviations from axial symmetry only reduce penetration. In a lot with random deviations, even small average deviations may result in alarge dispersion. Variations in wall thickness along a slant height are also important, but their effect isnot as serious as those around a circumference. The requirement of axial symmetry extends also to the explosive charge and the assembly as well as to the cone.

2-86. Tolerances. Tolerances on the cone dimensions have been recommended for some designs. This does not imply that there exists a tolerance within which the cone performs properly and beyond which it does not perform properly. Any deviation of the cone from axial symmetry will, in the long run, result in a degradation in performance, either in average penetration or variability, or both. However, it must be remembered that extreme accuracy in cone manufacture is very expensive. The tolerance allowed must be a compromise between the desire for top performance from the round and what one is willing to pay for it.

For the 90-mm T108 round the following tolerances were recommended in the Aberdeen Proving Ground Memorandum Report:

0.001-in. maximum wall variation in transverse plane;
0,005-in. maximum wall variation in longitudinal plane;
0.003-in. maximum waviness.

These were simple copper cones, 45" apex angle, 2 3/4 -in. diameter, 0.062-in. wall thickness.

Recommended tolerances for wall thickness of the blank for 57-mm and 105-rnm cones (fluted cones, rotated) are given in paragraph 2-138. The 57-mm liners are about 1 11/16-in. diameter, 0.054-in. thickness; the 105-mm, 3 1/4-in. diameter, 0,100-in. thickness. These tolerances seem a little more liberal than those quoted for the 90-mm. Since the latter were determined from firings with inaccuracies of 0.004 in. to 0.005 in. of wall thickness, it might be possible to bring the 90-mm tolerances in line with the 57-mm and 105-mm. General observations of miscellaneous tests indicate that, for 3/4 -in. diameter, 45° electroformed cones, variation in wall thickness should be less than 0.001 in. for 0.025-in. thick cones and not more than 0.002 in. for 0.050-in. thick cones.

One might expect tolerances on wall thickness to increase with the thickness. Experience with small cones (3/4-in. diameter) has indicated that a 4 percent inaccuracy on 0.025-in. cones is more damaging than a 4 percent inaccuracy on 0.050-in. cones. However, this may be due to the ease with which thin cones become out-of-round due to handling. It was found that particular care must be taken with 0.015-in. cones to prevent this. The writer has seen no discussion about whether large-diameter cones require smaller percent tolerances than small-diameter cones, though this question merits consideration.

2-87. The Effect of Design Parameters on Penetration. The effect of standoff,, cone thickness, and cone angle will be presented in the form of graphs. These curves are based on data published by various groups of investigators working at different places and different times. Under these conditions, differences in results are to be expected. It is to be remembered that cones available in the early days of shaped charge investigation were of relatively poor manufacture. As the importance of :accuracy became known and methods of manufacture improved, the quality of cones improved.;this resulted in increased average penetrations and smaller dispersions, especially at the longer standoffs.

This subject is most readily divided into early work and recent work. Where cyclotol and pentolite are quoted as the explosive, the usual compositions, 60/40 RDX/TNT for cyclotol and 50/50 PETN/TNT for pentolite, were used, unless otherwise stated.

2-88. Early Work With Simple Cones. Figure 2-15 shows penetration of steel cones into mild steel targets for 30" cone angle and various cone thicknesses. This work was done in the early part of the war and the quality of the cones was probably not too good. Cones were, generally, 1 5/8-in. diameter, cast in 1 5/8-in. unconfined pentolite charges of 4 to 5 in. length, and fired statically at zero obliquity. Each point was the average of five shots, except where shown on the graph. Curves were drawn by eye, with some attempt made to keep all curves of one family of the same general form. For most of this work, the dispersion, especially at long standoffs, was large.

The general characteristic of penetration-standoff curves for cones of early manufacture is a maximum penetration at a small optimum standoff, the penetration decreasing sharply for larger standoffs. The optimum standoff increases with the cone angle.

Similar curves can be plotted for larger angle cones from the data given in the NDRC Division 8 interim reports and the Du Pont reports. From these data, the maximum penetration for any standoff is plotted in figure 2-16 for the different cone angles. Figure 2-17 shows the optimum standoff and thickness as a function of the cone angle. Since steel is not regarded favorably as a liner material, these data are of limited usefulness, but material shortages in an all-out war may force the use of steel again. The data do show that standoff, cone thickness, and cone angle are interrelated, and that if any of these is changed it may be necessary to change the others for best results. For example, a comparison of cones of different angle may be biased, unless the thickness is also changed.

Figures 2-18 and 2-19 show penetration as a function of standoff for 45" copper cones. Here there is a distinct difference between the two sets of data at the longer standoffs. The reason for this difference is not known positively, but it is probably due to a difference in quality of the cones, since recent work with cones of high accuracy tends to confirm the Du Pont data.

Figures 2-20 and 2-21 give the results for aluminum cones. They show the characteristic

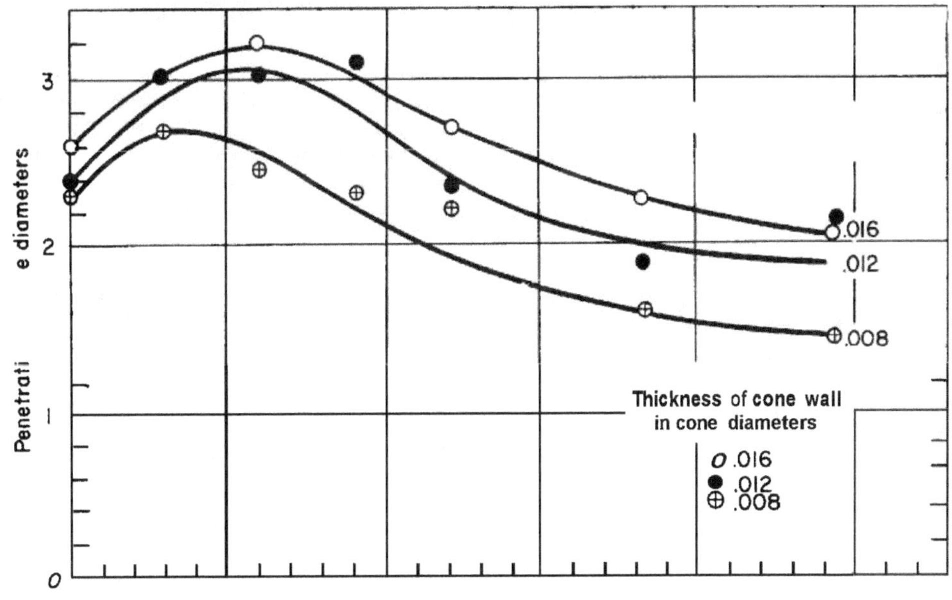

property of aluminum: the fact that penetration holds up well at long standoff. Considering its low density, one might expect that aluminum liners should be thick as compared to steel or copper, but figure 2-20 does not show any advantage for the thicker liners.

Figures 2-22 and 2-23 show the performance of zinc and lead cones, neither of which are of considerable importance at present. However, as mentioned previously, the Naval Ordnance Test Station has recently reported, informally, excellent results from a castable zinc alloy.

2-89. Recent Work With Simple Cones. Figures 2-24 through 2-27 show penetration as a function of standoff for copper, steel, and aluminum cones of a constant thickness for cone angles of 22°, 44°, 66°, and 88". The tendency of aluminum to maintain its penetration with increasing standoff is evident, as is also the tendency for optimum standoff to increase with cone angle. These curves do not necessarily show optimum results, since thickness may not be the best for some angles.

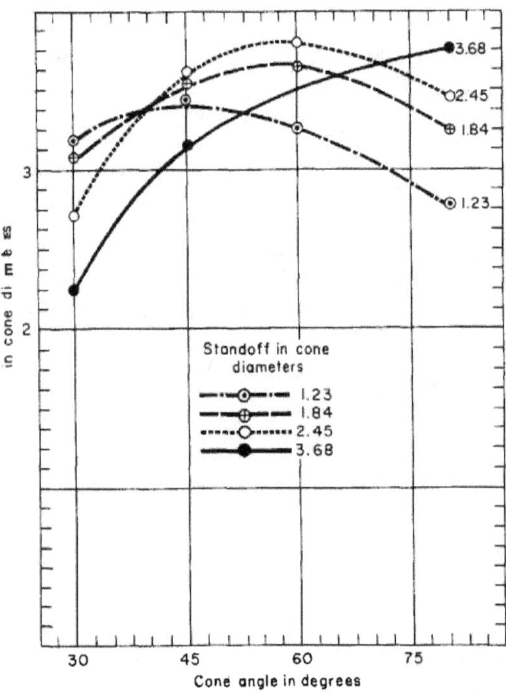

Figure 2-16. Penetration versus cone angle for steel cones and mild steel targets

Figures 2-28 through 2-30 show results obtained under conditions very different from those for the previous data. These charges were fired in shell bodies or cases closely simulating shell bodies. Thus, the explosive charge was short, in comparison with its diameter, and fairly heavily confined. The accuracy of the cones was very good. Figure 2-28 shows a very good penetration, fairly flat penetration-standoff curve, and a long optimum standoff. Figure 2-29 shows an optimum cone thickness of 3 percent, which is somewhat heavier than that for unconfined charges. This is in agreement with the general observation that, if the charge diameter is increased or the charge is confined, the thickness of the cone should be increased for optimum penetration. Figure 2-30 gives the results of varying the cone angle under different conditions of constant explosive loading. The results are, therefore, not of general application. A penetration-standoff curve obtained from firings at the Ballistic Research Laboratories is given in figure 2-31. These were drawn, 105-mm cones of good accuracy, confined in shell cases and fired against mild steel targets. Penetrations were unusually good and held up well at long standoff.

Figures 2-32 through 2-34 give the results of recent firings at the Ballistic Research Laboratories. These were small cones, 0.750-in. inside diameter, of electroformed copper machined on the outside to about 0.0005-in. tolerance. They were fired in unconfined pentolite charges of a diameter 20 percent greater than the cone diameter, and of a sufficient length (two cone heights above the apex) to insure that penetration was not restricted by short charge length. The cone thickness was much greater than that usually used. Figure 2-32 gives penetration as a function of standoff for 30" cones. Similar curves were obtained for 20°, 46°, and 60° cones. The results of most interest are those for the small-angle cones, which gave excellent penetration and a large optimum standoff. This is at variance with previous results for steel cones of smaller wall thickness and with charges the same diameter as the cone, for which both optimum standoff and maximum penetration decrease with cone angle (compare figures 2-16 and 2-17 with figures 2-33 and 2-34). However, figure 2-15 does show the penetration increasing with cone thickness even up to the maximum thickness fired. Figure 2-33 shows that the cone thickness is not critical for 46" cones; this is not true for other angles, especially 20" cones. Figure 2-'34 shows maximum penetration as a function of cone angle. It is thought that the differences in optimum standoff given are not of any significance, since no definite trend is shown and the penetration standoff curves are fairly flat. The optimum thickness for 46" seems a misfit. It is possible that a trend in thickness might be shown if intermediate thicknesses between 0.033 and 0.066 had been used.

2-90. Bimetallic Cones and Nonconical Shapes. The term "bimetallic cones" will-be applied to cones consisting of an inner layer of one metal with an outer layer of another metal. It does not include cones the apex end of which is of one metal, the base end of another metal. In most cases the composite cone consists of two separate cones nesting in intimate contact. However, in the case of copper-clad steel, the two metals are bonded together. Table 2-10 lists penetrations into mild steel targets by 45" cones 1.63 in. in diameter. For comparison, results from single metal cones are given in each case.

Nonconical shapes include hemispheres and spherical caps, trumpets, and combinations.

Table 2-10

Cone Thickness				Standoff (c.d.)	Penetration (c.d.)
Outside (in.)	Center (in.)	Inside (in.)	Total (c.d.)		
Steel 0.017	...	Aluminum 0.040	0.035	7.4	2.6
...	Aluminum	...	0.036	7.4	2.6
Steel 0.036	...	Aluminum 0.036	0.044	5.5	3.5
...	Aluminum	...	0.036	5.5	2.9
Steel 0.018*	...	Copper 0.012	0.018	0.9	4.0
Copper 0.005*	Steel 0.018	Copper 0.018	0.025	1.0	4.0
...	Copper	...	0.024	1.0	4.3
Copper 0.010**	Steel 0.029	Copper 0.025	0.031	1.2	4.3
...	Copper	...	0.031	1.2	4.6
Steel 0.036	...	Copper 0.036	0.044	2.5	4.7
...	Copper	...	0.030	1.4	5.0
Steel 0.018	...	Cadmium 0.006	0.015	0.9	3.0
...	Steel	...	0.015	1.3	3.5

*Copper-clad
**Copper-clad, 42°, 2.07-in. dia.

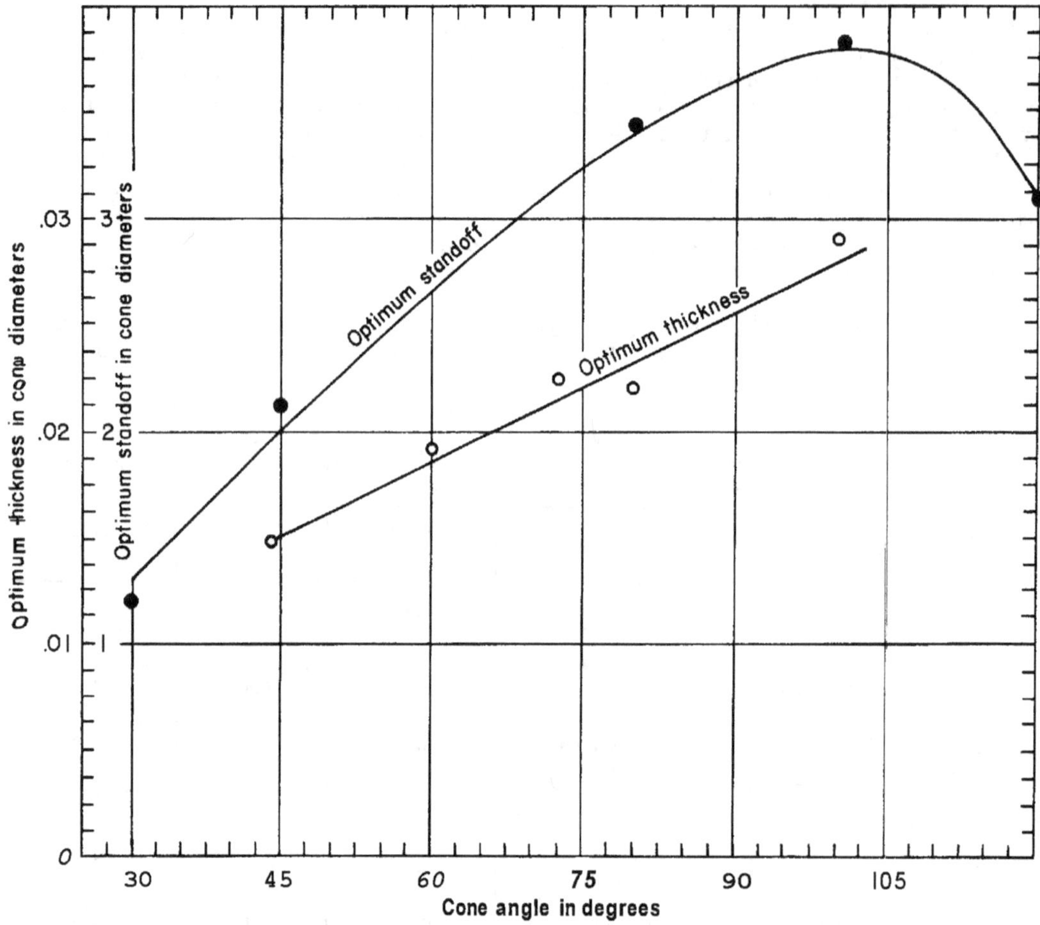

Figure 2-17. Optimum thickness and optimum standoff versus cone angle for steel cones and mild steel targets

Radiographs indicate that hemispheres do not collapse, with the formation of a jet, as do cones; they turn inside out before collapsing, the whole liner being projected as a stream of particles. Spherical caps (segments) are fragmented and projected as a cluster of particles which may be more or less focused, depending on the curvature. Some work has been done with spherical segments, especially by the British. The results were, generally, poorer than those from hemispheres.

Interest in double-angle cones has been revived recently, due largely to certain advantages shown for the French 73-mm round. Early firings of double-angle cones, in which the change from one angle to the other was made abruptly, did not show any increase in penetration for the double angle. In the French 73-mm round, the change from one angle to the other was made smoothly and the liner wall was tapered. This round gave peak performance at normally available standoffs.

An almost infinite variety of combinations is possible. Complete coverage is not warranted, since, generally speaking, the penetrations achieved from them are inferior to those from cones.

2-91. The Effect of Tapered Walls on Penetration. The British suggested that, since the thickness of liners should scale as the diameter, a cone would logically be thicker at the base

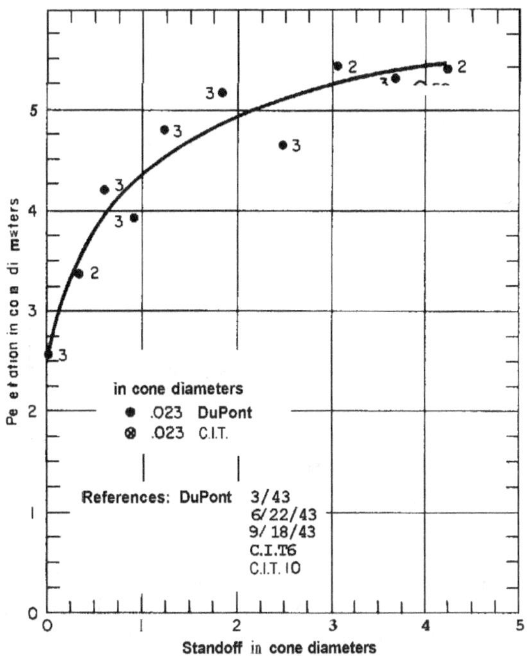

Figure 2-18. Penetration versus standoff for 45° copper cones and mild steel targets

than at the apex. This idea has been followed up by several groups of investigators.

In general, it does not seem likely that any very appreciable improvement in performance

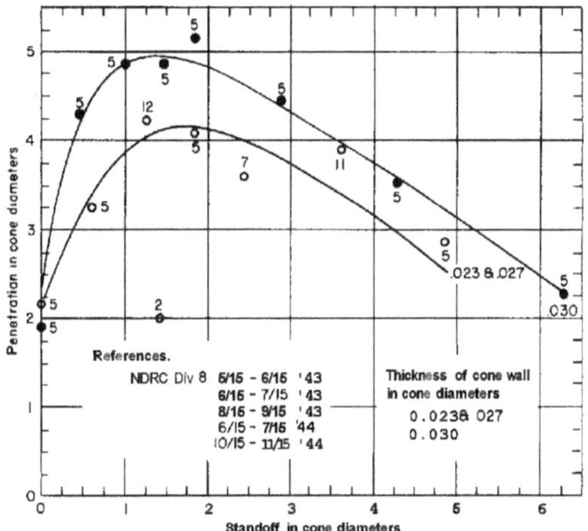

Figure 2-19. Penetration versus standoff for 45° copper cones and mild steel targets (no confinement)

can be obtained by the use of tapered-wall cones. However, it is possible that if the right combination of thickness and taper can be found, improved results may be obtained.

2-92. Wires and Other Obstructions Within the Cavity. A very large number of tests have been conducted to find the effect of wires, coils, simulated firing pins, and so on, placed within the cavity of the cone or on the axis in front of the cone. Practically all of these items were in

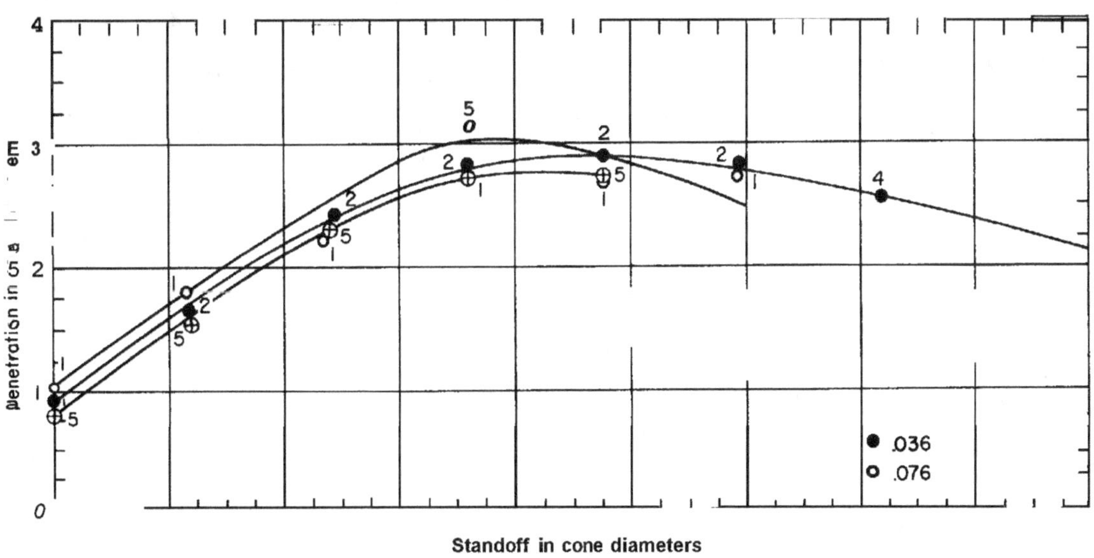

Figure 2-20. Penetration versus standoff for 45° aluminum cones and mild steel targets

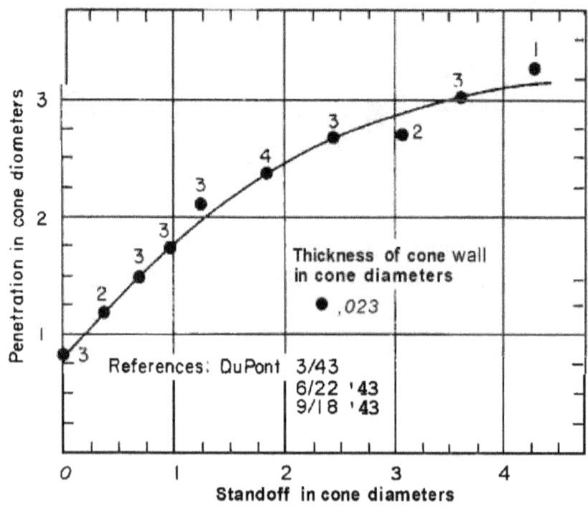

Figure 2-21. Penetration versus standoff for 45° aluminum cones and mild steel targets

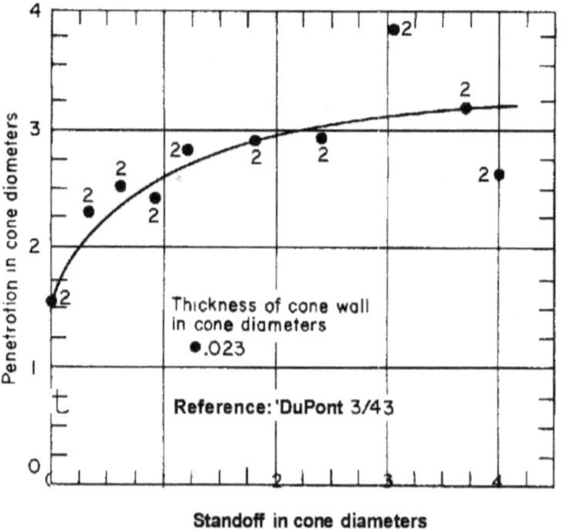

Figure 2-22. Penetration versus standoff for 45° zinc cones and mild steel targets

connection with fuzing. All such obstructions almost invariably cause very serious impairment of the penetration, often as much as 50 percent.

2-93. Flanges. The effect of the base flange of the cone on the jet formation is somewhat curious. The data given in table 2-11 were reported by Du Pont workers for M9A1 steel cones 45°, 1.63-in. base diameter, flange 2.0-in. diameter, unconfined.

Table 2-11

Dia. of explosive (in.)	Penetration, mild steel 1-in. standoff (in.)	No. rounds
1.63	5.45	4
1.75	5.70	5
1.88	4.50	5
2.00	4.00	5

As the diameter of the explosive is increased beyond the cone diameter, there is a slight increase in penetration, followed by a decrease. Similar effects have been noted by other experimenters. The Carnegie Institute of Technology explains this effect as follows. The velocities and the velocity gradients along a jet are quite sensitive to the times of arrival of the release wave at the liner. Since the release wave is initiated at the charge boundary, any change in the geometry will consequently cause a change in the velocity and in the velocity gradient. With the larger diameter charge, which has an explosive belt, the major portion of the release wave is initiated on the lateral surface, but a small portion is initiated along the base of the explosive shoulder during the later stages of the cone collapse process. This small portion of the release wave produces a greater gradient in the velocities of the rear elements of the jet, which contain a large fraction of the jet mass. To be of benefit, the magnitude of this gradient should be neither too great nor too small. If it is too large, this portion of the jet will break up quite rapidly and become ineffective. If it is too small, proper

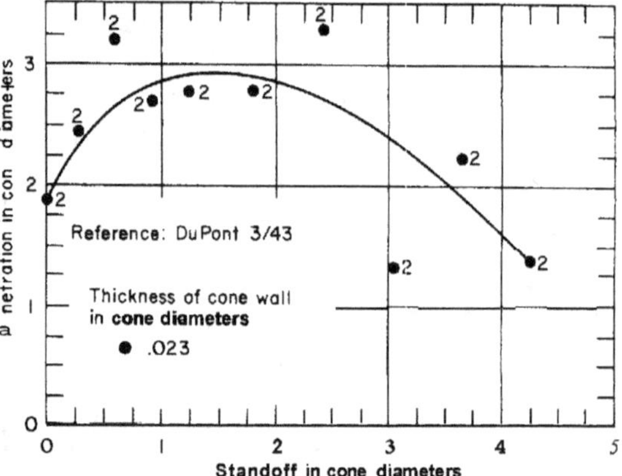

Figure 2-23. Penetration versus standoff for 45° lead cones and mild steel targets

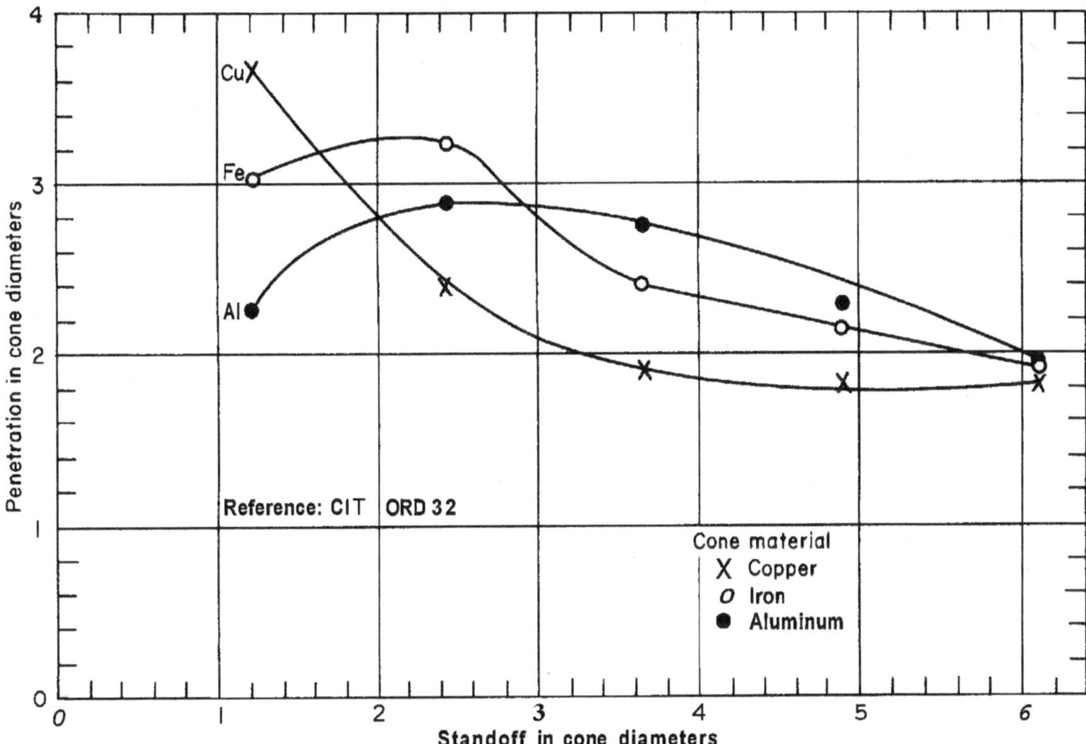

Figure 2-24. Penetration versus standoff for 22°, 0.023 c.d. thick cones and mild steel targets

lengthening will not be achieved for efficient penetration. This, in essence, explains the observed optimum charge diameter for a liner of given base dimension.

2-94. Effect of Spit-Back (Flash-Back) Tubes. For some types of fuzing, a small tube, called a spit-back tube, is attached to the apex end of the cone, extending away from the cavity. The portion of the apex inside the spit-back tube is removed. For M9A1 steel cones in unconfined charges, the presence of the spit-back tube caused little change in penetration, or a slight decrease. For copper liners in confined charges, there was no change, or an increase up to 20 percent.

2-95. Effect of Annealing. Results of tests to determine the effects of annealing and of hardening steel cones show that the penetration from drawn M9A1 cones is not changed by annealing, but that the penetration becomes progressively less as the hardness is increased. Drawn copper cones show no change with annealing. Results at the Ballistic Research Laboratories show that cast beryllium copper cones, whose normal penetration was low, were improved by annealing; that electroformed copper cones were not affected by annealing, except that when the annealing temperature was increased to 1,400°F the cones blistered on the inner surface, with a decrease in penetration; and that cones made by a shear forming process, which worked the metal so severely that its structure was impaired, were improved by annealing.

THE UNFUZED WARHEAD

2-96. Introduction. To accomplish its mission satisfactorily a projectile must be capable of defeating its assigned target, and of flying accurately enough to assure a high probability of striking the target with the first or second round fired. While flight of the projectile is properly the problem of the exterior ballistician and the destructive capacity that of the terminal ballistician, the requirements of accuracy and of destructive capacity are so often at variance that the designer is compelled to make compromises and to attempt to arrive at the best overall balance of accuracy and destructive potential. The effect of various design parameters will be discussed here only in relation to the limitations imposed upon the designer by the realities of practical shell design. An effort will be made to trace the development of a typical warhead and to point out the general areas where successful compromises have been made.

2-97. Selection of Weapon Type and Size. The design for a specific type of shaped charge missile begins with the performance specifications for the weapon system. The range, accuracy, and mobility requirements determine the velocity of the projectile and the type of weapon required, while the maximum penetration determines the minimum size of liner and charge. Table 2-12 contains typical accuracy, range, and velocity data for existing shaped charge projectiles. The range shown in the table is that which offers a fairly high probability for a first-round hit on a small target (such as a tank) and is not to be confused with the maximum range of the missile.

The caliber or size of the missile depends upon its required destructive capacity, its peak acceleration, and the type of weapon from which it will be fired. Although perforation of the target is a necessary condition for the defeat of a target, it is not sufficient. Unfortunately, very little is known about the effect of shaped charge design parameters on the extent of damage beyond penetrable armor. The work which has been reported is insufficient for establishing adequate criteria for shaped charge effective-

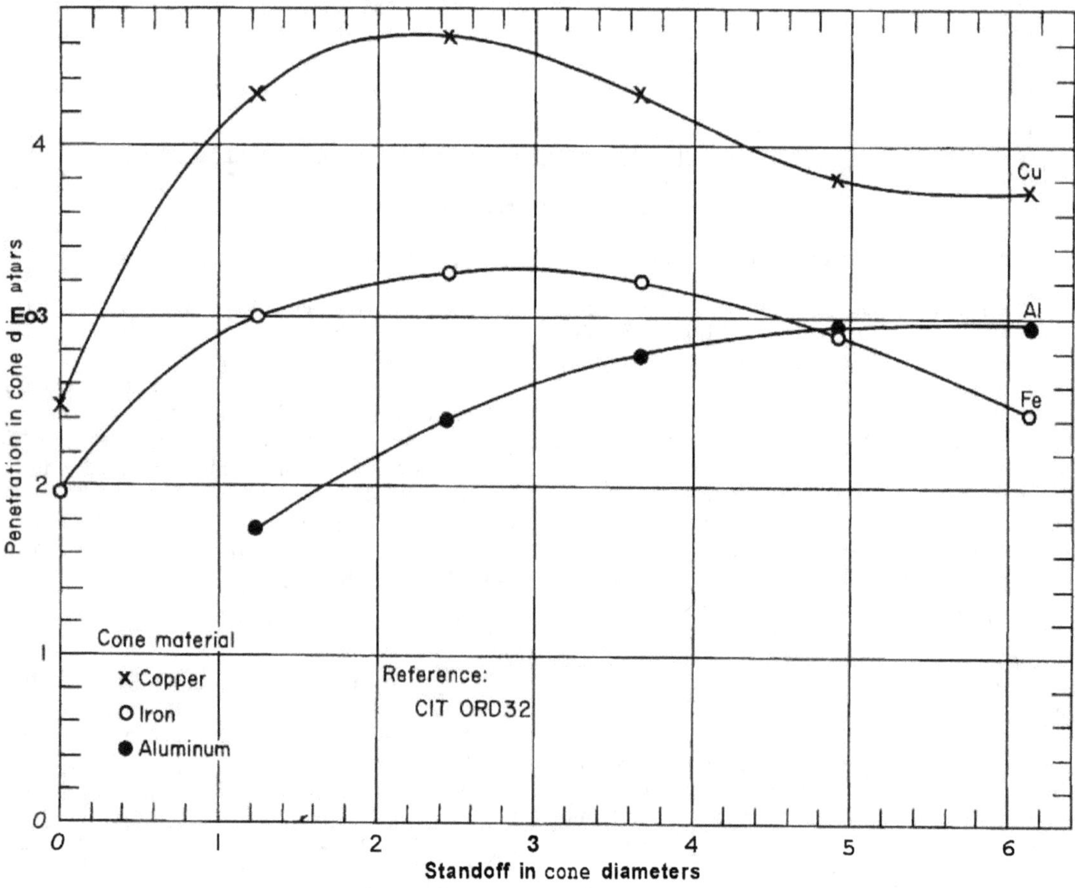

Figure 2-25. Penetration versus standoff for 44°, 0.023 c.d. thick cones and mild steel targets

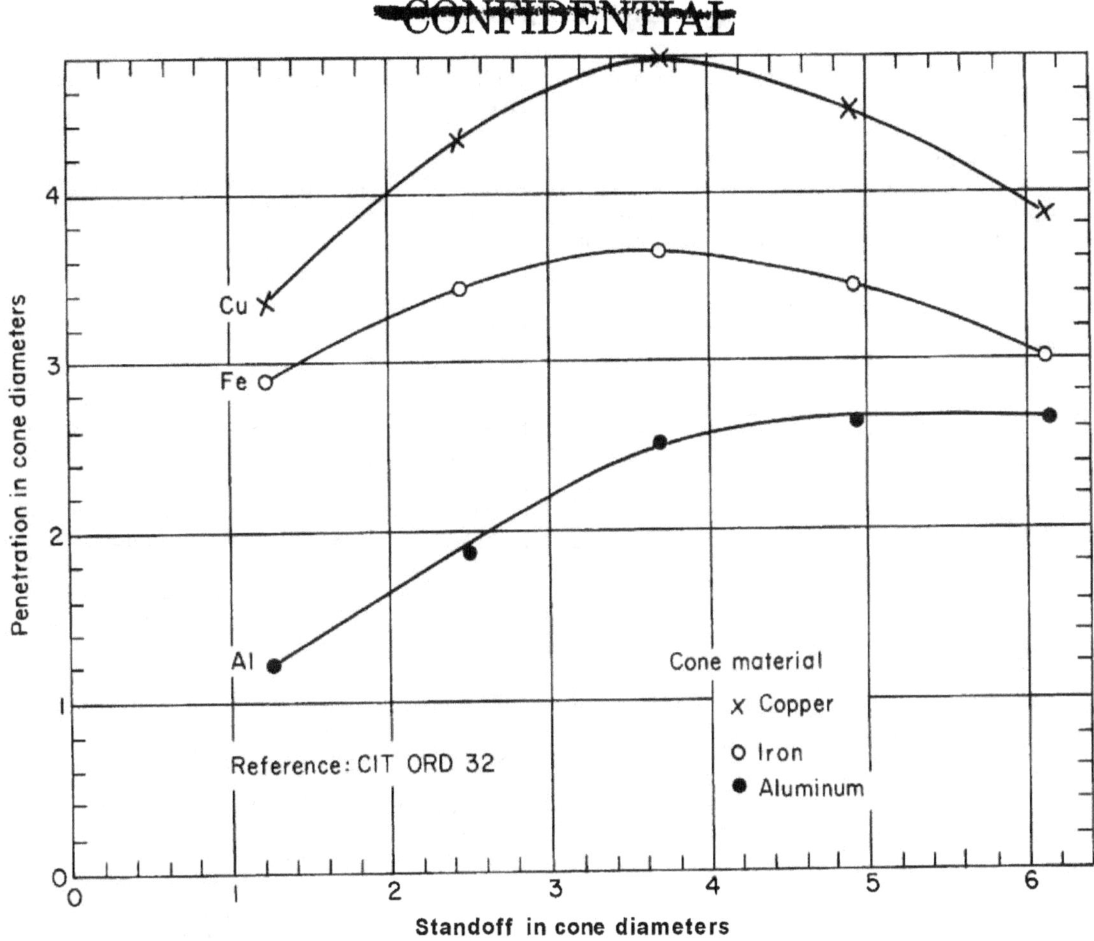

Figure 2-26. Penetration versus standoff for 66°, 0.023 c.d. thick cones and mild steel targets

Table 2-12

Type	Range (yards)	Velocity (fps)	Typical ballistic accuracy (probable error)
Bazookas and grenades	100 to 400	150 to 250	± 2 mils
Rockets (fixed launchers)	1,000 to 4,000	1,000 to 4,400	± 1 to 2 mils
Recoilless rifles	500 to 2,000	500 to 2,200	± 0.5 mil
Guns and howitzers	1,000 to 2,000	1,000 to 4,000	± 0.15 to 0.30 mil

ness. It has become customary, therefore, to evaluate shaped charge effectiveness on the basis of depth of penetration and relative hole volume, and to trust that a provision for some arbitrary "residual" penetration — usually 2 in. of homogeneous armor — will be enough to assure the defeat of the target. Additional studies of shaped charge effectiveness should certainly prove to be fruitful avenues for futher research.

The maximum thickness of the armor to be penetrated, without provision for any "residual" penetration, quite clearly establishes the minimum diameter of the liner and charge. Based upon present standards of shaped charge performance, the minimum diameter of an unrotated copper cone (D in.) required to penetrate a given thickness of armor (T in.) 90 percent of the time, without provision for residual damage effect, may be estimated quite well by equation (7).

$$D = \frac{T+2}{5} \text{ in. homogeneous armor, BHN 300} \qquad (7)$$

Table 2-13 shows the penetration level above which 90 percent of the rounds would fall for cones and charges of various diameter.

Table 2-13

D (in.)	T (in. armor)	Approximate projectile caliber (mm)
2.5	10.5	75
2.6	11.0	
2.7	11.5	
2.8	12.0	
2.9	12.5	
3.0	13.0	90
3.1	13.5	
3.2	14.0	
3.3	14.5	
3.4	15.0	
3.5	15.5	105
3.6	16.0	
3.7	16.5	
3.8	17.0	
3.9	17.5	
4.0	18.0	
4.1	18.5	120
4.2	19.0	
4.3	19.5	
4.5	20.0	

After selecting the type of weapon and velocity from the accuracy and portability requirements, and the minimum size of cone and charge from equation (1) and table 2-12, the caliber of the missile may be determined from the thickness of the projectile walls required to withstand the stresses of firing.

Before proceeding with a more quantitative discussion of the effect of the various design parameters on the penetrating potential of the charge, we summarize the present state of the development of the hypothetical projectile. The type of weapon, peak gun pressure, acceleration forces, muzzle velocity, projectile caliber, cone size, projectile wall thickness, and the material of construction of the projectile body have been tentatively defined. All have been fixed as a result of a consideration of the specifications for weapon weight, weapon accuracy and range, and by the projectile penetration requirements. The type, shape, material, wall thickness, apex angle and method of mounting of the cone, the amount and distribution of the explosive, the size and positioning of the booster, provision for the fuze, and the manufacturing precision required for obtaining the shaped charge performance predicted in table 2-13 remain to be determined.

2-98. Consideration of Liner Parameters. Although the effect of various liner parameters — shape, material, wall thickness, cone angle, and so on — are described in detail in paragraphs 2-80 through 2-95, the projectile designer is not free to treat these parameters independently. He must be guided in his choice by the projectile parameters fixed by the requirements for projectile accuracy. It is therefore quite appropriate to treat each of these parameters here in order to illustrate the manner in which a judicious choice of each of these parameters may be related to the boundary considerations of the projectile design.

2-99. Standoff Distance. In a real projectile the effective standoff distance is determined by the length of the ogive, the velocity of the projectile, and the fuze functioning time. Although the optimum standoff distance for a well-made conical liner is usually more than four cone diameters, the actual standoff is usually limited to from one to three cone diameters by aerodynamic considerations involved in ogive shape and size. However, this standoff may be enough to permit the attainment of about 90 percent of the penetration expected at optimum standoff. The shorter standoff has advantages in certain instances. For example, if the shell must be rotated at some low spin rate, 10 to 15 rps, in order to achieve projectile accuracy, the optimum standoff may be reduced from four to two cone diameters. Also, if the enemy employs spaced armor in an effort to reduce the efficiency of shaped charge projectiles, the spaced armor itself may provide the increased effective standoff required for maximum penetration.

2-100. Charge Length. The length of the projectile body, and hence of the charge, is most frequently limited by aerodynamic performance and projectile weight specifications. In general, the penetration and the hole volume obtained increase with increasing charge length and reach a maximum at about 2 or 2.5 charge diameters for heavily confined charges or four charge diameters for lightly confined or unconfined charges (refer to paragraph 2-115). In most cases involving rockets or projectiles a charge length of two charge diameters can be provided, and this is sufficient if the charges are subjected to close quality control during manufacture and loading. The usual effect of reduced charge length is a lowered average penetration, reduced hole volume, and an increased number of rounds with below-normal penetration.

2-101. Charge Shape. Existing shaped charge designs usually have one of the shapes shown in figure 2-35. Although each can be made to perform satisfactorily, (a) has the advantages of somewhat greater ease of manufacture, high explosive loading, and blast effect (because of the larger amount of explosive); (b) and (c) are sometimes necessitated by the requirements for accuracy. There is some slight evidence that a tapered charge has a shorter optimum standoff and a slightly lower maximum penetration than the cylindrical charge. The greater amount of explosive in the cylindrical charge makes it more valuable than the tapered charge for the secondary effects of blast and fragmentation damage.

2-102. Selection of Liner Material. Although depth of penetration is not the only criterion for judging the maximum damage to the target, there is only limited information available as to the relative damage beyond the target caused by target penetration by liners of different materials. Such information as there is indicates that the relative damage decreases in the order aluminum, steel, and copper. The relative penetrating ability of various materials is described in paragraph **2-84,** and has been the subject of many investigations. If the type of weapon is such that the caliber of the projectile overmatches the penetration requirement for the defeat of the prospective target, a most desirable circumstance, it may be possible to select aluminum or steel in preference to copper. But if, as is most frequently the case, the penetration requirement taxes the penetrating

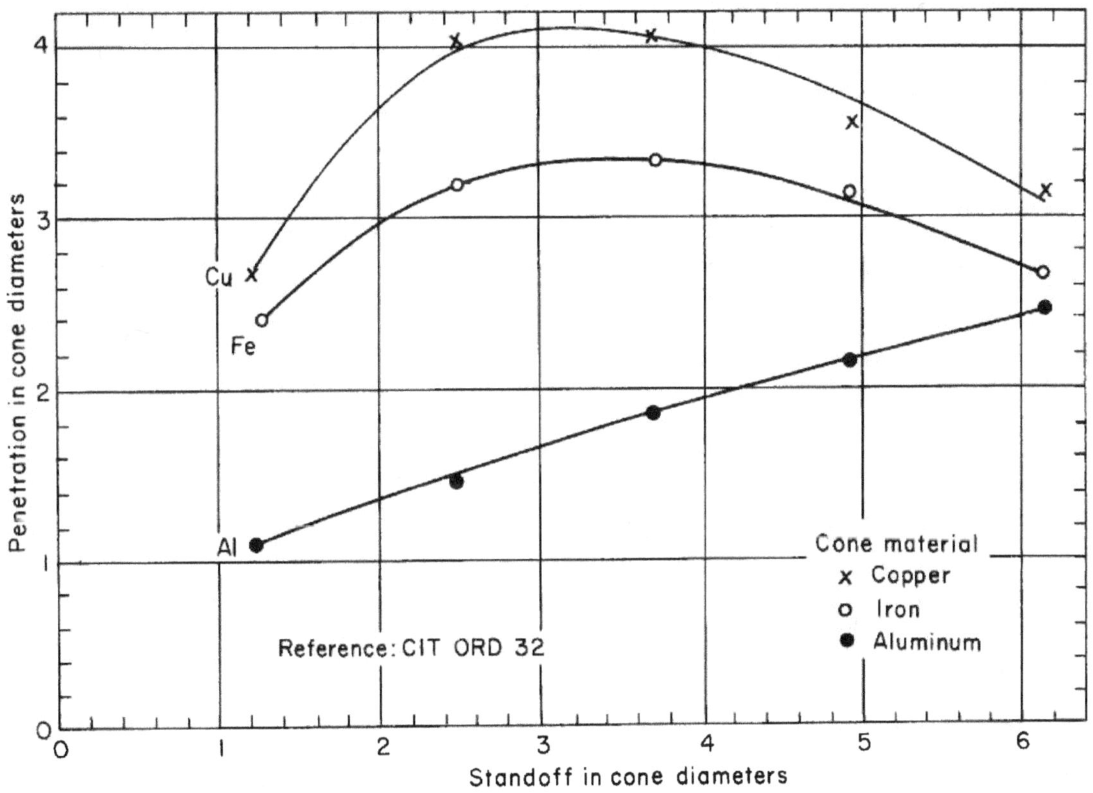

Figure 2-27. Penetration versus standoff for 88°, 0.023 c.d. thick cones and mild steel targets

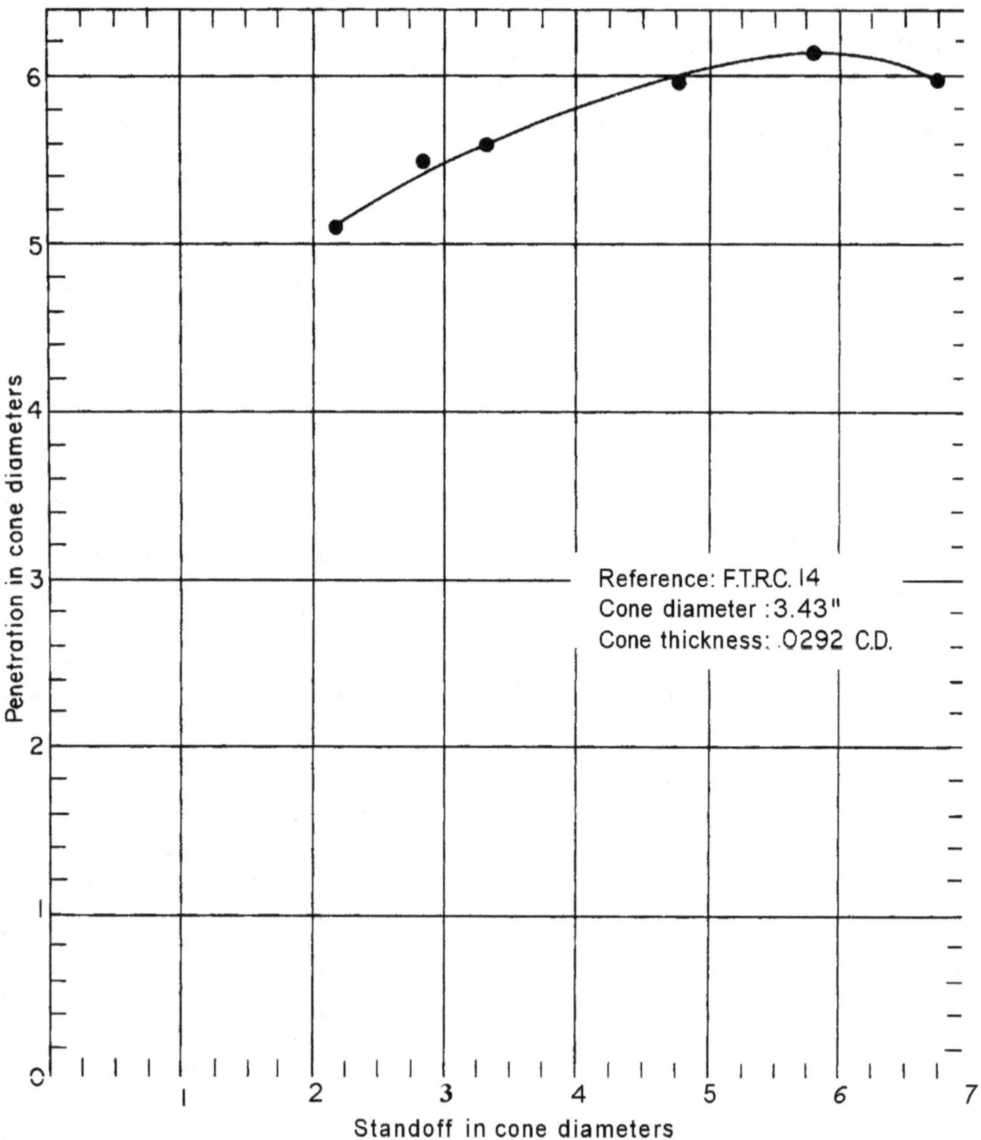

Figure 2-28. Penetration versus standoff for 45° copper cones

ability of the projectile, only copper can be considered seriously. In this case aluminum sleeves or bimetal cones should be considered.

Having reached a decision as to the type of cone material to be used, it is necessary to specify the composition or alloy. Although only very scanty evidence can be cited to support this, it is the considered opinion of those closely associated with the art of shaped charge design that the purity of the material, or the type of alloy to be specified, should be that which has the greatest ductility. This conclusion is deduced from the fact that potential depth of penetration is governed by the length and density of the jet. The density of the jet is, of course, characteristic of the type of material used and is only slightly influenced by impurities and alloying ingredients. The length of the jet is determined not only by the size of the cone but

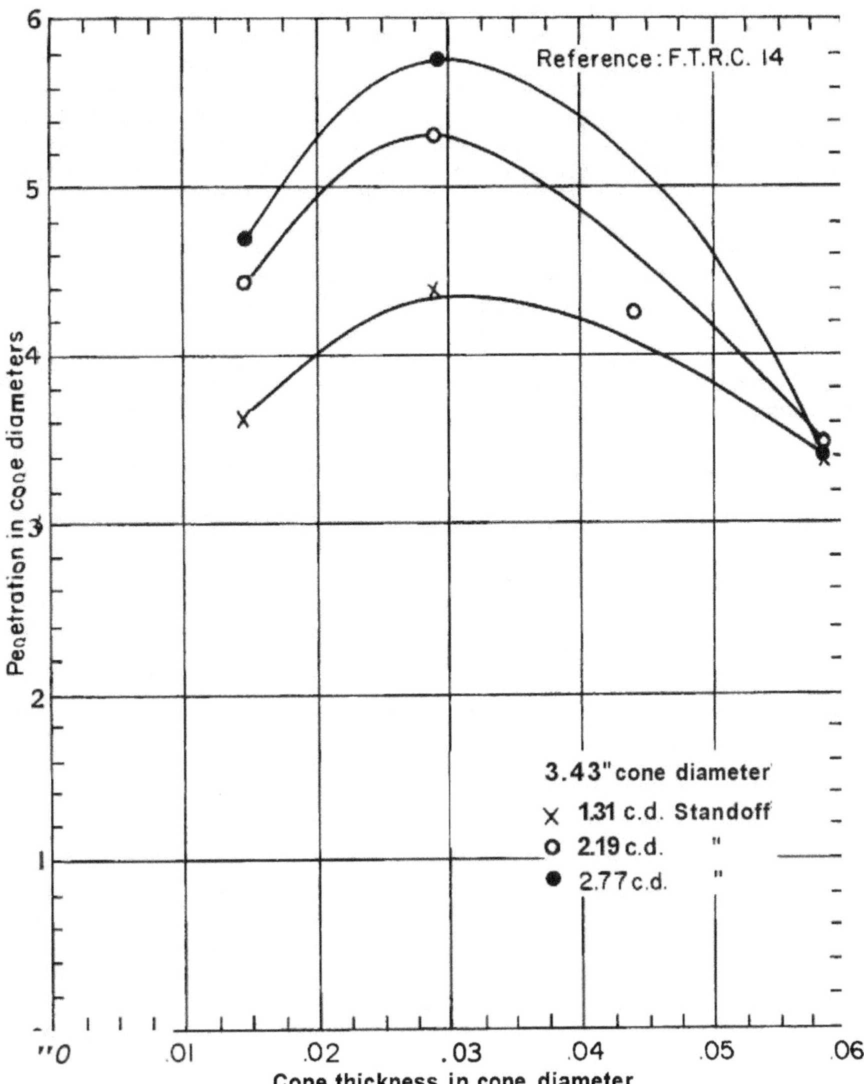

Figure 2-29. Cone thickness versus penetration for 45° copper cones

also by the velocity gradient resulting from the design of the charge, and by the ability of the ductility of the jet material to support the velocity gradient, during elongation of the jet, without rupturing. This effective jet ductility is, of course, dependent upon the inherent ductility, the strength, and the melting point of the material. Much more work has to be done before the influence of these factors is understood. At this time, however, the best choice of material for shaped charge liners is believed to be oxygen-free electrolytic copper.

2-103. Liner Shape. Liners of many different geometric shapes have been tested for penetrating efficiency (paragraphs 2-90 and 2-91), but experience in the United States seems to indicate that the best and most consistent results can be obtained with conical liners of appropriate apex angle and wall thickness. Some very recent data indicate that double-angle conical liners may offer certain advantages, but the performance of these liners has not yet been determined sufficiently for a complete evaluation of their true worth.

2-104. Cone Wall Thickness. For each type of cone material, standoff, projectile wall confinement, type of explosive, shape of charge, and apex angle there is an optimum wall thickness. From the practical consideration of projectile design, however, projectile confinement and cone apex angle are most determining.

Figure 2-36 shows reasonable values of liner wall thickness for copper cones with apex angles between 40" and 45° plotted as a function of the confinement. As a guide for liners of different apex angles, or for shapes other than conical, an approximately correct wall thickness may be obtained by maintaining the thickness constant in the axial direction (figures 2-17 and 2-29).

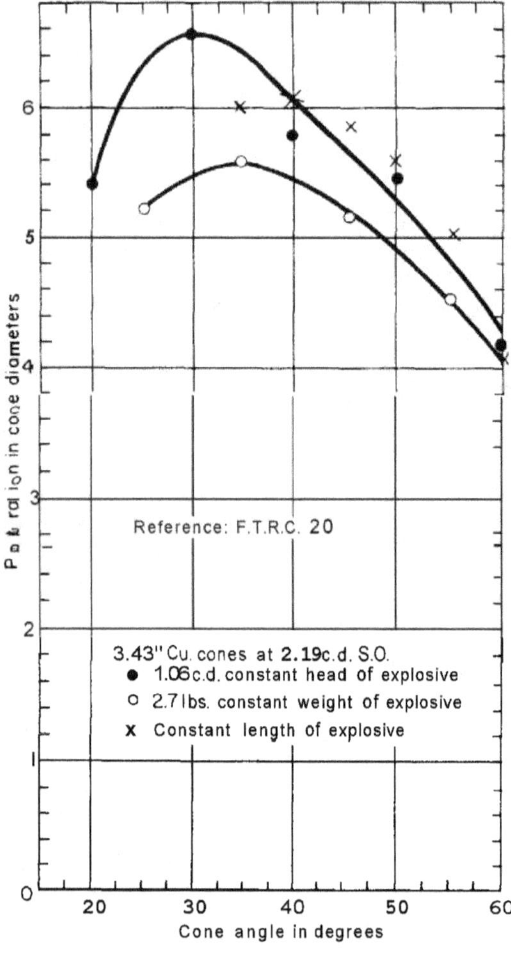

Figure 2-30. Penetration versus cone angle at constant standoff

Curves of penetration vs. wall thickness are frequently unsymmetrical. A thicker wall is generally to be preferred over a thinner wall. The performance of the latter is typified by an excessive variability from charge to charge, the former by good reproducibility with only a tolerable decrease in penetration. It would, therefore, seem to be good practice to select a wall thickness about 5 percent greater than the optimum to assure that in production the wall thickness will not be less than optimum.

Cones with tapered wall thickness have been studied from time to time (paragraph 2-91). Though more work in this field is desirable, the available evidence indicates that tapered walls offer slight, if any, real improvement in the performance of conical liners. The data do show, however, that rather wide tolerances may be placed on the variation in wall thickness between apex and base without reducing penetration, provided the wall thickness is held constant at each transverse section of the cone.

For liners of shapes other than conical (double-angle, hemispheres, trumpets, and so on) the observation that optimum wall thickness depends upon the inclination of the surface indicates that in such cases tapered wall cones may be advantageous.

2-105. Cone Apex Angle. The choice of cone angle is quite important, both from a performance and a manufacturing point of view. Data are available which show that optimum standoff increases with increased apex angle up to about 65°; optimum standoff then decreases as the apex angle is increased. However, the optimum standoff is also dependent upon the cone material, wall thickness, and charge length (figures 2-24 through 2-27, and 2-34). As with most other cone parameters, the effect of apex angle becomes less important as the spin rate of the projectile increases. For example, at 0 rps a 45°, 3.4-in. copper cone penetrates 3 in. deeper than a 60" cone of the same wall thickness, but at 45 rps the difference is less than 1 in.

The penetrating performance of small apex cones (20") is characterized by lowered efficiency and increased deviation of scatter of the data. It is probable, however, that this merely reflects the difficulty of manufacturing good cones of very small angle (figure 2-34). With the precision of modern manufacturing methods

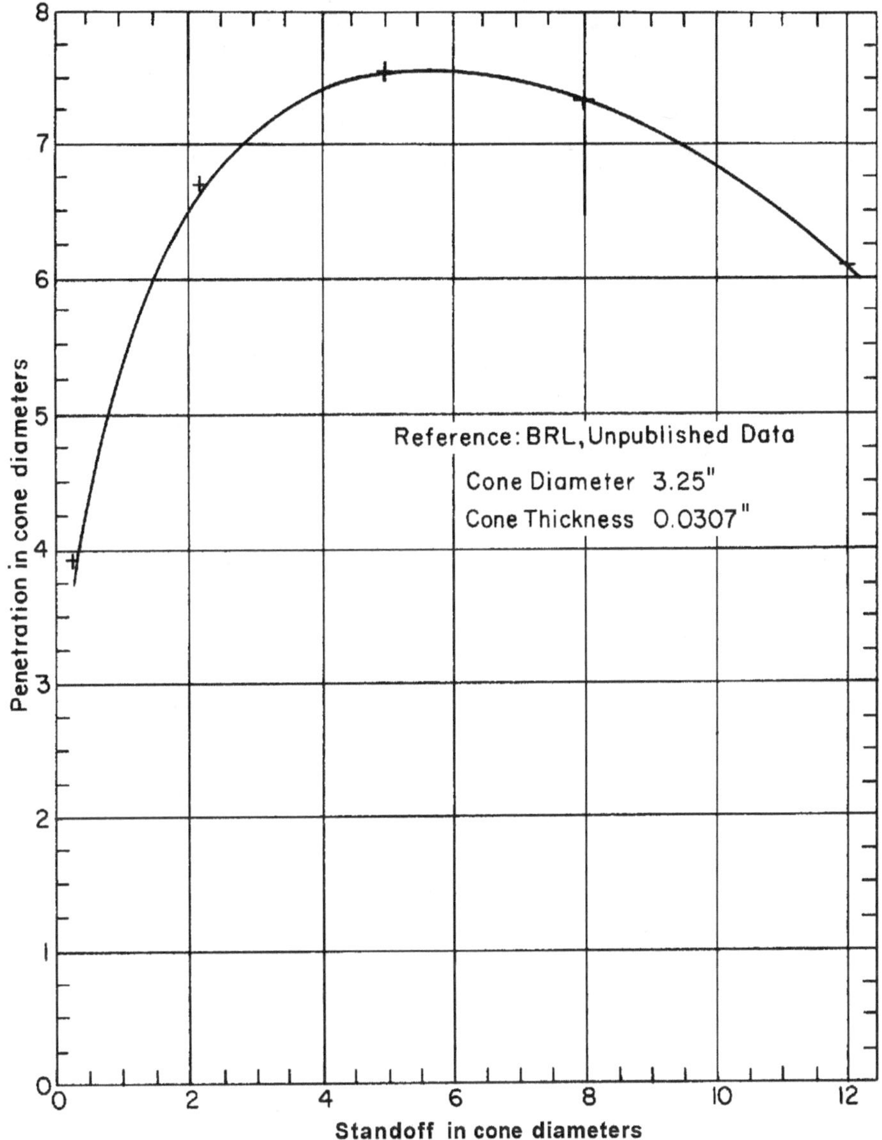

Figure 2-31. Penetration uersus standoff for 45° copper cones confined in shell bodies

the optimum cone angle for projectiles with copper cones is close to 40" or 45°. In some cases best penetration performance has been observed with 20" (figure 2-34) cones, and in others with 60" cones. As a first choice a cone angle of either 40" or 45" may be selected, which will give good performance in projectiles with an ogive length of 2 calibers.

2-106. <u>Sharp Apex Versus Spit-Back (Flash-Back) Tube Cones.</u> In most reported cases involving copper cones where charges differing only in the presence or absence of a spit-back tube have been compared, equal or slightly better penetration is obtained with a spit-back tube. There is no effect upon optimum standoff, rotation, or optimum wall thickness. In addition to the usually better performance of spit-back tube cones, it is easier to manufacture cones with a short spit-back tube section and to maintain close tolerances than it is with a sharp apex cone, and less difficulty is encountered in obtaining sound charges when spit-back tube cones are used.

It is standard practice to specify hard-drawn copper tubing with a wall thickness of 0.060 to 0.065 in. for spit-back tubes. The effect of tube diameter has received only limited attention, but satisfactory results have been obtained with tubes having a diameter between 20 and 30 percent of that of the cone.

Little information is available on the effect of method of attachment of spit-back tubes to cones. The tubes may be integral with the cone, or may be attached by means of soft solder, brazing, buttress threads, cementing, or crimping. Although all methods have been used only the use of an integral mounting, buttress threads, cementing, or crimping are both relatively inexpensive and not subject to warpage as a result of application of heat to the cones. In any soldering or cementing operation great care must be exercised to see that any excess material is removed from the tube and cone. Even a small quantity of cement smeared on one side of a cone has been shown to be enough to reduce penetration by 40 percent.

2-107. Method of Attaching the Liner. Four different methods of cone attachment are commonly employed: (1) the cone flange is crimped between ogive and body flanges (M9A2 grenade); (2) it is brazed or cemented (M28A2 or T205 rockets); (3) the cone flange is registered in the body and clamped firmly by a threaded ring or ogive (M67, T108, T184 HEAT projectiles); (4) the cone is pressed into the ogive and held in place by a locking groove (T119, T138 projectiles). Although each method has certain advantages in manufacture, the last two methods have performed well consistently and may be used satisfactorily.

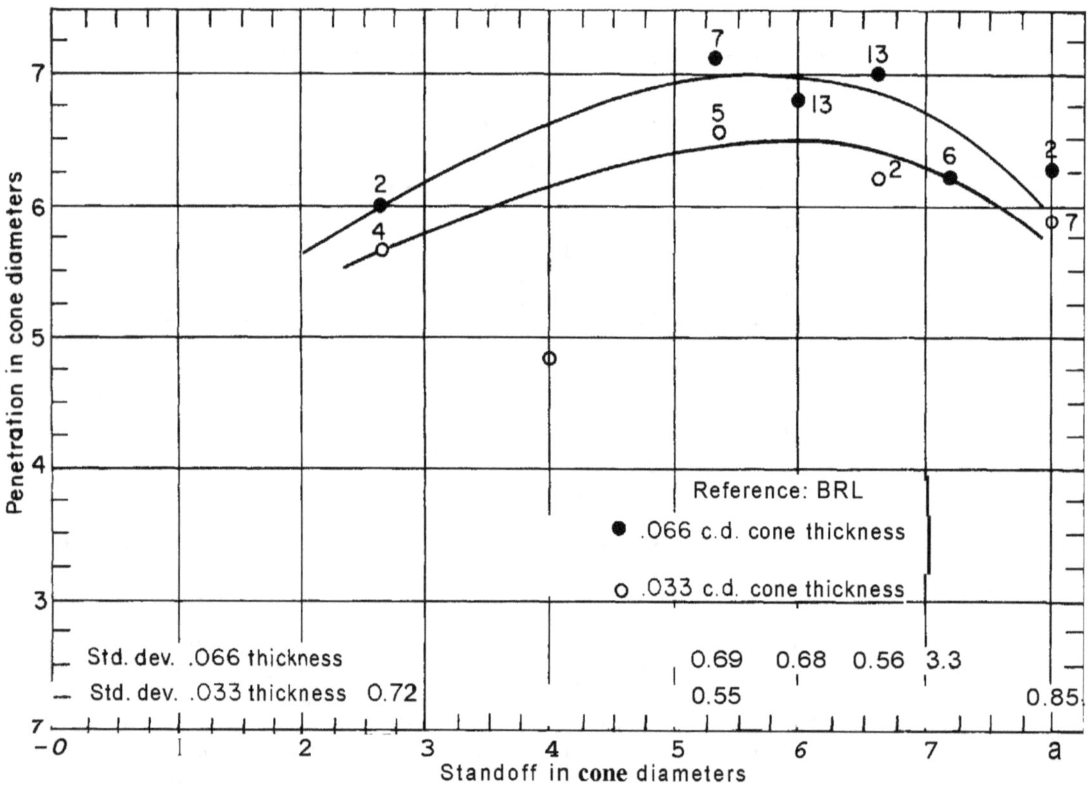

Figure 2-32. Penetration for 30° electroformed copper cones into mild steel targets

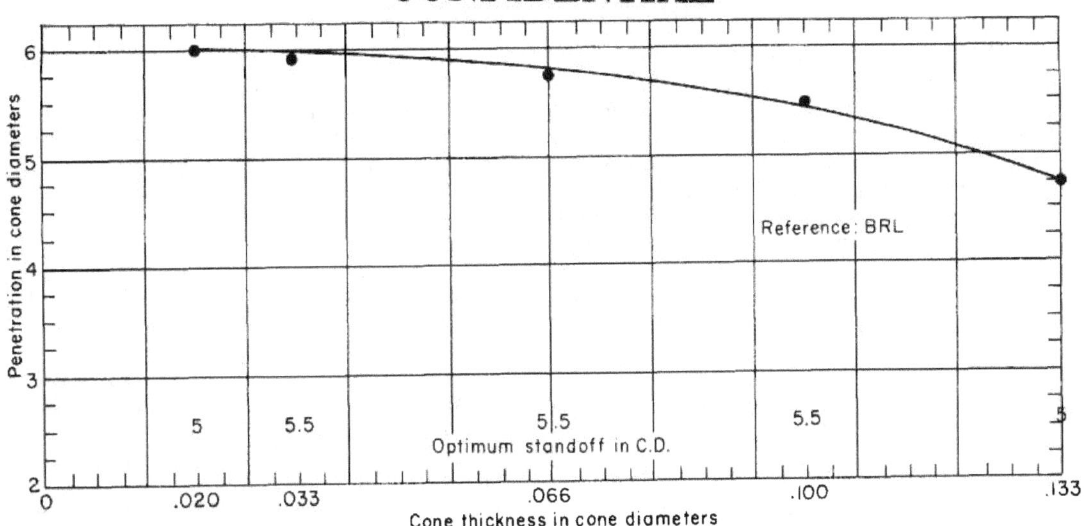

Figure 2-33. Maximum penetration into mild steel targets of optimum standoff versus cone thickness for 46° electroformed copper cones

2-108. Alinement of Cone and Charge. For best and most reproducible performance, the axis of the charge and cone should coincide. In actual practice, however, the axes may be parallel but displaced, or may not be parallel. A large number of experiments have been described in which the importance of these variables is treated. The importance of extremely careful control over this type of imperfection cannot be overemphasized.

Figure 2-34. Maximum penetration into mild steel targets at optimum standoff versus cone angle for electroformed copper cones

Tilt of the liner results in a reduced average penetration. The lowered average is the result of a larger number of "poor" shots. There are some good shots, even with angles of tilt as high as 2.0°, but in general the average penetration is reduced by 50 percent when the cone is tilted 1°, about 20 percent at 0.5°, and 10 percent at 0.3°, and a difference in the spread and average penetration can be detected between tilts of 0.05° and 0.15°.

The second type of misalinement, in which the cone charge axes are parallel but offset slightly, must be controlled just as carefully. In one experiment, an offset of only 0.015 in. (1 percent of the base diameter) reduced the penetration approximately 20 percent.

From the standpoint of manufacturing, however, it is not difficult to maintain the coincidence of

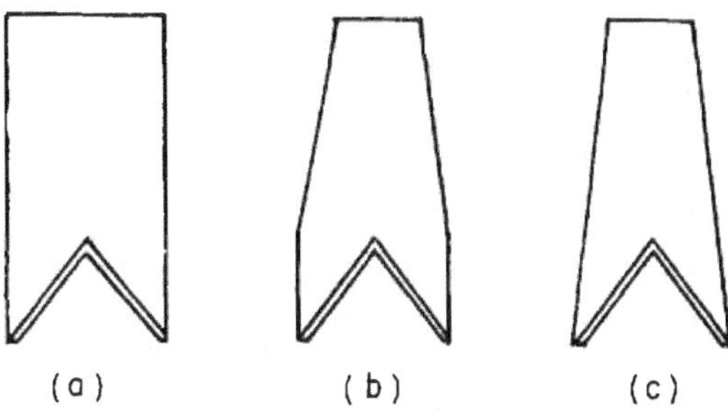

Figure 2-35. Typical charge shapes

the charge and liner within 0.010 in., provided the cones are properly registered and clamped in place. It is much more difficult to maintain alinement in brazed, welded, or cemented assemblies. However, regardless of the method of cone attachment and the care exercised in maintaining proper alinement, it is very important to be able to inspect the alinement after the cone and charge are assembled. Every effort should be made to avoid blind assemblies of the projectile.

2-109. Boostering of the Charge. The size, shape, location, and alinement of the booster have all been studied. In most cases electric detonators have been used to initiate the booster. In one experiment the thickness of the booster was varied from 0 to 1 in. without any indication of a detrimental effect upon performance, and it was concluded that the detonator was sufficiently powerful to initiate the charge. In a real projectile, however, the detonator is enclosed in a rotor in the fuze, and even though a tetryl lead may be employed, the probability of being able to initiate the charge satisfactorily without a

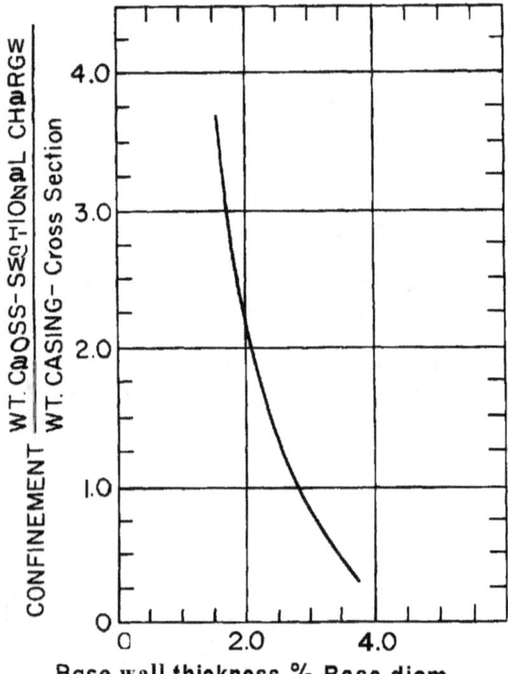

Figure 2-36. Confinement versus cone wall thickness

booster is not high. Nevertheless, experience with several projectiles has shown that the booster does not need to be large; a pellet 1 in. in diameter and 0.4 in. high appears sufficient for 3.5-in. charges (T208 E7 base element).

The effect on penetration of the "head" of explosive, or the distance between the booster and the apex of the cone, has been examined. The head of explosive required seems to vary with the "order" of the initiation. If the maincharge is satisfactorily initiated in a symmetrical fashion, the booster may be placed directly above the liner. If, however, the initiation is borderline, satisfactory performance will be obtained only if the booster is from one to two cone diameters above the cone. If the detonator and booster are adequate, it is believed that satisfactory shaped charge effect will be obtained if the booster is not less than 1 in. above the cone. It does seem likely, however, that the effect of misalinement of the cone will become increasingly severe as the booster is moved toward the cone; therefore, the booster should be placed as far rearward in the charge as other design considerations permit.

Eccentric initiation of the charge has been studied extensively. For point initiation it has been shown that the detonation wave front is essentially spherical with the detonator at the center of curvature. If the detonator is moved off center a decrease in penetration is observed, but the effect is relatively small. Placing the detonator 0.5 in. off center, in a charge length of two cone diameters, resulted in a loss in penetration of 20 percent. Since it is not difficult to hold booster and detonator alinement to within 0.060 in., off-axis initiation is not an anticipated problem with electric or magnetic fuzes, but some difficulty might be experienced with a spit-back type fuze. In the latter case initiation at a point 0.5 in. off center can occur unless care is taken in assembly of the spitter cone.

2-110. Confinement. The relationships between cone wall thickness and projectile wall thickness were described in paragraph 2-104. There are, however, other effects of confinement that are of considerable interest to the designer. Increasing the confinement greatly increases the hole volume. This effect is noted whether the confinement is provided by increased wall thickness or by a "belt" of explosive. The presence of explosion products at high pressure within the explosive belt retards the expansion of the products in much the same manner as does a steel casing. The "confining" effect of different inert materials is, of course, proportional to their density.

In early experiments with charges of diameter larger than that of the cone, a significant effect was noted in those cases where the cones were flanged. It was observed that when an explosive belt is in contact with the flat flange of a cone the penetration was lower than when the flange was removed. The loss in penetration was considerably greater when the charges were heavily confined. With the typically heavy confinement of a 105-mm projectile a loss in penetration of 48 percent resulted when a 0.10-in. flange was backed by explosive. Recently, the results of an extensive study of the effect of confinement on the performance of flanged and unflanged cones have become available (paragraph 2-93). From these data the author drew the following conclusions.

a. The addition of a small explosive belt obtained by increasing the charge diameter from 1.63 to 2.00 in. produces approximately the same effect on penetration and hole volume as the addition of 0.25 in. of steel confinement.

b. When heavy base confinement has been added to the 2-in. charge, the penetration is decreased about 27 percent.

c. The addition of both lateral and heavy base confinement to the 2-in. charge causes a drastic reduction of about 45 percent in penetration performance.

d. When the larger charge is confined laterally, the presence of the flange causes a relatively small, but significant, decrease in penetration, as compared with a similarly confined charge lined with a deflanged cone.

e. The hole volume produced by the 2-in. charge is increased by about 50 percent when lateral confinement of 0.25-in. steel is used (compared with the 100 percent increase which occurs with the 1.63-in. charges); boundary conditions at the base of the charge have little or no effect on hole volume in spite of the large changes in depth of penetration.

This experiment illustrates how an apparently superficial change in charge design can cause profound changes in charge performance. While it is possible to explain these changes satisfactorily in the light of fundamental information, and to predict qualitatively what might have been expected, great care should be exercised in designing experiments so as to be sure that the variable studied may be honestly evaluated.

2-111. Internal Ogive Shape. The internal shape of a conical or tangent ogive does not interfere with the normal collapse process of the shaped charge liner. However, a number of HEAT shells now being developed for the Ordnance Corps employ a tee, boom, or spike ogive, which can reduce penetration greatly. Ogives of this shape are of interest because they have a low lift and, therefore, smaller restoring moments are required for projectile stability. While such ogives do have a much higher drag than conical or tangent ogives at projectile velocities up to 2,000 fps, the advantage of lower drag possessed by the latter is much less marked at velocities of 3,500 and 4,000 fps.

The effect of internal tee configuration on shaped charge effect has been given a great deal of attention. Figure 2-37 shows six of the many configurations which have been tested, and also the penetrations each of these booms permit. A consideration of designs A to F discloses two important design requirements: (1) a free space not less than 0.6 cone diameters must be provided in front of the cone, and (2) the bore of the boom must be as large as the maximum diameter of the slug. It seems clear that near the base of the cone the collapsing elements follow a forward curved path. Cone collapse is not complete until the cone has moved forward a distance of nearly one cone diameter and useful jet elements are formed during the time the slug is moving forward a second cone diameter. If the bore of the boom is not at least as large as the major diameter of the slug, the jet will be pinched off when the slug jams in the bore and a portion of the potential penetration will be lost. The tests reported above were static tests. It is reasonable to suspect that in dynamic firings the boom may be jammed rearward toward the cone by the impact velocity, and that this will reduce the effective free space. Therefore, some additional free space must be provided, and the actual amount required will probably depend upon the maximum impact velocity of the projectile.

THE EXPLOSIVE COMPONENT OF SHAPED CHARGES

2-112. Introduction. The shaped charge effect depends upon the pressure impulse of a detonated explosive to accelerate the liner walls in the collapse process, which produces the jet. The explosive is therefore fundamental to the phenomenon and it is essential that charge parameters be carefully selected. This means that proper distribution, initiation, and explosive, or an adequate compromise of these factors, be made.

2-113. Effect of Explosive on Performance. Considerable experience has been gained, from which it is generally possible to make adequate shaped charge designs. The effect of compromises with the ideal design can also be estimated reasonably well. However, the problems of explosives in shaped charges have not all been solved. Conditions arise wherein minor variations cause an appreciable performance change, which can be attributed only to the explosive. Small modifications in charge preparation technique, or a change in explosive distribution about the liner, may affect the penetrations significantly. The exact bearing nonuniformity of the explosive charge has on performance requires further investigation. Proper shaping of the detonation wave in the explosive has shown promise of large increase in penetration performance but has introduced additional difficulties, which must be overcome before it can be considered seriously for application.

Practically all studies of explosives in shaped charges have been experimental. This does not mean, however, that the basic studies have been neglected. Detonation theory is being actively pursued, as is also the study of explosive-metal interactions. Direct applications of these research sutdies are being carried out by the

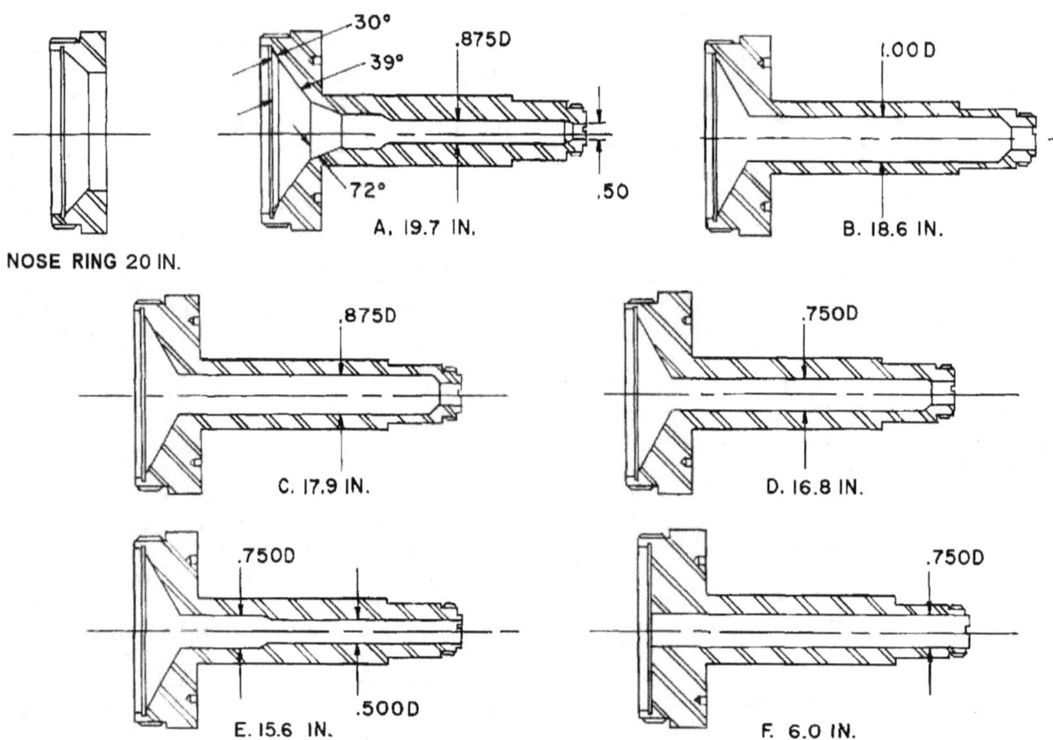

Figure 2-37. Tee configurations

Carnegie Institute of Technology group in their work on a release wave theory as applied to liner collapse.

2-114. Effect of Different Types of Explosives. In shaped charge development considerable work has been performed on standard charges to evaluate the relative performance of various explosives. Only a few of these have found ordnance application. Some explosives with low detonation pressures are marginal and form very poor jets, but no upper limit to performance has been found; that is, as the detonation pressure is increased the penetrations increase. The trend in explosives research is the development of new compounds with higher detonation velocities and pressures. Thus small additional improvement in performance might be anticipated.

Table 2-14 lists various high explosives, with their properties and shaped charge penetrations. The list is not complete, and some are unacceptable for wide application because of sensitivity, compatability, stability, or production difficulties. The densities given are those actually obtained in the charges used for penetration comparisons. The detonation velocities given are computed on the basis of experimentally derived density-detonation velocity slope data for the explosives. Sensitivities are taken from impact studies at Naval Ordnance Laboratory only to avoid introducing calibration constants for different testing machines. The penetrations are from Naval Ordnance Laboratory work on point-initiated, unconfined charges 4.0 in. in height, 1.63 in. in diameter, with M9A1 steel cones, and fired with 4.0-in. standoff into mild steel plates. A few similar explosive comparisons performed at Du Pont's Eastern Laboratories are also given. This furnishes a reasonable comparison of different explosives.

Formulas or correlations relating penetrations and cavity volumes to parameters of the explosive have been developed, but they do not take into account the properties of the liner or the nature of jet formation and penetration processes, and hence have limited usefulness.

2-115. Explosive Distribution. The distribution of the explosive about the liner and the type of initiation used to detonate the charge have a very marked influence on the performance of a shaped charge. Distribution as discussed in this paragraph is concerned only with the geometrical arrangement of the explosive. Inhomogeneities or variations from a uniform charge will be considered later. The parameters which describe the explosive geometry for cylindrical or near cylindrical charges are height and diameter. The dependence of performance on the distribution is closely related to the manner in which it controls the pressure impulse delivered to the liner walls.

It is not possible to generalize much on the effect of the explosive distribution parameters without first defining certain supplementary conditions. If one takes an unconfined, point-initiated charge, the mean penetration will increase with increasing charge height. Penetration is very sensitive for heights up to several cone diameters, after which it shows only small changes with increase of the explosive column. However, it is still observable at lengths up to six or seven cone diameters. Figure 2-38 is indicative of the normal behavior of penetration as a function of charge height under the conditions previously enumerated. Actually varying the length of explosive above the liner apex affects the shape and magnitude of the high pressure region in the explosive reaction zone and also varies slightly the direction of the wave front which interacts with the liner, especially at short charge heights.

Under similar conditions the effect of varying the explosive-to-liner diameter ratio results in a penetration relation as shown in figure 2-39. Varying the explosive diameter with a fixed-cone diameter results in a performance similar to that for changes in the confinement wall thickness.

The hole volume increases with increasing length, as well as with increasing diameter of explosive, within the range normally observed. A limiting value is approached and, of course, it becomes more difficult to observe the smaller increases, which are hidden by the spread in the data.

Although the preceding paragraphs would indicate a relatively simple correlation for performance with explosive length and diameter, in reality it is a complex problem. It should be noted that the results presented were for the simplest case and under restricted conditions. The shape and magnitude of these curves might

Table 2-14

Explosive	Composition	Density (gm per cc)	Detonation velocity (m per sec)	Sensitivity (drop height in cm)	Penetration (in.)	
					NOL	Other
TNT	Trinitrotoluene	1.60	6,980	162	4.25	(du Pont) 4.2
Composition B	59.5 39.5 1.0 RDX TNT Wax	1.69	7,930	60.4	6.17	(du Pont) 6.2
Composition A*	91 9 RDX Wax	1.60	8,230	58.8	6.22	...
50/50 Pentolite	50 50 PETN TNT	1.65	7,560	23.5	5.56	(du Pont) 5.5
HBX-1	40 38 17 5 RDX TNT Al D-2	1.72	7,350	95.7	5.16	...
PTX-2	43.2 28.0 28.8 RDXPETNTNT	1.68	8,000	...	6.57	...
70/30 Tetrytol	70 30 Tetryl TNT	1.64	7,310	...	5.12	(du Pont) 65/35/Tetrytol/TNT 5.0
70/30 Cyclotol	70 30 RDX TNT	1.70	8,100	37.6	6.27	...
75/25 Cyclotol	75 25 RDX TNT	1.71	8,160	...	6.56	...
75/25 Octol	75 25 HMX TNT	1.78	8,350	45.6	6.98	...
77/23 Octol	77 23 HMX TNT	1.81	8,440	...	7.45	...

*Pressed (all other explosives listed were cast).
Du Pont charges were similar except chg. height was 6.0 in.

be greatly changed by any one of the large number of variables not considered in the discussion up to this point. In general, unless experimental results are available for the particular situation at hand, it is difficult to predict the effect of variation of charge height or explosive diameter with any certainty. This is also true if the shape or contour of the explosive charge deviates from cylindrical symmetry.

2-116. Initiation of Explosive. Plane-wave or peripheral initiation, which shape the detonation wave, may change the penetrations obtained. Figure 2-40 compares point, plane-wave, and peripheral initiated standard charges for different charge heights. These results are for steel liners. Limited tests for other liner materials indicate an increase in penetrations with peripheral initiation, but the percentage improvement varies considerably with the material of the liner. These special results are given to indicate what may be achieved with proper wave shaping. Penetrations from point or plane-wave initiation are fairly reproducible. However, small asymmetries anywhere in the system will produce large variability with peripheral initiation. Application of such an initiation system must be made with caution if increased penetrations are to be achieved. Experience with a large number of shots under well-controlled conditions has shown that increased penetration for steel-cone lined charges and peripheral initiation is real. However, the large increase (25 to 30 percent) reported here has been shown to depend critically on the liner used. Furthermore, the cavity volume may be reduced by as much as one half.

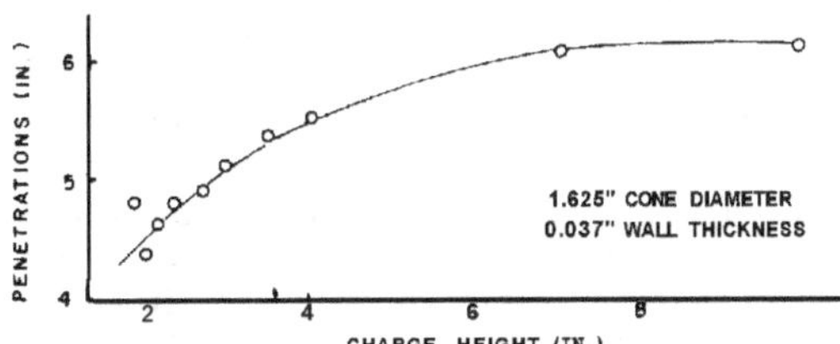

Figure 2-38. Shaped charge penetrations as a function of charge height for M9A1 steel cones and 50/50 pentolite explosive

2-117. Charge Preparation. In charge preparation the problem is to produce an explosive loading which will result in maximum shaped charge performance. The method should insure (1) uniformity of the explosive, (2) axial symmetry, and (3) maximum density. Radial uniformity and axial symmetry are highly important to jet formation, and small deviations from these conditions may produce a significant decrease in mean penetration. Maximum densities are required to obtain the highest possible detonation pressures and, hence, largest penetrations. Lack of uniformity in the charge does not always result in poorer performance. Increased penetrations have been reported when composition and density gradients along the charge length were changed inadvertently. Increased penetrations have also been reported that could be attributed to charge imperfections in the form of axial pipes that produced some shaping of the detonation wave.

2-118. Charge Imperfections. Experiments with liquid explosives, which may be considered

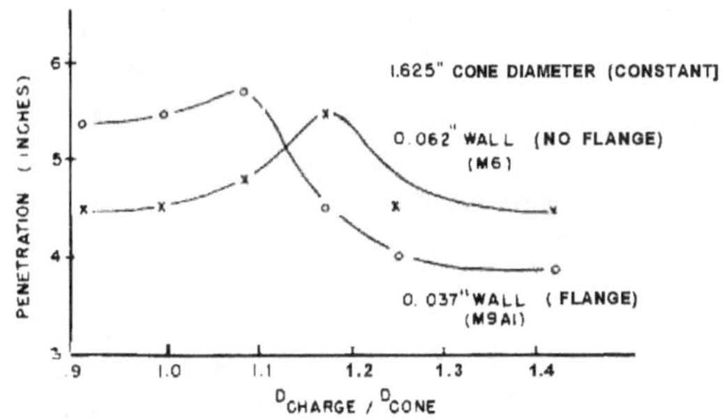

Figure 2-39. Shaped charge penetrations as function of charge diameter (unconfined)

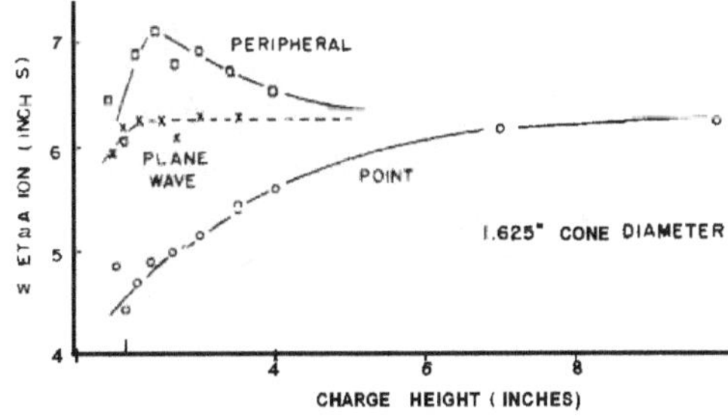

Figure 2-40. Shaped charge penetrations as function of initiation for M9A1 steel cones and 50/50 pentolite explosive

to be perfectly homogeneous, indicate that the variability of results with solid explosives may be decreased greatly by improving homogeneity of the charge. Inhomogeneities in a multicomponent explosive, such as Comp B, due to the unsymmetrical segregation of one of the constituents, could account for approximately half of the variability of results, the other half being attributable to variations in metal parts. Voids and bubbles in the vicinity of the liner may cause marked reduction in penetration. Small bubbles or voids, distributed throughout the charge, lower the loading density and result in decreased detonation pressure. Consideration of the effect of detonation pressure indicates that a 2 percent loss in pressure may produce as much as 5 percent variation in penetration.

FUZES FOR SHAPED CHARGE MISSILES

2-119. Fuzing of High-Velocity Rounds. A 90-mm, fin-stabilized, gun-fired projectile may travel at a velocity of some 2,500 fps. The distance from the nose of the round to the location of the detonator is approximately 1 1/2 ft. This means that, if the round is to be detonated a very short time after the nose contacts the target (the time being limited arbitrarily by the requirement that the nose must collapse not more than 1/4 in. before the initiation of explosion), the initiation must be started in 8 μsec, because that is the time required for the shell to travel 1/4 in. If the detonator requires 6 μsec to detonate after receiving the signal, it follows that the information must travel from the tip of the shell to the detonator in 2 μsec. This immediately rules out any mechanical means of transmitting the information from the front to the base of the shell. Under these conditions an electrical fuze must be used. Several electrical methods have been tried. One is to use a power supply, such as a battery, a switch in the nose (which may be a simple double shell), and a detonator at the base (with an appropriate arming system). The second, which is really a modification of the first, is to use a source of electrical energy that is inert until firing, such as a simple impulse generator that charges a capacitor on firing. This capacitor can be made to hold its charge for the duration of the flight of the projectile, and can be discharged by a switch as in the previous case. This last approach was used in the first model of the T208 but was abandoned in favor of the simplicity of the piezoelectric generator. A third possible electrical method is to use a generator, located in the nose of the projectile, that is energized by the impact with the target; either an electromagnetic or electrostatic device can be used.

2-120. Fuzing of Low-Velocity Rounds. In the case of a subsonic round, such as the 3 1/2-in. rocket grenade or, even better, the T37 rifle grenade, the requirements for speed of initiation of the explosion are far less stringent. A rifle grenade travels at some 150 fps. Again, if one permits the round to deform [illegible] in. before setting off the high-explosive charge, the time available is 140 μsec, and a mechanical transmission of information from the nose to the rear element becomes at least theoretically possible. Two general methods are employed to provide mechanical transmission. One is the so-called "spit-back" (flash-back) fuze, where a small shaped charge explosive in the nose of the round is initiated by a percussion primer and fires a jet backwards through a passage provided in the main charge into a base booster. Since the velocity of such a small jet is very high, this provides an extremely rapid method of transmitting the trigger action from the front to the rear. In spite of the very high jet velocity, this method depends upon a clear path from the shaped charge in the nose to the booster; a condition not always satisfied due to misalinement of metal parts or deformation of the fuze upon impact. Another approach used in rocket grenades is to have an inertia weight located at the base of the round. When the round contacts a target, it decelerates, and the inertia weight slides forward and fires a percussion cap. The disadvantage of this type of fuze is that it is inherently slow and that the shell is required to have a very rigid nose section, so as to prevent collapse while the fuze is going through its triggering cycle.

THE EFFECT OF ROTATION UPON SHAPED CHARGE JETS

2-121. The Deterioration Process. The development of a triple-flash X-ray system for studying jets from large rotated charges has clarified the details of the deterioration of the jet. The sequence of events as the rotational frequency increases is shown in figures 2-41 through 2-44, which show the effects of increasing rotation upon the jet from a 105-mm copper liner. The deterioration process can be broken down into the following distinct steps.

a. The jet, which is normally continuous when unrotated, begins to break up into separate pieces along its length.

b. As the rotational frequency increases, the cross section of the jet starts to deviate more and more from a uniform circular shape and shows evidence of deformation into a ribbon-like structure.

c. There is finally a definite bifurcation or separation of the jet into two essentially parallel jets, with each jet broken into separate pieces. When the bifurcation first appears, the two portions of the bifurcated jet generally seem to lie in a plane of bifurcation.

d. Increasing rotational frequency causes the plane of bifurcation to be distorted into a helical surface.

The bifurcation in the jet appears to be associated with a critical frequency that depends on the caliber. Thus, bifurcations have not been seen in jets from 105-mm charges rotated at 15 rps, whereas all jets from 105-mm charges rotated at 45 rps show bifurcation, as do most jets from 105-mm charges rotated at 30 rps.

The incidence of bifurcation is clearly associated with the steepening portion of the penetration fall-off curves (see figure 2-45). Finally, the plateau region associated with the highest spin frequencies indicates that the later modifications of the bifurcation process contribute very little to further reduction in penetration. It was originally conjectured that the original bifurcation was perhaps followed by bifurcation of each of the new portions of the jet. This has not been ruled out, but the observations on the target plate upon which this was based can also be explained by the distortion of the plane of bifurcation into a helical surface.

2-122. Theory. It was pointed out by Tuck in 1943 that rotation could result in a malformed jet. The vector addition of the rotational velocity and the collapse velocity of any element results in a velocity which has a direction tangent to a circle whose radius (r') is dependent upon the velocities and the cone geometry. This would result in a hollow jet, and could cause a drastic decrease in penetration if r' became large enough. On this basis, Birkhoff estimated that a 3-in. diameter liner would show appreciable deterioration at 100 rps.

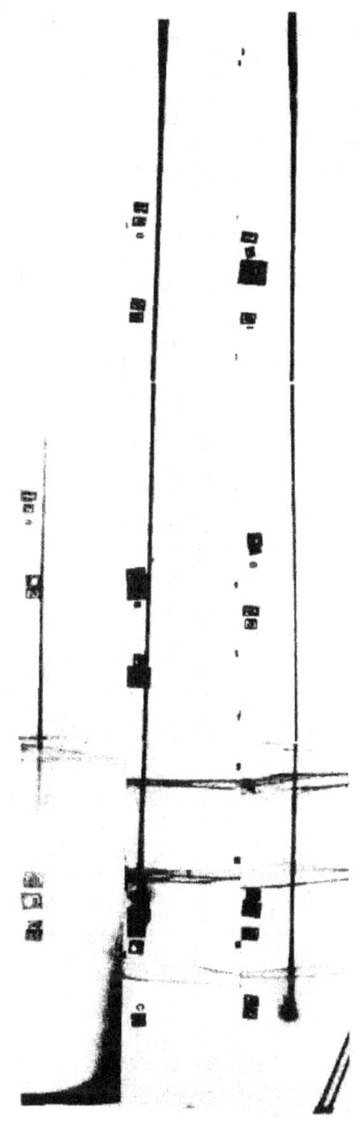

Figure 2-41. Effect of rotational frequency upon the jet from a 105-mm copper liner

Birkhoff, using a different approach which neglects initial malformation of the jet, has estimated the decrease in the penetration from a given element of a properly formed jet caused by the increase of the cross-sectional area resulting from the expansion of the jet resulting from rotation. The magnitude of the force causing this expansion can be appreciated by considering the extremely high rotational velocities of the jet, which result from the conservation of the angular momentum of the cone.

2-123. Scaling Under Rotation. For the scaling of results of rotated shaped charges, several theories have been advanced. The simplest of them scales penetration in cone diameters against ωd where ω is the angular velocity of the projectile, and d is the cone diameter. Figure 2-46 is a plot of this type. This scaling law can be expected to hold only if the other parameters (such as standoff and cone thickness) are also scaled. It has not been too well verified in the high spin frequency range.

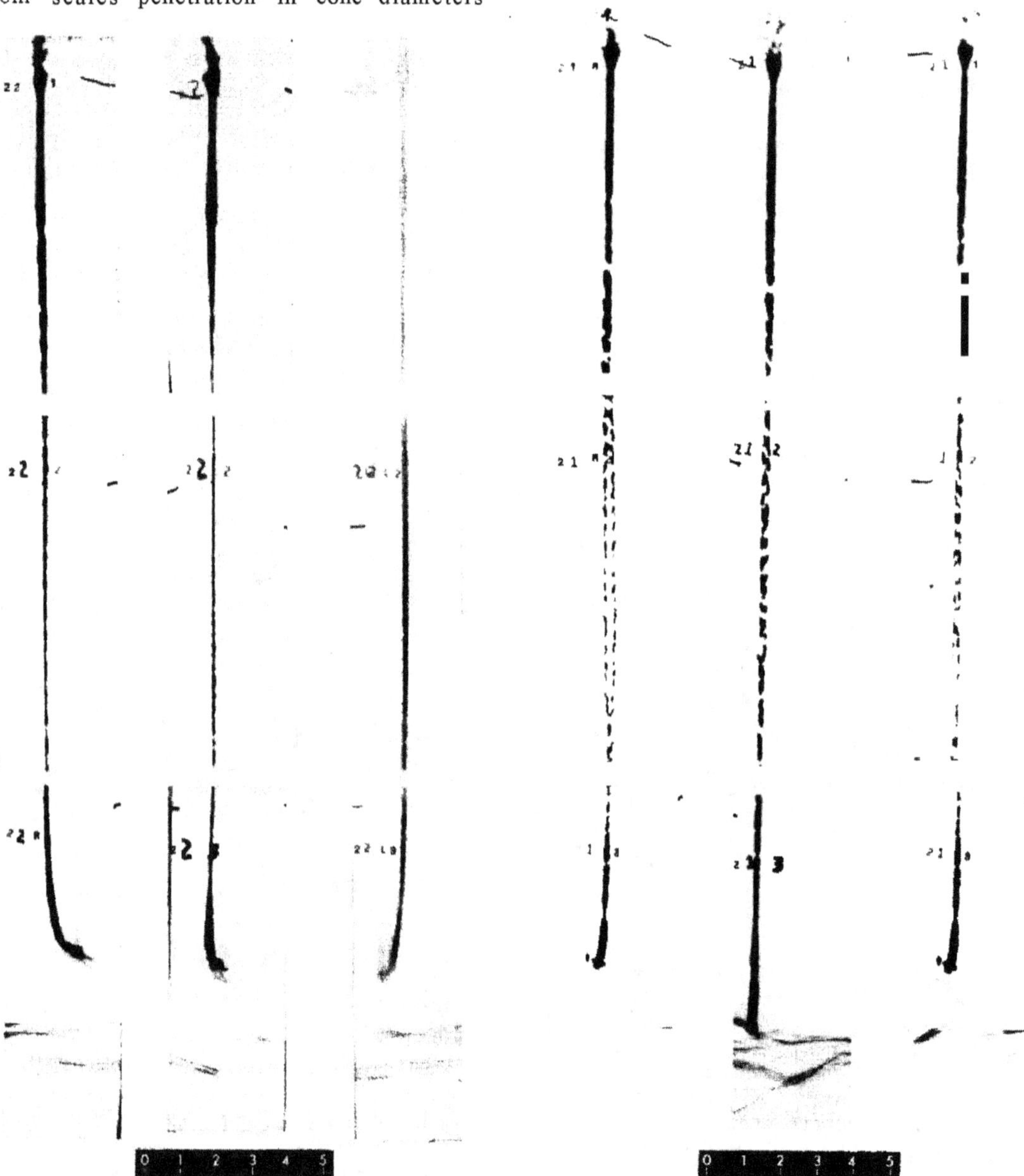

Figure 2-42. Effect of rotational frequency upon the jet from a 105-mm copper liner

Figure 2-43. Effect of rotational frequency upon the jet from a 105-mm copper liner

For the designer, the use of ωd as a scaling variable for predicting the effect of rotation upon penetration appears to be the best available basis over the range 0 to 100 rps and 57-mm to 105-mm calibers. For higher spin frequencies it is still the best guide, but experimental verifications are recommended as a check on predictions. It is expected that experiments currently under way will eliminate the uncertainties that exist at the highest values of ωd.

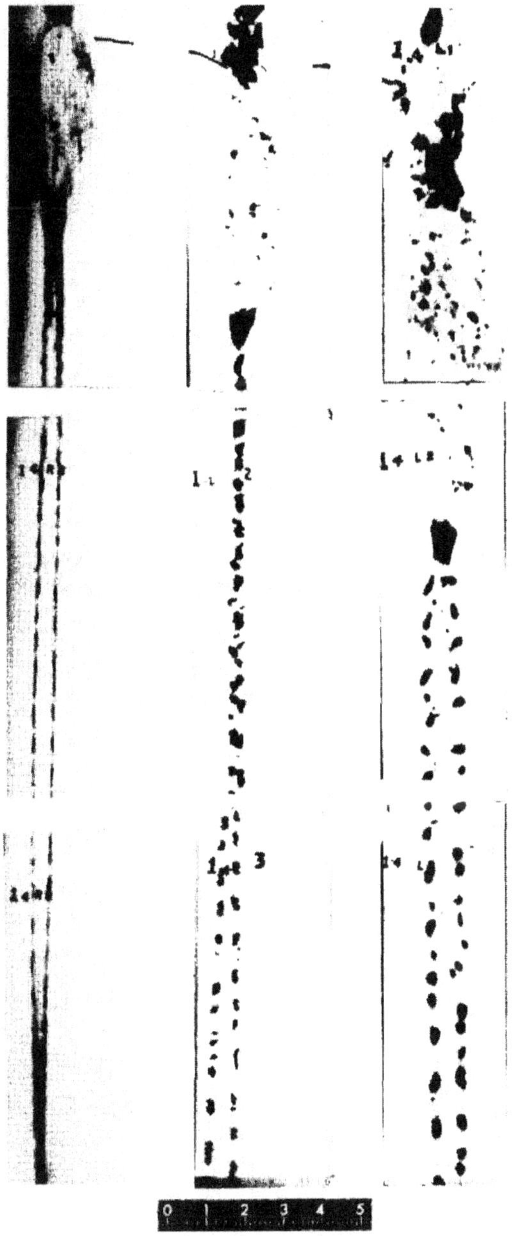

Figure 2-44. Effect of rotational frequency upon the jet from a 105-mm copper liner

2-124. <u>The Effect of Cone Angle on Penetration Under Rotation.</u> Theoretical consideration of the effect of cone angle on jet rotational velocity indicates that small-angle cones will be more sensitive to rotation. The experimental data obtained during the war by the Office of Scientific Research and Development on the effects of cone angle are not easy to interpret because of large experimental dispersions. However, the general conclusions drawn by their investigators are essentially as follows.

a. At short standoff the larger angle liners show little deterioration as a result of rotation. However, since their unrotated penetration is relatively poor, this is of little practical value.

b. Because of the increased effective standoff (due to the increased cone height) of a small-angle cone, it is more seriously affected at a given external standoff and its penetration is therefore not appreciably better than that of a large-angle cone.

These conclusions are unfortunately not as specific as would be desired by a designer. Additional experimental data have been obtained by the Ballistic Research Laboratories, using 105-mm charges of a given fixed height at a standoff of 7 1/2 in., a value near the common

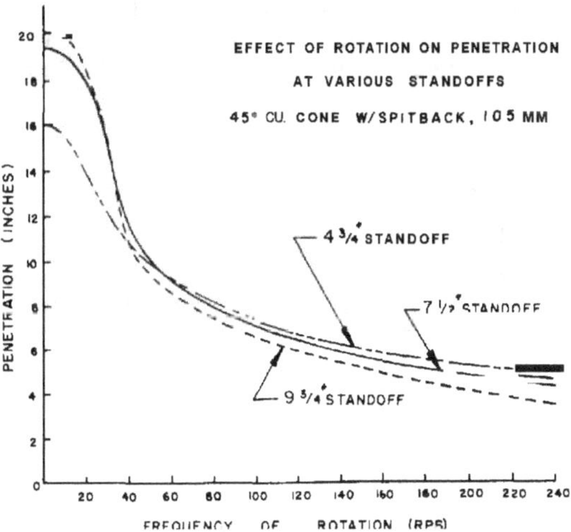

Figure 2-45. Effect of rotation on penetration at various standoffs, 45° copper cone with spit-back, 105 mm

built-in ammunition standoff (~ 2.3 cone diameters). These data are shown plotted in figures 2-47 and 2-48. In these curves comparisons are made of the unrotated penetration, and the penetration at 45 rps, as a function of cone angle. The results within the range of variables so far explored clearly indicate an increased sensitivity of small-angle cones to deterioration by rotation. From the practical viewpoint of the designer, from this experiments the best cone angle at 45 rps appears to be about 45°, even though the smaller cone angles have better unrotated performance and the larger cone angles have reduced sensitivity to rotation. However, caution must be used in extrapolating to other conditions. The experiments, which are being continued, will cover a much larger range of the variable ω and the standoff for various cone angles.

Under standoff conditions normally existing for ammunition, the designer can expect to find small-angle cones more sensitive to rotational deterioration than large-angle cones. There is not sufficient good information on the cone angle effect at large standoffs.

2-125. <u>The Effect of Liner Thickness on Penetration Under Rotation.</u> There have been very few rotation experiments reported involving liner thickness as a variable. Those conducted up to the present time seem to confirm the theoretical expectation that over the range of thickness studied the effect of thickness is not of major importance, and that the penetration performance of a uniform conical liner under rotation is, within the precision of the experiments, essentially unaffected by thickness.

Improved experiments, with extra care taken to reduce dispersion, will be required to establish the existence and magnitude of the thickness effect in rotation. Separate experiments are required to ascertain the contribution, if any, of

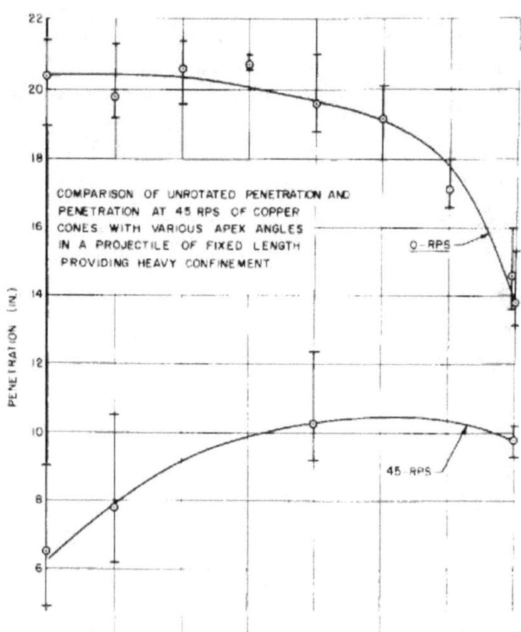

Figure 2-47. Comparison of unrotated penetration and penetration at 45 rps of copper cones with various apex angles in a projectile of fixed length providing heavy confinement

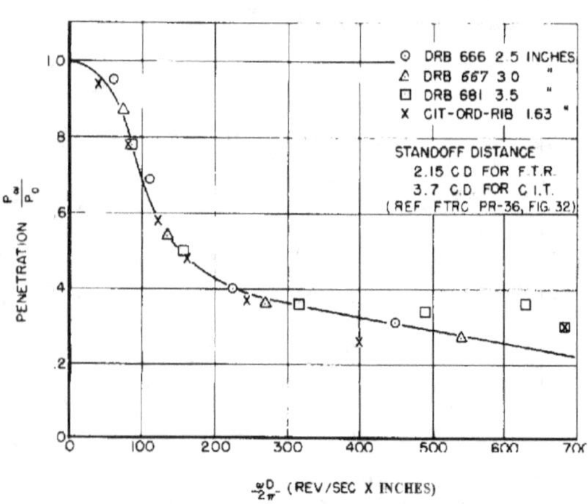

Figure 2-46. Effect of rotation on penetration

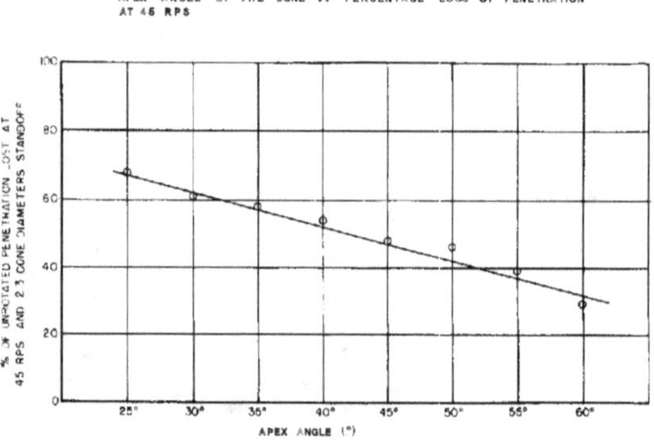

Figure 2-48. Percentage of unrotated penetration loss versus apex angle

the liner thickness to changes in jet stability to breakup. Such experiments are under way at the Ballistic Research Laboratories and at Carnegie Institute of Technology, but at the time of this review there are no definitive results. The best course for the designer at this time is to treat the thickness variable as though it has no effect on rotational penetration; the best performance under unrotated conditions should determine the thickness.

2-126. The Effect of Standoff on Penetration Under Rotation. Recently there has been completed at the Ballistic Research Laboratories a very comprehensive experimental study of the effect of rotation and standoff on the penetration of heavily confined 105-mm drawn-copper liners. The most useful way to summarize this study is to present the experimental results in graphical form; these are shown in figure 2-49.

These results can be considered typical of good liners, since the unrotated performance of the basic liners compares favorably with the best results ever reported.

The conclusions, of value to the designer, that may be drawn from these results are as follows.

a. The penetration at a given standoff decreases monotonically as the rotational frequency increases.

b. The standoff corresponding to peak penetration decreases as the rotational frequency increases, until at the highest frequencies used (~ 240 rps) the optimum standoff is only a few inches.

c. At low rotational frequencies useful penetrations are obtainable even at the largest standoffs (42 in.) used. The implications of this result are important for the problem of defense by spaced armor.

2-127. The Effect of Liner Material on Penetration Under Rotation. Penetration experiments comparing various liner materials under rotation have been carried out by OSRD, by Firestone, and by Carnegie Institute of Technology. In addition, flash radiographic jet studies have been carried out by the Ballistic Research Laboratories. The penetration experiments generally lead to the conclusion that no material

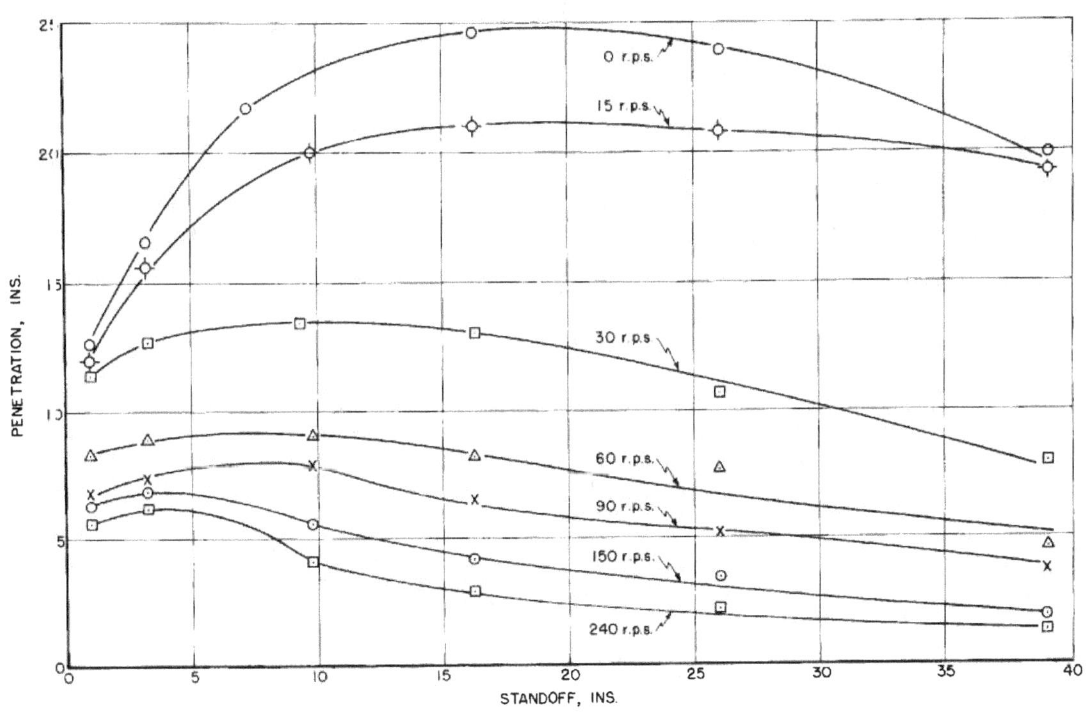

Figure 2-49. Effect of rotation and standoff on the penetration of heavily confined 105-mm drawn copper liners

studied so far offers any striking advantages over any other material insofar as rotational effects are concerned; even though the advantage of copper over other materials diminishes as the rate of rotation increases, the predominant role of the unrotated penetration still makes copper the proper choice for penetration purposes, according to the penetration experiments.

The flash radiographic studies by the Ballistic Research Laboratories have indicated a basis for expecting differences in the behavior of various materials due to the expected dependence of the critical frequency for bifurcation upon the physical properties of the materials. Such differences have actually been observed. These studies, however, have not yet progressed to the point where conclusions of value to a designer may be drawn. It may even turn out that the differences which seem to exist may be too small or may require the use of strategic material for their exploitation.

2-128. <u>The Effect of Liner Shape on Penetration Under Rotation.</u> It has been suggested by various investigators that trumpet-shaped liners might show increased resistance to deterioration by rotation. This view is based on the notion that since the trumpet liner is on the average closer to the axis of rotation than the equivalent cone of equal altitude, it ought to be affected less by rotation.

Experiments by the Carnegie Institute of Technology several years ago did not bear out such expectations. However, experiments which have been carried out at the Ballistic Research Laboratories using trumpet liners with peripheral initiation have indicated that one can indeed obtain reductions in the deterioration of the performance under rotation by means of a trumpet shape.

These experiments were for some time plagued by an inability to reproduce the experimental results. This difficulty has recently been traced by Lieberman to an inadvertently overlooked mechanical interference with the late collapse stages, which has since been eliminated. In addition, asymmetries in the explosive have also been shown by Lieberman to be of importance in hindering reproducibility.

A comparison of the most recent performance of peripherally initiated trumpets with the corresponding cones of 45" apex angle is shown in figure 2-50. The performance of electroformed trumpets (peripherally initiated) is compared with the best drawn conical liners available at the Ballistic Research Laboratories in the same caliber. It is quite evident that the peripherally initiated trumpets are resisting deterioration quite effectively. More complete coverage of the pertinent variables is still needed. (For instance, a truly valid comparison would require that liners of both shapes be made by the same process. However, the effect is sufficiently clear to warrant consideration of this system in applications involving lower rotational frequencies. This system may be considered competitive with fluted liners in this range, and may even have advantages, since there is no peaking of the penetration performance at a given rotational frequency, but rather a reduced deterioration, the performance improving monotonically as the rotational frequency decreases. The possibility of increased sensitivity to loading asymmetries is a disadvantage that must also be considered. It should, however, be possible to overcome this with careful loading techniques.

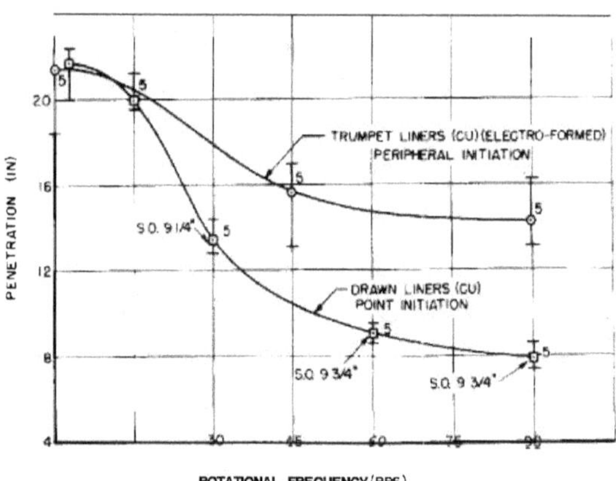

Figure 2-50. Comparison of penetrations obtained with point initiated conical liners and peripherally initiated trumpet liners

The flash radiographs of the jets shown in figures 2-51 and 2-52 bear out the increased resistance of this system to rotational deterioration.

In order to minimize the effects of rotation, it is logical to start the collapse as near the axis as possible; that is, by the use of a cylindrical liner. The earliest recorded experiments with cylindrical liners are those of the British. The group at the Ballistic Research Laboratories, unaware of such experiments, started a similar investigation in 1950. Since that time, investigators at Frankford Arsenal have also attacked the problem and have produced the wave shaping system, which so far has given the best penetration performance. This performance level and the reproducibility, however, have both been inadequate.

The major problems in the investigation of cylindrical liners are the following.

a. Devising a system whose unrotated performance will compare favorably with that of a conical liner in the same projectile.

b. Perfecting a wave shaping system which will be sufficiently reproducible to make experi-

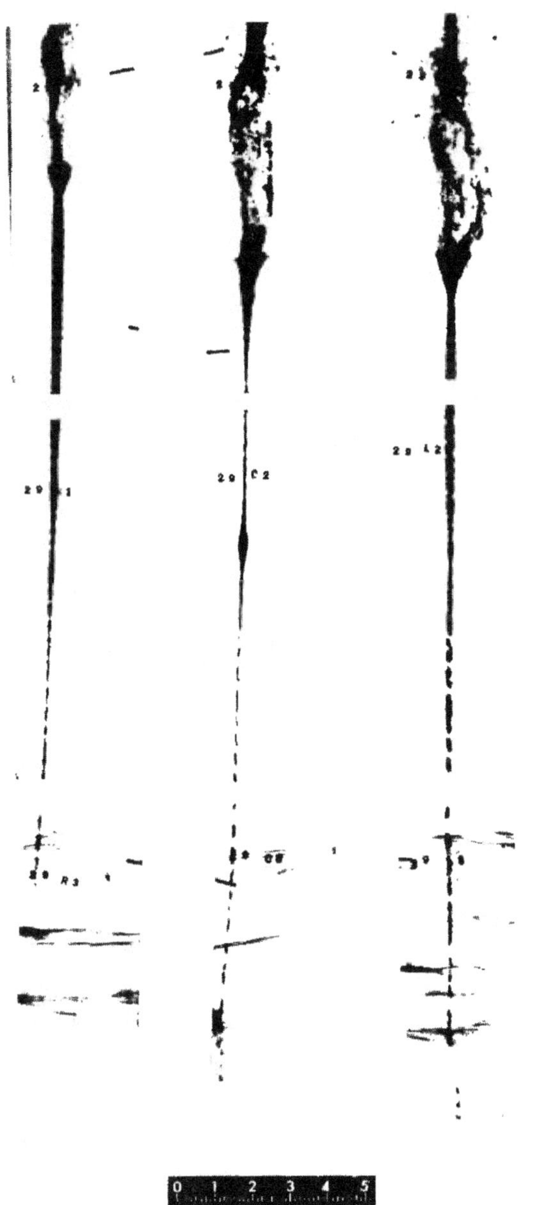

Figure 2-51. Jet radiographs

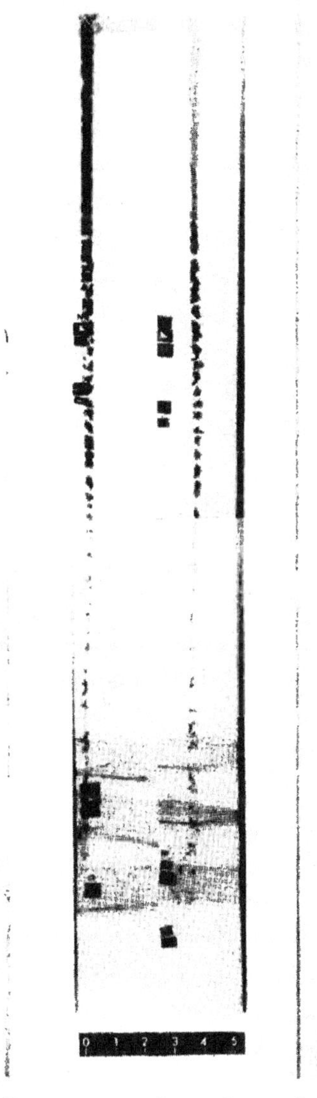

Figure 2-52. Jet radiographs

mental investigation of other parameters possible.

The advantages of a small-diameter cylindrical liner are the following.

a. The cylinder should exhibit a high ability to resist deterioration by rotation.

b. There is potential value in the possibility of making the penetration depend upon projectile length rather than projectile caliber.

c. The simplicity of the geometry should have advantages from the production viewpoint.

The possible disadvantages of such a liner are the following.

a. Very high precision will probably be required for the cylinder liner.

b. A wave shaping system is required, according to present designs, to get enough material into the jet to make the sizeable hole diameter essential for adequate lethality.

c. Present designs have, up to this time, given penetration performance no better than half that attainable in the same projectile with a cone.

Figure 2-53 shows the appearance of the jet from a 4-in. long cylindrical liner, of 1 in. interior diameter and 1.1 in. exterior diameter, in a heavily confined 105-mm body.

Figure 2-54 shows the hole made in mild steel by such a jet.

In summary, the designer should be aware of two developments, involving liner shapes, aimed at reducing sensitivity to rotational deterioration. Of the two, the system involving trumpets, with and without peripheral initiation, is much nearer realization and application than the system involving a cylindrical liner with a wave shaping device. Both of these systems should be distinguished from the fluted liners and other methods (discussed in paragraphs 2-129 through 2-141). The latter are more properly considered methods for actively overcoming the effects of rotation, while the systems discussed in this chapter are passive systems.

SPIN COMPENSATION

2-129. Fluted Liners. The most promising method of compensating for spin is the use of fluted liners. From the viewpoint of application, the best results that have been obtained to date are:

Figure 2-53. Jet radiographs

57-mm liners (charge diameter 1 5/8 in.)	4.0 charge diameters penetration at 360 rps
	5.0 charge diameters penetration at 180 rps
	4.7 charge diameters penetration at 250 rps
105-mm liners (charge diameter 3 1/4 in.)	6.2 charge diameters penetration at 50 rps
	4.5 charge diameters penetration at 85 rps

The potential performance of the 57-mm cones (as represented by smooth liners fired statically) is about 5.3 diameters penetration, and that of the 105-mm liners about 6.7 diameters, under appropriate conditions for comparison with the above. By interpolation from laboratory results, a penetration of 4.8 charge diameters should be readily obtainable from a 57-mm HEAT round

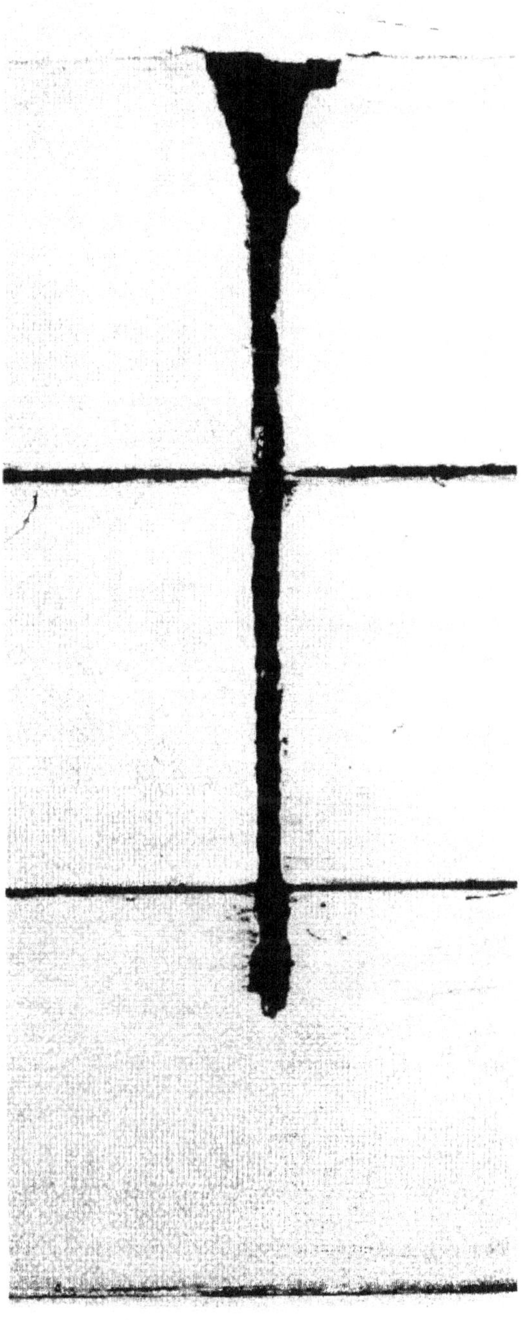

Figure 2-54. Jet penetration in mild steel

at its standard spin frequency of 210 rps. No liner has yet been tested that would provide very good spin compensation in standard 105-mm HEAT rounds (spin frequency about 200 rps). Although considerable development work is being done with fluted liners, no ammunition containing a fluted liner has been standardized.

2-130. Mechanism of Spin Compensation by Fluted Liners. It is now generally accepted that the detrimental effects of rotation are due to the requirements of conservation of angular momentum and the consequent tremendous rotational frequencies of the jet. In order to counteract this effect, it is obviously necessary that a tangential component of velocity be imparted to each element of the liner, by some means, that is equal in magnitude but opposite in direction to that set up by the initial spin of the liner. The simplest means of accomplishing this is to find a way of using the energy of the explosive to produce a countertorque on the liner.

The present concept of spin compensation is based on two phenomena that have been studied at the Carnegie Institute of Technology. One, sometimes called the "thick-thin" effect, is the observed dependence upon the thickness of the liner of the impulse delivered to a liner element by the product gases of detonation. The second, named the "transport" effect, is the dependence of the impulse delivered to the liner upon the angle at which the detonation products impinge on the liner. Both of these effects are strictly dynamic phenomena; that is, they are to be observed only in a rapidly flowing fluid. They represent departures from Archimedes' principle.

The thick-thin effect is represented graphically in figure 2-55. The curve shown was derived from the theory of shock waves and has been verified by experiment. A very similar result has also been obtained on the basis of gas kinetics. Application of the thick-thin effect to a fluted liner is also illustrated in figure 2-55. The impulse per unit area is always greater on the offset surface, since the thickness normal to that surface is greater. Furthermore, the impulse is directed along the surface normal. When the impulses delivered at all surface elements are resolved into radial and tangential components and summed, the total tangential component does not vanish, as in the case of a static fluid, but has a net resultant that produces a torque, in the direction shown, which can be used for spin compensation.

The transport effect can be represented simply by the equation

$$I_\theta = I_n\left(\frac{5 + \cos 2\theta}{6}\right)$$

where I_θ is the impulse delivered to unit area of a liner whose surface normal forms the angle θ with the direction of propagation of the detonation wave, and I_n is the impulse delivered in normal impact. This equation has also been derived from both shock theory and gas kinetics, and has been verified by experiment. It is significant in spin compensation because the angle e at which the detonation wave strikes the canted surface is generally (except in spiral flutes) less than for the offset surface. Thus, a net torque is produced in the direction opposite from that produced by the thick-thin effect.

By combining information gleaned from theoretical considerations, basic experiments, and observations with fluted liners, the following conclusions can be reached.

a. The phenomena responsible for spin compensation (that is, the thick-thin effect and the transport effect) are second-order in magnitude compared with the overall effect of an explosion on an inert liner.

b. The two effects are of approximately the same magnitude, but are opposite in direction under the conditions thus far studied experimentally. They are largely independent of one another and can be varied separately. Consequently, they are competitive, and either one can be made dominant by appropriate design, leading to the possibility of reversals in direction of spin compensation.

2-131. General Experimental Results With Fluted Liners. The essential effect of fluting is to introduce an angular impulse in the collapsing liner that, under appropriate conditions, can be made to compensate for the angular momentum due to initial spin. Thus, a fluted liner spun at its designed optimum frequency produces a jet exactly like that produced by an equivalent smooth liner fired statically. When fired statically, the fluted liner produces a dispersed jet like that from a rotated smooth liner. A set of flash radiographs (figure 2-56) taken at the Ballistic Research Laboratories illustrates this.

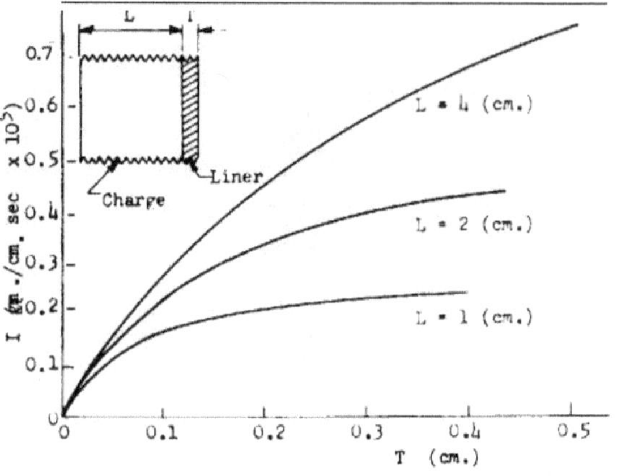

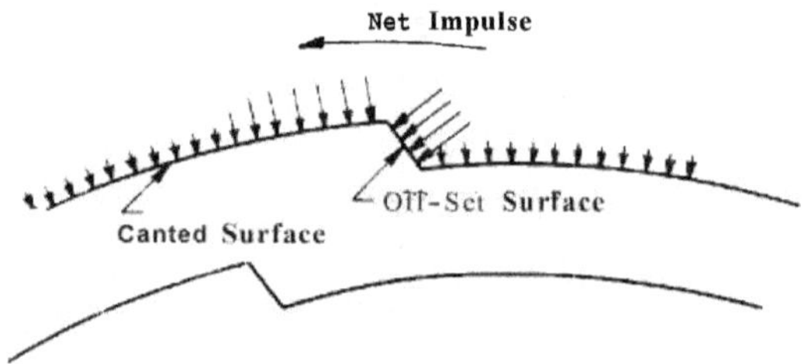

Figure 2-55. Illustration of the thick-thin effect and its application to a fluted liner. Due to the variation with liner thickness of impulse per unit area delivered by the explosion products to the liner (see upper figure), the impulse delivered to an element of a fluted liner depends upon the shape of the liner in the neighborhood of the element. The greater the thickness of the liner element (measured normal to the outer surface) the greater the impulse delivered to it. The result is illustrated schematically in the lower figure, where the lengths of the arrows roughly represent the magnitudes of the impulses. Because of the nonuniformity of the impulse, there is in general a net force tending to rotate the configuration in the direction of the curved arrow. The illustration, of course, oversimplifies the application, but conveys the general idea.

Further evidence of the effects of compensation is shown in the plots of figure 2-57, where experimental points and curves are given for the

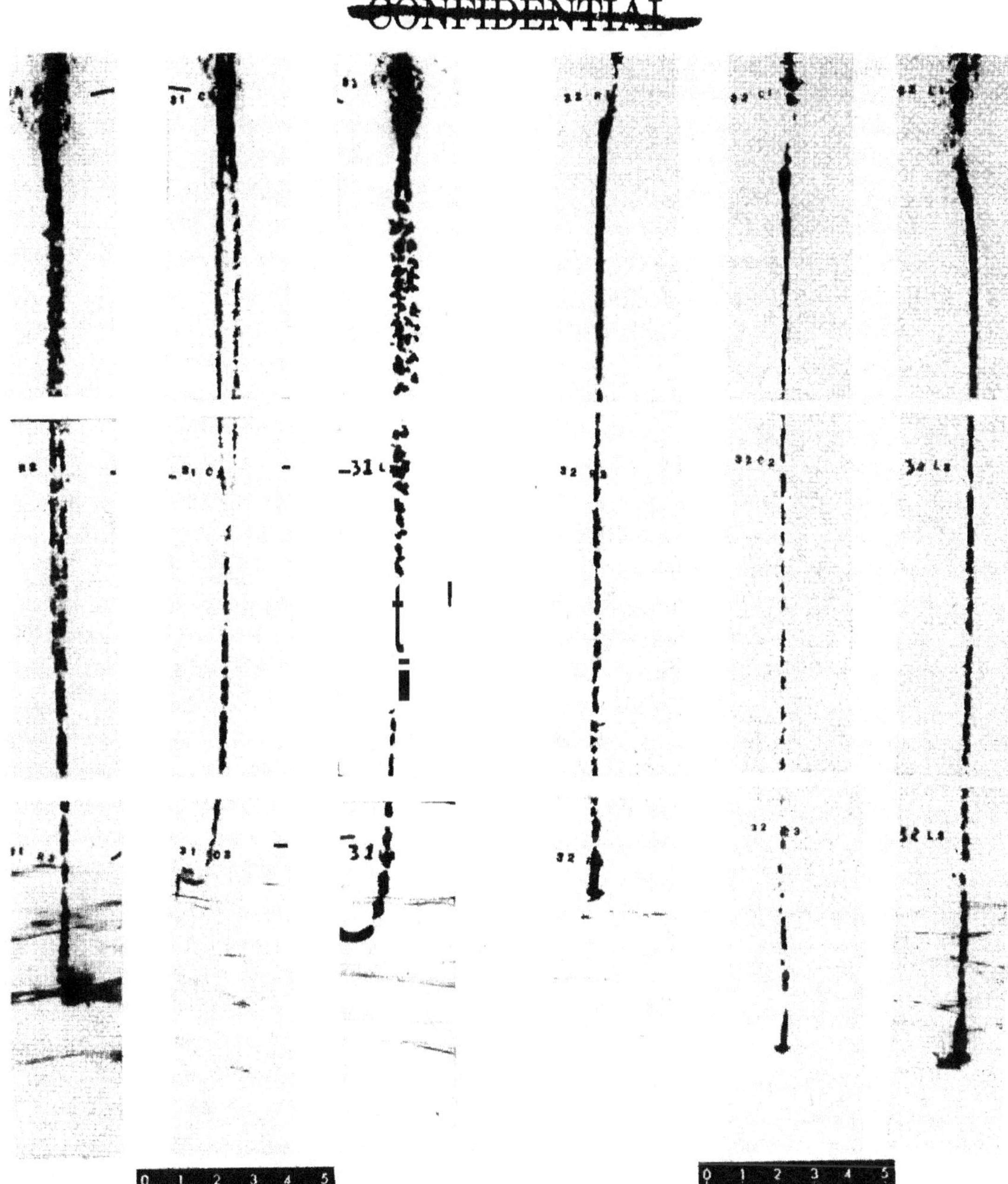

Figure 2-56. Radiographs showing effect of fluted liner

depth of penetration as a function of rotational frequency for smooth and fluted 57-mm and 105-mm liners. It is evident that the behavior of the fluted liners as the frequency is changed is the same as that of the equivalent smooth liners, except that the maximum penetration is obtained with the fluted liners at some rotational frequency other than zero. This optimum frequency is determined by the design of the flutes, wall thickness, and so on. The cases shown are typical. For the 105-mm liners, the fluted liners gave a higher average penetration at their optimum frequency than was obtained with the statically fired smooth liners. (The ostensible increase in penetration can be explained by reduction in average wall thickness of a cone initially thicker than optimum caused by machining of flutes; this is not a typical characteristic, of course.) With the 57-mm liners, the penetration at optimum frequency by the fluted liners is somewhat less than that obtained with statically fired smooth liners, but it will be noted that the optimum frequency is 250 rps (well above the rate of spin of a standard 57-mm HEAT shell). The experimental points shown on the plots also illustrate that the variability in performance with a satisfactorily made fluted liner is no greater than the variability of the equivalent smooth liners.

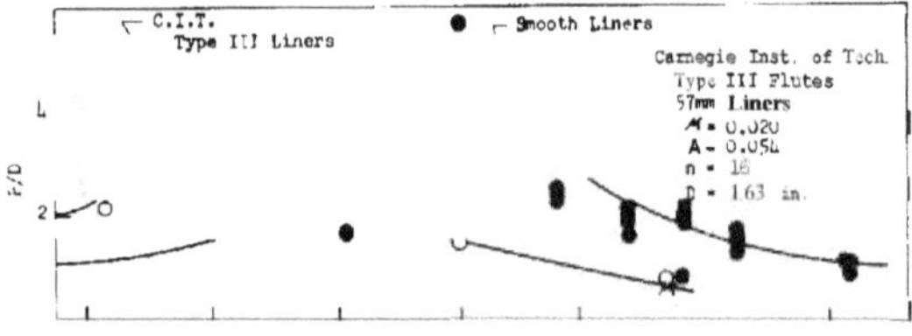

2-132. Scaling Relations for fluted liners are not yet well established. Theoretical considerations based on modeling laws lead one to expect that, for liners and charges that are geometrically similar in all respects, the optimum frequency (that is, the frequency at which the highest degree of compensation is obtained) should vary as $\frac{1}{d}$, d being the charge diameter. The only experimental evidence available at present is obtained by comparison of results with liners of different sizes that are not really scaled replicas. Early comparisons of this sort seemed to indicate that ν_0 was more nearly proportional to $\frac{1}{d^2}$, but this has since been contradicted by work at both Firestone and Carnegie Institute of Technology. At present, it appears that $\nu_0 \propto \frac{1}{d^n}$ with n slightly larger than unity. The uncertainty of the experimental comparisons is such that the departure from the theoretical expectations is not certain. Consequently, through the remainder of the discussion of spin compensation, the theoretical scaling relation $\nu_0 \propto \frac{1}{d}$ will be adopted. Definitive scaling tests are being carried out at the time of writing, but have not been completed.

It must be noted that even the most favorable scaling law that can be anticipated raises a great deal of difficulty in obtaining compensation at standard spin rates with large liners. In order to obtain compensation in a 105-mm shell, the ratio of flute depth to charge diameter or to liner wall thickness must be about twice as great as that required to obtain compensation at the same spin rate in a 57-mm shell. With flute designs that have been tested to date, it has not been possible, for this reason, to achieve a useful depth of penetration with the larger liners at standard spin rates.

With regard to the possibility of achieving compensation at very high spin rates, it can only be stated now that there has appeared no essential limit to the attainable optimum frequencies. It is certain, however, that the difficulties will increase rapidly as the optimum frequency sought increases. Certainly, the simple designs of flute and the relatively liberal tolerances used to date must be altered. A practical limit to the attainable compensation frequency exists, certainly, but its magnitude cannot be estimated from present information. It seems entirely possible that compensation can be achieved at values of $\nu_0 d$ as high as 1,000 in. per sec without requiring impractical designs.

2-133. Specific Experimental Results With Fluted Liners. Five distinct types of flutes have been tested to date. They are illustrated in figure 2-58 and given designations that will be used throughout the following discussion. Class I and Class II flutes are formed between one fluted metal die (male for Class I, female for Class II) and a rubber-padded smooth mate. The undulating flute formed by the padded tool characterizes both types. Class III flutes are formed between matching fluted metal dies and Classes IV and V between one fluted die (female for Class IV, male for Class V) and a smooth metal mate. As will be seen presently, quite different results are obtained with the various types of flutes.

All significant flutes tested thus far have been made so that (at least nominally) their depths increased linearly with cone radius; hence the flute depth can, at least nominally, be represented by

$$a = \mu R$$

where μ is a constant for each cone. The designs have been limited so that any can be nominally described by the five design parameters illustrated in figure 2-59, and defined as follows.

$\mu = \dfrac{a}{R}$

$\lambda = \dfrac{T}{R}$

n = number of flutes

ψ = angle between flute offset and radius through its root

δ = angle of indexing (when matching fluted tools are used)

where

a = flute depth

R = pitch radius of liner element

T = wall thickness of blank before fluting.

Tests have been carried out with all types of liners, but the Type III liner is by far the most promising.

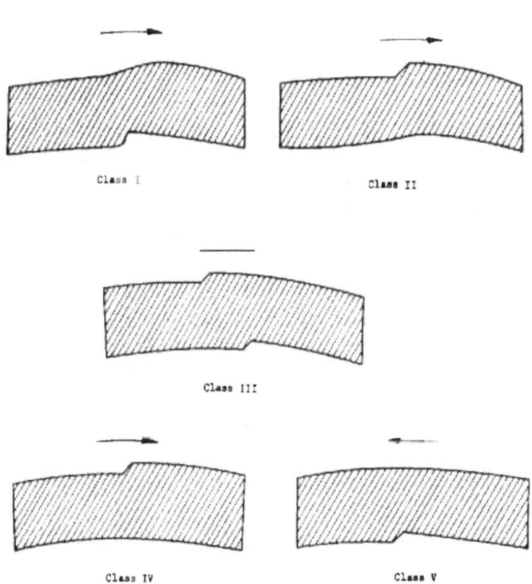

Figure 2-58. Profiles typifying five general classes of flute design. Arrows indicate direction of compensative impulse for small numbers of flutes; for Class III flutes, direction of compensation depends on index angle as well as on number of flutes.

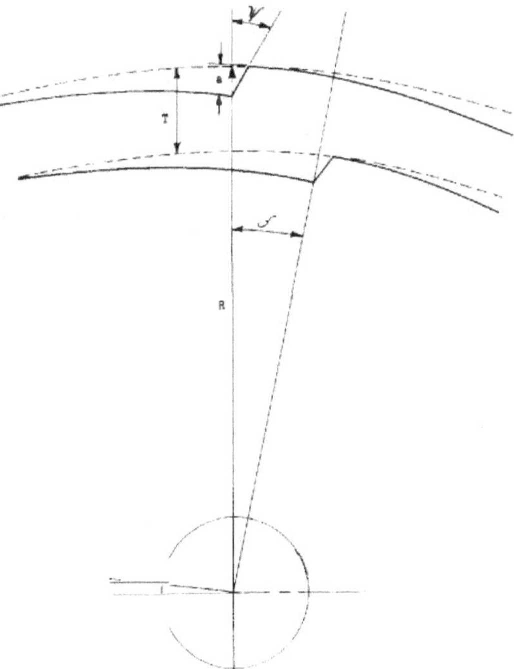

Figure 2-59. Definition of design parameters for fluted cones

2-134. The Behavior of Type III Flutes is best described by considering series of tests in which only the index angle has been varied. The results of a series of tests with 1 5/8-in. charges containing liners with 16 flutes of maximum depth 0.015 in. are illustrated in figure 2-60.

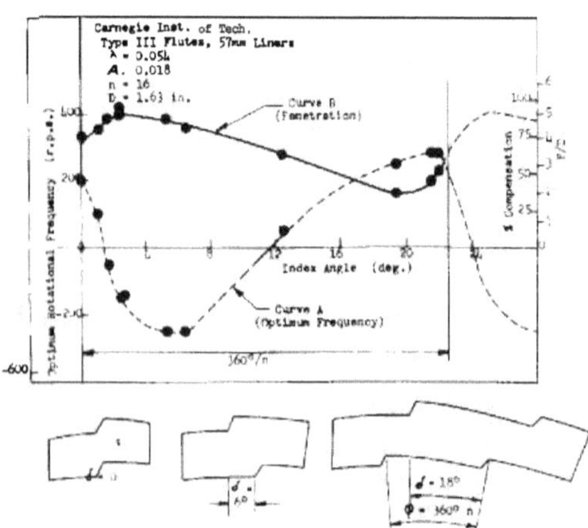

Figure 2-60. Variation of optimum frequency (cuvveA) and the penetvation at optimum frequency for a specific group of liners with Type III flutes. Sketches illustrate appearance of liner profiles for several index angles. Variations in penetration are due to varied degrees of necking of the flute profile. Sketches illustrate change in flute contour with varying index angle.

The relation between optimum frequency and index angle is, of course, cyclic, repeating itself at intervals of 360/n degrees —that is, at 22 1/2 degree intervals for the case illustrated.

2-135. Variation of Indexing. It is evident from the plot that variations in indexing alone, and the attendant changes in relative magnitude of the competing mechanisms of compensation, cause drastic variations in optimum frequency. The range covered in the experiment illustrated is from +275 rps to -250 rps — that is, a range of 525 rps. Of especial academic interest are the index angles 1 1/2 and 11 1/2°, at which the competing mechanisms exactly balance and produce zero optimum frequency. Of more practical interest are the indexings -1/2 (or +22) and 6°, where the largest (absolute values) optimum frequencies were obtained. It is evident that there is a 'definite preference, on the basis of penetrating ability, for the 6" index angle.

The results illustrated in figure 2-60 have now been substantiated by tests with 105-mm liners that are approximate scaled models of the 57-mm liners used in the original tests. The results of the larger scale tests are compared in figure 2-61 with the original; the close agreement between the two sets of observations is evident.

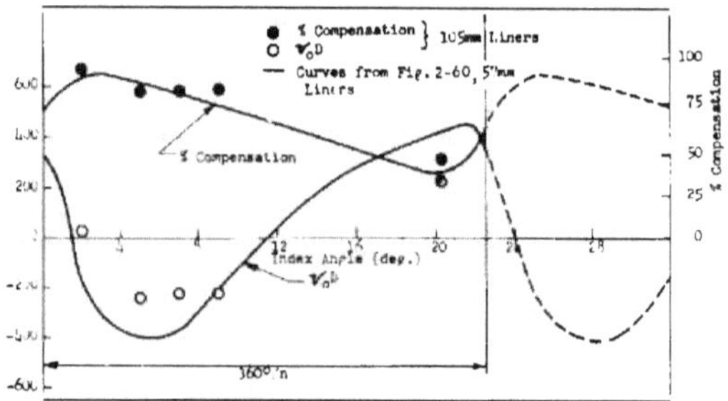

Figure 2-61. Comparison of Firestone tests with 105-mm liners having Class III flutes with results on 57-mm liners. Coordinate axes are normalized to permit combination of data. Since the two sets of liners and charges were not accurate scaled models, exact quantitative agreement is not to be expected. The close qualitative similarity substantiates the original C.I,T. observations.

2-136. Variation of Thickness. A brief series of tests has also been completed with liners formed with the same dies used in the experiment illustrated in figure 2-60, but with liners of approximately 50 percent greater wall thickness (0.063 in. instead of 0.045 in.). The results are compared in figure 2-62 with those of the original tests. While the general features of the two sets of observations are very similar, there is some evidence that the ratio of the optimum frequencies obtained with the two different wall thicknesses varies with the index angle and that the indexing at which zero compensation is observed may also depend upon wall thickness. If these indications are substantiated by further tests, it will mean that the behavior of Type III flutes cannot be reduced to simple empirical relations.

Still a third set of experiments is illustrated in figure 2-63. In these tests, the wall thickness of the liners was the same as that used in the original tests (that is, 0.045 in,), but two different flute depths were used, one shallower, and one deeper, than in the original series. Both sets of observations exhibit general features similar to those of the first tests, although there is some unsubstantiated evidence of variations in the indexing that produce zero compensation and of a nonlinear relation between optimum frequency and flute depth for a given index angle.

The behavior of Type III flutes is even more complicated than that of the four types formed with single dies. Variations in the additional variable δ produce complex changes in the geometry of the liner and, consequently, in the shock interactions that affect spin compensation. The very considerable technical advantages of the Type III flute, which are discussed more fully later, more than recompense for their more complex behavior.

2-137. Effect on Penetrating Power. The fluting of a liner can also affect its potential penetrating power quite drastically. Even though the impulses involved in compensation at the spin rates attained thus far are too small to affect appreciably the basic character of the cone collapse or of the jet formed (that is, specifically, one does not expect an appreciable change in the distribution of energy in the jet), it is quite evident from both Firestone and Carnegie Institute of Technology experiments that mechanical strength effects to a very large degree govern the depth of penetration obtained at optimum frequency. This is

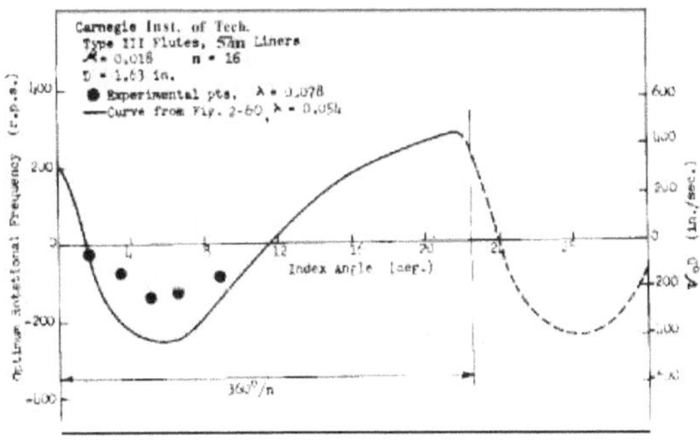

Figure 2-62. Comparison of optimum frequencies at various index angles obtained with liners of 0.063-in. wall thickness and with 0.045-in. wall thickness. The experimental points representing the thickev liners are tentative and subject to slight changes pending completion of gaging analysis.

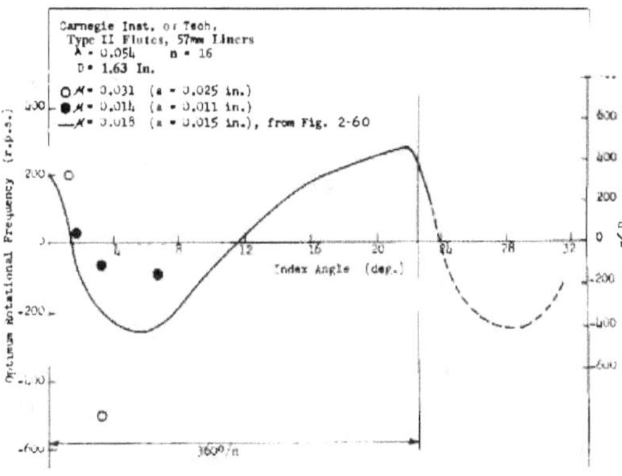

Figure 2-63. Comparison of optimum frequencies at various index angles for three flute depths. The experimental points are tentative and subject to slight changes pending completion of gaging analysis.

to be expected of liners having linear flutes (that is, $\mu = \frac{a}{R}$ constant along the flute), since all theoretical and experimental evidence indicates that the ideal flute is far from linear. With a linear or any other nonideal flute, the various elements of the liner tend to compensate at different frequencies rather than at a common frequency of rotation. So long as the natural frequencies of adjoining elements are not too different, or so long as the liner wall is sufficiently strong to resist the tendency for relative rotation of the elements, this causes no serious difficulty. But if the liner is badly necked in the fluting, the strains set up by such a situation cause the liner to rupture, instead of collapsing coherently and forming a jet. Theory, however, cannot be made to yield a usable design. The task of determining the ideal form of $\frac{a}{R}$ as a function of position on the liner must for the present be an empirical one. It has been undertaken, but no reportable results are available as yet.

The experimental evidence of deterioration in penetration, due to mechanical strength effects, is best illustrated by means of correlation between the depth of penetration at optimum frequency and the minimum thickness of the wall of the fluted liner. (Usually the minimum thickness is found near the base of the flute, where necking occurs.) Figure 2-64 shows plots for both Carnegie Institute of Technology and Firestone data of correlations between the

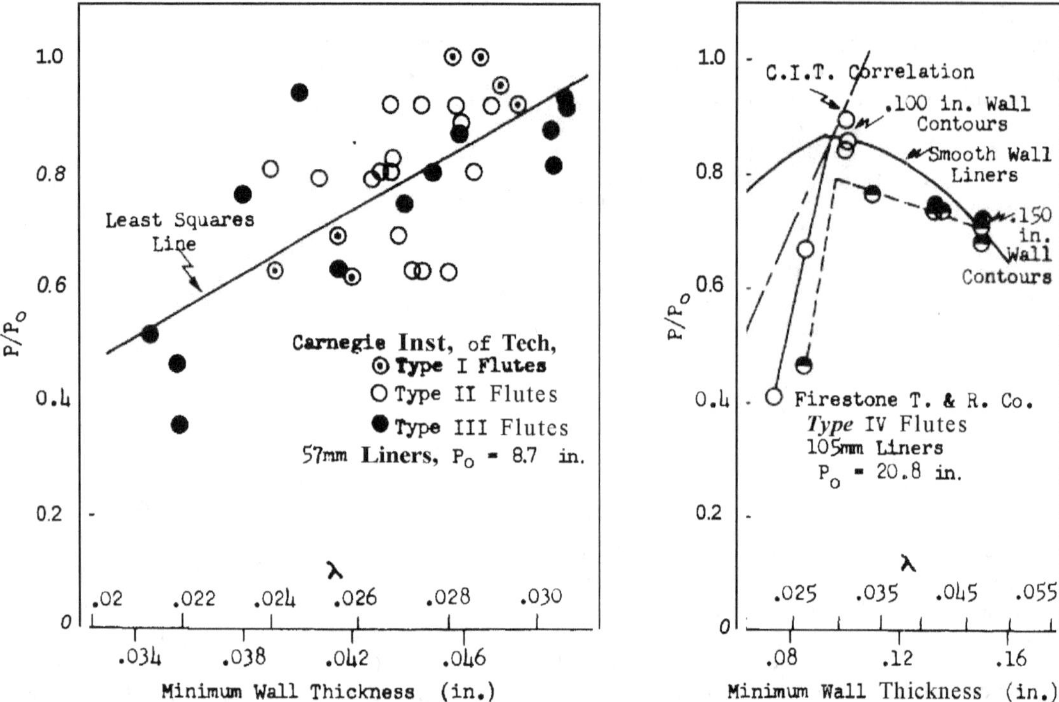

Figure 2-64. Correlation between penetration obtained at optimum fvequency of rotation and the minimum wall thickness across aflute profile. The observed penetrations have been divided by the penetration obtained from optimum smooth liners fired statically. The large-scale plot of C.I.T. data on the left shous a great deal of scatter but a correlation is evident. It has been assumed that all three types of flutes included can reasonably be treated as a single statistical population in this treatment. On the right, the C.I.T. cowelation line is shown with the correlation of Fivestone data. The agreement is as good as can be expected in view of the scatter of the C.I.T. data and the relative scarcity of Fivestone data. It is tempting to conclude that all four types offlutes follow the same correlation, and that linear scaling laws hold. (P_0 indicates the average Penetration by the best available smooth liners of each size.)

maximum penetration observed and the minimum wall thickness. It is evident that there is a critical value of the minimum wall thickness. When the thickness falls below this value, the penetration falls off very rapidly with decreasing thickness. There is, of course, asecondary correlation between flute depth and maximum penetration for any homologous series of cones, because increasing flute depth inevitably produces more pronounced necking of the liner wall and decreases the minimum wall thickness. Analysis shows that the primary correlation is that with minimum wall thickness, however.

One of the most interesting observations of this sort has been made in connection with the indexing tests described earlier. If the two curves P_{ν_0} versus δ and ν_0 versus δ shown in figure 2-60 are used to eliminate δ, the plot of P_{ν_0} versus ν_0 shown in figure 2-65 is obtained. Such a correlation is of practical interest, although as pointed out above it does not represent any fundamental relationship (these same liners are included in the general correlation between P_{ν_0} and minimum wall thickness shown in figure 2-64. Figure 2-65 shows that for Type III flutes, within the limits of optimum frequency fixed by the values of μ, A, n, and so on, used, one can obtain a given magnitude of optimum frequency by four different indexings. But, because the different indexings result in different degrees of necking in the liner wall, different penetrations are Obtained, so that there is a clearly optimum choice of indexing for overall performance. Caution must be used in generalizing from figure 2-65.

It applies only to a specific group of liners and does not in any way represent limitations on either the optimum frequency or the depth of penetration that can be obtained with other designs.

While compromises between wall thickness and flute depth can provide suitable combinations of optimum frequency and maximum penetration temporarily, the ultimate solution is to eliminate the influence of mechanical strength effects by use of appropriate nonlinear flutes.

2-138. Variability in Performance of Fluted Liners and Tolerances Required. For the types of liners tested to date by Carnegie Institute of Technology and Firestone, tolerances of the magnitude given in table **2-15** yield performances that are not appreciably more erratic than those of equivalent smooth liners.

Table 2-15

Parameter	Permissible tolerances (nominal)	
	57-mm	105-mm
a	± 0.001 in.	± 0.002 in.
T	± 0.002 in.	± 0.004 in.
$\frac{360°}{n}$*	± 15 min	± 15 min
ψ	± 2 deg	± 2 deg
δ	± 15 min	± 15 min

*The angle subtended at the axis by each flute.

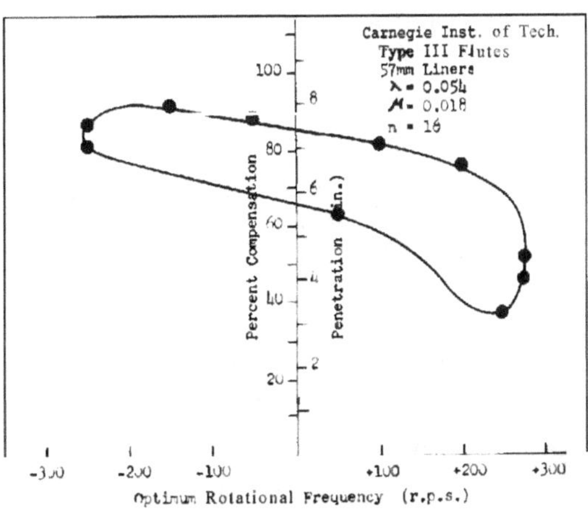

Figure 2-65. Correlation between penetration at optimum frequency and the optimum frequency of rotation for a particular group of liners having Type III flutes, the index angle being varied. For any (absolute magnitude) frequency up to 250 rps, any one of four index angles can be used. Only the one having the greatest minimum wall thickness is of practical interest, however, because it yields better penetration than the others. Consequently, only those liners corresponding to the upper left branch of the above Plot (1 1/2≤δ≤6 degrees) are of interest. Since liners of this series follow the correlation of figure 2-60, it is possible to predict, within limits, the range of index angles that are of interest in a series of this sort. The range will, of course, depend on the scientific values of λ, μ, etc. used.

The tolerances quoted are somewhat conservative. Quite respectable performance could be expected with somewhat more liberal figures. In the case of the 57-mm cones, especially in the early tests, the variations in test pieces have been much larger than the tolerances given. There seems no reason, however, why reasonably careful techniques should not be capable of providing pieces well within the specifications.

The shapes of the flutes must also be consistent, of course, although it is difficult to give quantitative tolerances for shape. Other parameters should be kept to the same tolerances that have been established for smooth liners.

2-139. Methods for Manufacturing Fluted Liners. One of the chief obstacles in early research on spin compensation was the procurement of suitable test pieces. The difficulties have now been very largely overcome, although manufacturers undertaking the task of producing fluted liners for the first time sometimes experience recurrence of the old troubles.

The only significant method of manufacture used to date is pressing. Firestone has made extensive use of machined liners of Type IV design, which are entirely satisfactory for laboratory purposes for that one design only. The method is entirely unsuited to quantity production. Die casting has been considered many times, but thus far appears unlikely to yield pieces of adequate homogeneity or dimensional stability. It is improbable that die casting could compete with pressing on an economical basis, in any event. Electrodeposition of liners with flutes has also been considered, but has never been attempted. There is little chance that such a procedure would be competitive, either economically or in quality of product.

2-140. <u>Means of Spin Compensation Other than Fluted Liners.</u>

a. The use of spiral detonation guides (variously called "lawnmowers" and "spiral staircases") to guide the detonation wave in sections along separate spiral paths has been tested by Firestone and Carnegie Institute of Technology. The technique introduces a component of impulse in the appropriate direction for spin compensation. Experiments indicate that the principle operates as expected, but that the optimum frequencies attainable without excessive loss of penetrating power are relatively low.

b. A technique of fluting the explosive adjacent to a smooth liner has been tested by Carnegie Institute of Technology. The flutings were similar to those ordinarily used on the liner. Spin compensation was observed, as expected, but the procedure seems unlikely to afford as great benefits as do the fluted liners.

c. Very recently, Firestone has carried out a series of tests with smooth liners made by a "shear-forming" process. The liners were very well made insofar as dimensional characteristics are concerned, and the charges were of standard design. The shear-forming process, however, tends to produce spiral deformations of the liner, even though they may be invisible on superficial inspection. The deformations may be manifested in the final product by asymmetric variations in density, metallurgical properties, or dimensions. Liners made by this process had previously been tested statically at Ballistic Research Laboratories and found to perform poorly; these results had prompted an analysis by Pugh that may have some bearing on the Firestone results.

Firestone fired shear-formed liners of 90-mm size at several different spin rates, obtaining the penetration depth versus frequency plot shown in figure 2-66. The liners showed a definite and, for liners of the size, fairly large optimum frequency. The tendency of these liners to compensate for spin is presumably due to the asymmetries caused by the method of manufacture. Much more experimental work

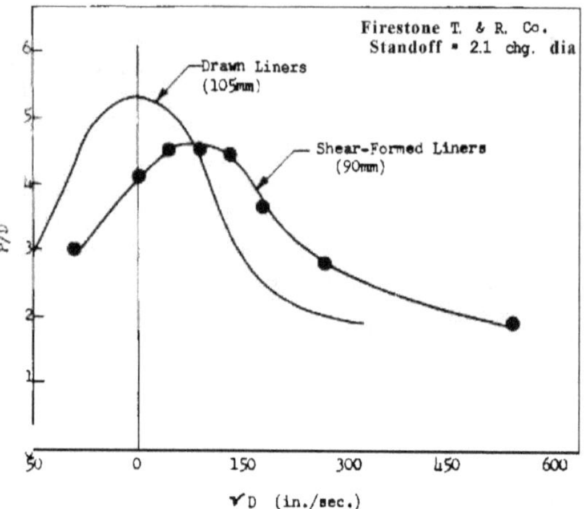

Figure 2-66. Experimental observations by Firestone showing spin compensation by smooth liners made by shear-forming process

must be done before the practical significance of this observation can be evaluated.

2-141. <u>Spin Compensation (by Use of Fluted Liners) Compared With Other Methods of Eliminating Spin Degradation.</u> For standard small-caliber weapons (57-mm and probably 75-mm), there is little room for argument concerning the best method of eliminating spin degradation. Designs are already available and tested for fluted liners that will provide performance, at the standard spin-rate of the 57-mm HEAT round, essentially equal to that obtained with smooth liners fired statically. The development of a comparable liner for 75-mm rounds should be relatively simple. There is no comparison so far as simplicity and low cost of application are concerned between fluted liners and the other methods available.

For larger weapons (90-mm, 105-mm, and larger), the situation is quite different. In the special case of the 105-mm BAT weapon, Firestone has found the slow-spin T138 round with a fluted liner to be the best solution. Static, smooth-cone performance is obtained at spin rates up to about 60 rps with available fluted liner designs, and shells have been designed that afford sufficient accuracy at spin rates below 60 rps. The design is not a satisfactory solution to the general problem, however, because it requires a special rifle. It is necessary to find a shell design that can be fired from guns currently in the field or in production.

There are three alternative exterior ballistic solutions to the problem under consideration. The concept of using peripheral jet engines on the shell to stop its spin before the target is reached has not yet been adequately tested. Firestone, in conjunction with Picatinny Arsenal, has some tests planned, but computations of the torque needed offer little hope that this method will prove practical.

The use of bearing-mounted charges that permit part of the shell to spin, for stability, while the charge itself spins very slowly, if at all, has been proven practical by Firestone and Frankford Arsenal and by Midwest Research. Two designs have been used — one, a tandem arrangement, in which the front part spins; the other, a concentric arrangement in which the outer shell casing spins about a relatively slow-spinning core containing the charge. Either design seems capable of providing charge spin rates of the order of 50 rps *or* less when fired from currently standard rifles. Clearly, a fluted liner is needed to compensate for even those low spin rates in large shells, and it immediately becomes critical to know whether the spin frequency of the charge is consistent.

A number of laboratories, both industrial and military, have been working on fin-stabilized rounds. Both long-boom and folding-fin models have been designed that can be fired from standard weapons with apparently satisfactory accuracy. Even these have a small amount of spin imparted to them (ordinarily less than 20 rps) to improve interior ballistics, so a fluted cone might be desirable for large caliber shells.

A comparison of the bearing-mounted charge with the fin-stabilized shells is very difficult at present, because of the limited experience with bearing-mounted projectiles. It seems certain that they must weigh more than a standard spin-stabilized projectile of similar caliber and that the concentric design demands the use of a smaller-than-normal liner. It is not at all certain, however, that they will weigh more or be more costly than finned rounds. The fin-stabilized shells involve some complications in design and firing, but would appear to offer the better solution of the two for larger caliber rounds.

It seems clear, however, that any technique for reducing the rate of spin of the charge can provide only an interim solution. Once a fluted liner can be designed to compensate for the frequency of spin of a standard round of any given caliber, it immediately provides the most convenient and economical solution to spin degradation for that caliber.

DEFEAT OF SHAPED CHARGE WEAPONS

2-142. Defeat of Shaped Charge Weapons. Of all possible means of defeating shaped charges, the most promising found to date consists of a combination of glass and steel armor. The glass may be in the form of plates, blocks, or large balls, possibly in conjunction with a suitable shock absorbing material. This means of protection has the advantage of low density and hence low overall weight, in addition to utilizing the abnormal stopping power of glass, which has not been approached by any other method of passive defense. Titanium displays abnormal stopping power to a much smaller extent. Recent results suggest that explosive pellets or linear-shaped charges can provide a very high degree of protection under certain circumstances, but these last two methods have not been developed to the point where they can be considered practical.

TERMINAL BALLISTIC EFFECTIVENESS OF SHAPED CHARGES AGAINST TANKS

2-143. The Criterion of Shaped Charge Effectiveness must in the final evaluation be its ability to defeat an armored vehicle — not just to perforate its armor. The first element to be considered in the criterion is that of perforation; the second element is damage after a perforation. Perforation of a tank's armor is not the same as perforation of an idealized target. Much of the target presented by a tank is irregular and nonhomogeneous in composition, so that predictions of the effect of perforation cannot be made from flat plate data. Damage does not occur to a tank just because a perforation of the armored envelope occurs. The perforation must cause further destruction inside. This damage depends upon where the perforation occurs and the residual damaging effect of the jet after the perforation. Other factors such as stand-off, liner material, and so on, also are involved.

Usually, a tank is destroyed by releasing the forces that it carries within it. Thus, tank destruction depends to a great extent on ignition of the ammunition or fuel. However, other types of damage reduce tank effectiveness equally well, depending upon the circumstances of combat. For instance, if a tank is immobilized during a retreat, it is lost just as surely as though it had been burned. In the case of an attacking force, if the firepower of the tank is destroyed, the tank is no longer of use in the particular action.

Three categories of damage to a tank have been defined.

K Damage is damage that will cause the tank to be destroyed.

F Damage is damage causing complete or partial loss of the ability of the tank to fire its main armament and machine guns.

M Damage is damage causing immobilization of the tank.

2-144. Sources of Terminal Ballistic Data. There are three sources of data for terminal ballistic damage of shaped charges to tanks: historical data, terminal ballistic firings at tanks, and box tests. Each of these three methods has its importance.

2-145. Historical Data. The principal historical data compiled on the damage effectiveness of shaped charges were obtained by the Military Operations Research Unit in Great Britain. The data available are entirely for German infantry hand-fired weapons, such as the Panzerfaust and the Raketenpanzerbuchse, against U. S. and British tanks. This information is valuable in that it gives some idea of the points of impact, the ranges of engagement, and the crew casualty experiences in World War II. Table 2-16 shows the proportion of Sherman and Stuart tanks that burned completely and the proportion that were repairable, as a result of shaped charge attack.

Table 2-16

Total no. considered	No. burned	Percent burned	Percent casualties repairable
64	27	42	54

This table suggests that most of the tanks that were not burnt were repairable. Table 2-17 suggests that the rule of thumb that one man killed and one wounded for a perforating round is a good one for shaped charge rounds.

Table 2-17

No. of men	Percent	Percent	Percent
235	20	24	6

These are short-range weapons. Kills rarely occurred at ranges greater than 120 yards. The mean range of 227 allied casualties (all the collected data available on shaped charges from World War II) is 43.5 yards. The angular distribution of casualty producing attack data available for these short-range weapons is shown in table 2-18.

Table 2-18

No. of perforations	No. on front	No. on side	No. on rear
100	30	63	7

2-146. Terminal Ballistic Firings. The second and the most definitive source of damage information is the proving ground firing at a tank. Such firings have been carried out by the British, the Ballistic Research Laboratories, and other organizations from time to time.

The largest and most systematic program has been at the Ballistic Research Laboratories. The method of obtaining these data has been as follows. A fully equipped tank (usually a T26E4 or T26E5) is loaded with wooden crew members in each crew position and stowed with inert ammunition. A small amount of fuel is placed in the fuel tanks to operate the engine, so that it can be running when the tank is fired on. A round is then fired on a selected surface of the tank. The angles of fire usually considered are normal to front and side and 45" azimuth to front and side. Another angle of attack is at 45° elevation angle. The range of firing varies, depending on the round to be tested. For instance, the 90-mm T108 round was fired at 500 yards and the 3.5-in. rocket was fired at 100 yards. For the first firings the attempt is made to cover the tank with hits in a fairly uniform manner. However, after the nature of the damage of a particular round is generally understood, the rounds are fired at those surfaces of a tank where there is the greatest doubt about the damage.

As soon as a hit is obtained on the tank, two combat-experienced assessors go to the tank and examine the damage. So long as the tank is operable, operable components are checked (such as turret traverse, gun elevation, radio intercommunication, and so on). The damage is then assessed with a description of every item of damage to the tank. These descriptive assessments are then translated into numerical assessments, which have been determined by the assessors to be standard. A list of standard assessments of components is given in table 2-19. M, F, and K damage are defined in paragraph 2-143.

Table 2-19 is representative and is not inclusive of all the damage that could happen to such components as the electrical circuit, and so on.

The determination of a personnel kill is made from examination of the wooden dummies. Assessment of the possible damage that could have occurred from fuel or ammunition fires is made by observing where the hits occurred and correlating this with actual experiments carried out against these components separately. Fuel and ammunition are removed from the tank prior to firing, for practical reasons.

The descriptive and numerical assessments are the basic data for the analysis of tank vulnerability. The information is contained in this form in the firing records. British data are presented in a somewhat different form in that only the descriptive part of the damage is given.

2-147. Box Tests. A box is placed behind the armor plate. In this box are instruments to measure pressure and temperature. Usually there are witness plates to give an indication of scatter. The box appears to be an admirable way to obtain developmental data on shaped charge design. Its value will probably increase as correlation can be established between box measurements and tank damage.

2-148. Qualitative Description of Shaped Charge Damage. Fin-stabilized shaped charge rounds

Table 2-19
List of standard assessments

	Probability of overall damage to tank when destroyed		
	M	F	K
Ammunition			
Cases — main gun	...	...	1.00
HE projectile	...	...	1.00
Small arms stowed in turret bustle	...	.15	...
Small arms stowed in driving compartment	.20	.10	...
Small arms stowed near loader	...	.15	...
Grenade box	.30	.45	...
Personnel			
Commander	.30	.30	...
Gunner	...	.20	...
Loader	...	.15	...
Driver	.10	.10	...
Bow gunner	.10	...	...
Gun			
Main gun and breech	...	.80	...
Equilibrator	...	.80	...
Elevating and traversing mechanism	...	.80	...
Recoil mechanism	...	.80	...
Coaxial machine gun	...	.10	...
Bow machine gun	...	.10	...
Engine compartment			
Engine, transmission	1.00	...	...
Oil and coolant coolers	1.00	...	...
Fuel tanks	...	...	1.00
Battery	.40	.40	...
Fighting compartment			
Radio and intercommunication	.50	.10	...
Fire control — dependent on system	...	...	...
Driving controls	1.00	...	...
Heater (using liquid engine coolant)	1.00	...	...
Exterior components			
Front idler hub	.50	...	...
Track	1.00	...	...
Driving sprocket	1.00	...	...
Final drive	1.00	...	...
Track guides	.10	...	...

with copper liners, such as the 3.5-in. rocket, 90-mm T108, 2.75-in. FFAR, and 8-cm AR, will usually do damage (provided "sufficient residual penetration" is available) in a narrow cone along the path of the jet. Fragments can be expected to do damage to soft targets, such as personnel and communications equipment. The fragments are not likely to ignite ammunition. The jet will ignite the projectile propellant. The jet also will ignite gasoline by a perforation into the fuel tanks, either above or below the fuel level.

Diesel fuel is not nearly so easily ignitable as gasoline. Rounds with a large residual penetration have an appreciably better chance of igniting diesel fuel. Another effect in the diesel fuel firings is produced by container size. In the firings of the 3.5-in. rockets, small containers containing five gallons of diesel fuel were not ignited in 13 attempts.

Exactly what constitutes "sufficient residual penetration" cannot yet be specified. The amount of damaging power left in a shaped charge jet after a target perforation that is necessary to do damage will vary, depending on the point of entry into the tank. If "residual penetration" is acceptable as an index, the range of values that can be selected is probably greater than 1 in. and less than 3 in., to do the type of damage that is confined to a narrow path behind the perforation. A figure frequently used is 2.5-in. residual penetration.

The damage from shaped charges using liner materials other than copper is somewhat different. Materials such as steel or aluminum tend to cause more fragments to fly off the rear face of the armor, and thus fragment damage is more widespread than damage from copper cones. However, neither steel nor aluminum lined cones have as great a penetration as copper cones of the same diameter. Both steel and aluminum cone-shaped charges produce considerable pressure effects inside a tank upon perforation. The pressure from aluminum cones is apparently somewhat greater than from steel. Tests on animals placed in a tank fired on by a 5-in. shaped charge showed them to be unharmed unless hit by fragments. The approximate pressure measured by paper blast gages was of the order of 50 psi. This pressure did, however, tear off hatch doors and bend bulkheads within the tank.

Although shaped charges do not in general wreck a tank by their own energies, they are nearly equally as efficient as the kinetic energy rounds in igniting fuel and ammunition in the tank. Shaped charges are equally as good as kinetic energy rounds at knocking out the engine or transmission. They do not, however, assure a kill, when a perforation of the tank's armor occurs, any more than do kinetic energy rounds of the same caliber.

2-149. Target Characteristics. Tables 2-20, 2-21, and 2-22 contain information regarding the armor of various tanks and the effectiveness of several HEAT projectiles in penetrating this armor.

Table 2-20 gives the probability of encountering an obliquity of θ or less for various tanks, averaged over the expected angles of attack, if the attacking projectile strikes the presented surface of the tank in a random manner. In averaging, the distribution of angles of attack was considered to be either circular or in the form of a cardioid, as noted in the table. The circular distribution is approximately what would be expected in the case of attack by handheld AT weapons, and the cardioid distribution is what would be expected from mounted AT guns.

Table 2-21 shows the probability of HEAT projectiles encountering an equivalent armor thickness t_e or less, averaged over the expected angles of attack for various tanks, assuming that the projectiles strike the presented area of the tank in a random manner. An equivalent armor thickness is a thickness at 0° obliquity that gives the same protection as some other combination of thickness and obliquity. The distributions of attack angles are the same as used in table 2-20. Shielding by external components was not considered in the preparation of tables 2-20 and 2-21. However, the net effect of external components is to lower the values in table 2-21 by about 10 percent.

Table 2-22 gives the portion of the presented area of the armored parts of various tanks which can be penetrated from various angles of attack. Shielding by external components was not considered in the preparation of this table.

2-150. Methods of Data Analysis. The reduction techniques of damage data for tanks have

Table 2-20

Distribution of angles of obliquity (ground attack)
Probability of encountering obliquity θ or less

θ ,degrees)	M48 (circular)*	M48 (cardioid)†	T43 (circular)*	T43 (cardioid)†	JSU152 (cardioid)†	SU100 cardioid)†	T34/85 (cardioid)	JS III (cardioid)
0	0.03	0.03	0.05	0.05	...	...	...	...
10	0.05	0.05	0.06	0.06	0.09	0.08	0.06	...
20	0.06	0.06	0.07	0.07	0.31	0.11	0.13	0.08
30	0.17	0.16	0.16	0.16	0.39	0.31	0.23	0.27
40	0.24	0.24	0.27	0.24	0.53	0.43	0.38	0.32
50	0.43	0.40	0.46	0.43	0.69	0.56	0.49	0.4 1
60	0.59	0.58	0.64	0.56	0.89	0.79	0.71	0.68
70	0.77	0.77	0.84	0.82	0.95	0.96	0.86	0.88
80	0.95	0.95	0.95	0.95	0.99	0.99	0.95	0.98
90	1.00	1.00	1.00	1.00	1.00	1.00	1.00	1.00

*Circular (uniform) distribution of angles of attack considered.
† Cardioid distribution of angles of attack considered $f(\gamma) = \frac{1}{2\pi}(1 + \cos\gamma)$.

Table 2-21

Distribution of equivalent armor thickness with respect to ***HEAT*** *rounds (ground attack)*
Probability of encountering t_e or less*

t_e (in.)	M48 (circular)†	M48 (cardioid)‡	T43 (circular)†	T43 (cardioid)‡	T34/85 (cardioid)‡	JS III (cardioid)‡
0	...	...	...	...	...	...
2	0.18	0.07	0.06	0.04	0.22	...
4	0.48	0.37	0.41	0.31	0.62	0.16
6	0.63	0.57	0.59	0.48	0.72	0.23
8	0.72	0.68	0.68	0.62	0.76	0.46
10	0.74	0.72	0.72	0.66	0.78	0.67
12	0.76	0.76	0.74	0.71	0.78	0.72
14	0.77	0.77	0.75	0.72	0.78	0.74
16	0.77	0.77	0.75	0.73	0.78	0.74
18	0.77	0.77	0.76	0.73	0.78	0.74

*t_e = equivalent thickness at 0° for HEAT rounds (= thickness of armor x secant of the angle of obliquity) of armor measured from the normal plane.
† Circular distribution of attack angle.
‡ Cardioid distribution of attack angle.

Table 2-22

*Portion of presented area of hull and turret that is penetrable (ground attack)**

Angle of attack (degrees)	2.36-in. HEAT, M6A6			
	M48	T43	T34/85	JS III
0	0.06	0.22	0.72	0.08
30	0.29	0.16	0.70	0.00
60	0.59	0.66	0.69	0.00
90	0.74	0.73	0.75	0.61
120	0.65	0.70	0.77	0.62
150	0.58	0.53	0.64	0.01
180	0.74	0.76	...	...
	3.5-in. HEAT. M28A1			
	M48	T43	T34/85	JS III
0	0.60	0.63	0.75	0.73
30	0.67	0.55	0.79	0.70
60	0.72	0.75	0.77	0.70
90	0.86	0.84	0.76	0.73
120	0.71	0.79	0.79	0.74
150	0.74	0.76	0.79	0.73
180	0.75	0.78	...	...
	6.5-in. ATAR			
	M48	T43	T34/85	JS III
0	0.72	0.77	0.79	0.760
30	0.74	0.63	0.79	0.760
60	0.75	0.77	0.79	0.760
90	0.87	0.85	0.76	0.768
120	0.72	0.80	0.79	0,768
150	0.74	0.76	0.79	0.760
180	0.75	0.78	...	...

Angle of attack (degrees)	90-mm **HEAT,** T108			
	M48	T43	T34/85	JS III
0	0.60	0.53	0.75	0.702
30	0.67	0.52	0.79	0.637
60	0.72	0.74	0.76	0.623
90	0.86	0.84	0.76	0.713
120	0.71	0.79	0.79	0.737
150	0.74	0.74	0.75	0.629
180	0.75	0.78	...	...
	75-mm HEAT, M66			
	M48	T43	T34/85	JS III
0	0.02	0.05	0.53	0
30	0.07	0.06	0.53	0
60	0.29	0.37	0.61	0
90	0.44	0.56	0.74	0.504
120	0.34	0.50	0.60	0.171
150	0.35	0.33	0.13	0
180	0.64	0.67	...	...
	105-mm HEAT, M67			
	M48	T43	T34/85	JS III
0	0.05	0.10	0.60	0.063
30	0.20	0.15	0.62	0
60	0.55	0.53	0.61	0.099
90	0.65	0.63	0.75	0.596
120	0.63	0.62	0.63	0.576
150	0.53	0.51	0.53	0
180	0.72	0.74	...	...

*Remember that a penetration does not insure a kill.

not yet provided simple indices far the measure of the probability of a given round killing a tank. Such firings assemble basic data from firing on obsolete vehicles to provide an accurate estimate of vulnerability of new or proposed tanks that have not yet been fired upon. The problem here is how to use a small number of actual firings to give reliable overall damage probability estimates.

The technique of data reduction is influenced by the use to which the data are to be put. One use is the comparison of the effectiveness of specific weapons. A second is estimating the number of weapons needed to counter an enemy force. Assembly of data on relative kill probability on the particular target tanks used in experiments provides a reservoir of knowledge which the analyst must assimilate prior to making an estimate of a weapon's effectiveness or computing a kill probability against any target type.

Two principal methods are used for data reduction. These are called the "vulnerable area method" and the "distributed area method." The vulnerable area method is used when the target is small compared to the dispersion of hits on the target. The distributed area method is used when the dispersion of the hits is small compared to the size of the target. These methods and some approximations that have been made are discussed below.

2-151. The Distributed Area Method. The distributed area method will be described first, since the vulnerable area method is essentially a simplification of it. Consider the case when the probability of a hit being a kill by a projectile of high velocity and low dispersion is desired (such as the 90-mm T108 round against the JS III). Terminal ballistic damage data of this round on the T26E4 tank are assembled first. These data include both the numerical and descriptive assessments. Examination is made of the damage resulting from hits on components where damage is obtained only a part of the time, such as the suspension, the turret ring, the hull in front of driver's controls, and so on. For many other areas kill probability will depend only upon probability of perforating. Numerical assessments of damage for various types of rounds are compared to see if terminal ballistic damage after perforation is comparable (as are the 90-mm T108 and the 3.5-in. rocket). An examination of perforating and nonperforating hits is made to determine the reliability of fuze action of chemical energy rounds.

Vulnerability drawings of the target tank are prepared showing the arrangement of the interior components to the line of fire (see figure 67). Using an overlay grid, the probability of a hit, the chance of perforating, and probability of a perforation being a kill are entered into each square for a given point of aim.

The probability of a perforation being a kill is determined by estimating the fragment pattern and the expected damage from the jet. Reference to the qualitative description of damage from each round is used here. Numerical damage is computed by combining the damage from components lying in the path of the jet and fragments by the formula

$$P_M = 1 - (1 - P_a)\ (1 - P_b)\ (1 - P_c) \ldots\ldots (1 - P_k)$$

where P_M is the probability of M damage occuring, P_a is the percent of M damage resulting from a hit on component "a," and so on.

This calculation is carried out for M, F, and K damage for several views about the tank, and hit probability figures are varied for each range. By summation, the probability of an aimed round killing as a function of range r and azimuth e is obtained. These data can be combined with the expected angular and range frequency of attack to give an overall figure of the vulnerability of one tank to an antitank gun firing a certain round. Such values have been computed in the following table.

Table 2-23

Probability of killing JS III tank

Projectile	Kill category		
	M	F	K
90-mm HEAT T108	0.50	0.47	0.44
5-in. copper liner HC (assumed same dispersion as 90-mm T108)	0.66	0.66	0.60

All shots are aimed fire, without rangefinder, at center of the largest concentration of target vulnerable area.

Another calculation (table 2-24) made for the front of the JS III compares the 90-mm T108 HEAT round and the 105-mm BAT HEAT round fired at the JS III tank. The T108 round is fired under the conditions of table 2-23 above, and the BAT guns were fired in a salvo of two, using a spotting rifle for aim. No mismatch in the spotting rifle and the 105-mm rifles was assumed. (At present the mismatch is such that the values for the BAT rounds beyond 1,400 yards will not be appreciably higher than that for the T108 round.)

Table 2-24

Comparison of effect of two 105-mm BAT rounds with effect of 90-mm T108 round

Range (yards)	Probability of F damage on front of JS III	
	90-mm T108	105-mm BAT
500	0.33	0.33
1,000	0.10	0.25
1,500	0.04	0.16
2,000	0.02	0.10

2-152. The Vulnerable Area Method. The vulnerable area, which is the product of the hit probability on the presented area and the probability of a random hit on this area being a kill, is computed from the overlay of figure 2-67 merely by assuming a uniform hit probability in each square. It is assumed that the point of aim may be anywhere on the tank. There is no range effect to be considered for shaped charge ammunition. Many of the present day shaped charge rounds have sufficient dispersion for hits to be considered in this manner. Several calculations have been carried out on tank vulnerability using this method, and are included in table 2-25. Probabilities are given in terms either of vulnerable area or the probability of a random hit being a kill, the vulnerable area being the latter probability multiplied by the presented area of the tank.

To convert from vulnerable area to probability of a kill, the total presented area (table 2-26) is needed.

Table 2-25

Vulnerable area of JS III tank to ground attack by shaped charge rounds (ft^2)

Angle of attack	Round								
	6.5-in. ATAR			5-in. copper liner			90-mm T108		
	M	F	K	M	F	K	M	F	K
Front	39	34	21	32	23	14	5	6	3
60°	77	51	31	68	37	23	30	20	6
Side	75	54	33	76	47	2'7	35	28	11
120"	74	51	30	73	42	24	41	24	14

Table 2-26

Presented area of JS III tank (ft^2)

Angle of attack	Ultimate penetrable area	Total area including suspension
Front	34	62
60°	82	142
Side	84	137
120"	86	146

The probability of a random hit causing a kill averaged over the expected angles of attack, $f(Y) = \frac{1}{2\pi}(1 + \cos y)$, is given in the following table and may be compared with table 2-23, which gives the same figure for aimed fire averaged over the expected ranges of engagement.

Table 2-27

Probability of random hit falling on total presented area of JS III, causing kill

Projectile	Kill category		
	M	F	K
90-mm T108	0.18	0.12	0.05
5-in. HC (copper lined)	0.49	0.30	0.17
6.5-in. ATAR (steel lined)	0.55	0.41	0.26

Comparison of tables 2-23 and 2-27 shows the requirement for larger shaped charge rounds when inaccurate fire is to be used.

Table 2-28 gives a summary of vulnerable areas on the T34/85 Russian tank to the 2.36-in. and 3.5-in. rockets.

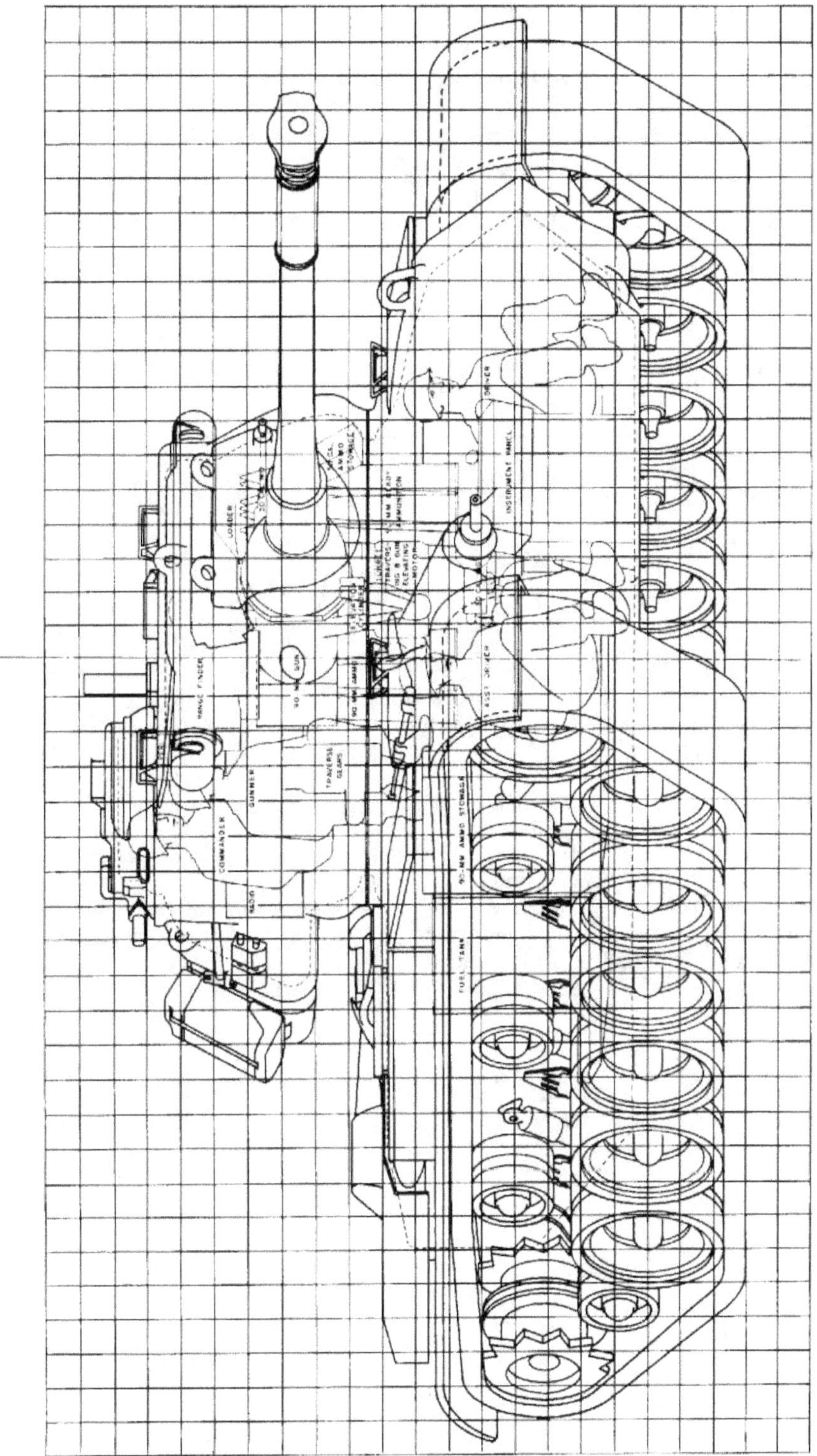

Figure 2-67. Vulnerability drawing of tank

Table 2-28

Vulnerable area of T34/85 (ft^2)

Angle of attack	Presented area (ft^2)	2.36-in. HEAT M6 M	2.36-in. HEAT M6 F	2.36-in. HEAT M6 K	3.5-in. HEAT M28 M	3.5-in. HEAT M28 F	3.5-in. HEAT M28 K
Ground attack							
Front	45	7	5	4	9	6	5
45°	95	18	11	8	20	13	9
Side	97	36	17	15	37	21	16
135°	95	35	14	11	40	17	14
Rear	45	23	5	5	25	6	6
30" Air attack							
Front	87	19	17	15	21	19	16
45°	135	30	16	14	35	21	17
Side	140	38	18	16	45	23	19
135°	135	45	17	14	51	20	17
Rear	87	41	7	6	46	9	7

Table 2-29 gives the vulnerable area of the M26 to the 3.5-in. rocket.

Some calculations using an approximation of the vulnerable area technique have been made using the product of the probability of a perforation averaged over the expected angles of attack and the probability of a kill in the unarmored components of the tank averaged over the expected angles of attack. Calculations were made for the average of the front and sides of the tanks only. These calculations approximate the probability of a random hit being a kill on the ultimate penetrable area, which is the penetrable area a tank presents to a round of infinite penetration.

Table 2-29

Vulnerable area of MZ6 to 3.5-in. rocket (ft^2)

Angle of attack	Kill category M	Kill category F	Kill category K
Front	7	7	2
30°	18	11	6
60"	29	16	11
Side	36	23	17
120°	33	16	11
150°	27	8	5
180°	22	2	1

A rough check of the consistency of these approximations can be made by comparing tables 2-26 and 2-27. This check shows that the values for the 90-mm T108 versus the JS III given in table 2-27 should be approximately half of those shown in table 2-30 which is approximately so.

2-153. Evaluation of Present Methods of Analysis. The present methods of analysis are not completely satisfactory. Fairly reliable estimates of the probability of a hit causing a kill can be made but the method is tedious. Additional data are needed to reduce the subjective elements of damage assessment. Further reduction of existing damage data should help.

The computation of vulnerability by considering the chance of a kill after perforation on each small area is cumbersome. However, it is reasonably reliable, and until a body of this reliable information is assembled, approximations must be viewed with suspicion.

Table 2-30

Approximate probability of a random hit on ultimate penetrable area of several Soviet armored vehicles giving a kill

Round	Tank			
	JS III	T34/85	JSU 152	SU 100
76-mm HEAT				
M	0.20	0.46	0.44	0.44
F	0.16	0.38	0.34	0.34
K	0.08	0.22	0.19	0.19
90-mm HEAT				
M	0.37	0.47	0.49	0.45
F	0.31	0.39	0.40	0.40
K	0.18	0.23	0.24	0.23
105-mm HEAT				
M	0.50	0.53	0.53	0.54
F	0.41	0.42	0.42	0.44
K	0.24	0.26	0.26	0.27
Energa rifle grenade				
M	0.09	0.38	...	...
F	0.08	0.31	...	...
K	0.04	0.17	...	...

The method of analysis does not yet accurately account for the transition point between inaccurate fire, where vulnerable areas can be used, and accurate fire, where the distributed area technique can be used. Where the gunner starts aiming at spots on a tank rather than the whole tank as a target is not known, and probably will depend to some extent on the training of the gunner.

Future analysis will be helped by witness plate data, such as the British have beenobtainingfor many years. However, there appears to be little likelihood that the vulnerability of a tank can be computed from syntheses of many tests made only with simulations of tanks. The analyst of vulnerability must never be misguided into the assumption that an exact measurement made of an assumed condition (such as the box tests represent) can be used to the exclusion of the inexact measurement of the real condition (vulnerability firings at vehicles).

FRAGMENTATION

INTRODUCTION

2-154. General. The purpose of any artillery projectile is the defeat of its intended target. The probability of accomplishing this mission depends on the characteristics of the weapon system, the damage potential of the projectile, and the resistance of the target. Not only must the projectile be accurate in its flight, but it must also be destructive enough to defeat the target. Ordinarily, to obtain the optimum design of a weapon, a compromise must be made between accuracy and damage potential. In the process of designing, the lethality is first determined, and then utilized in conjunction with other factors to obtain the optimum parameters.

In the succeeding paragraphs, some of the lethality criteria will be discussed, in addition to lethal area, fragmentation patterns, and shape of shell. Also, wound ballistics and the damage criteria of personnel targets, along with their optimum parameters, will be explained. Caliber, number and size of fragments, initial fragment velocity, target type and toughness, angle of fall, remaining shell velocity, degree of protection offered by the ground, location of burst, and similar considerations are all involved in determinations of the overall effectiveness of a weapon.

2-155. Nature of Fragmentation. A conventional artillery shell consists of a case, filler (high explosive), fuze, and detonator. When the shell detonates, the case first expands, and then breaks into fragments.

2-156. Fragmentation Process. Fragmentation and blast are the results of detonation of explosive missiles. The energy of the high-explosive gases is expended in projecting the fragments, expanding the gases, the shock wave, heating and deforming the fragments, and so on. The fragments are propelled at high velocity and, within a very short distance from the center of explosion, pass through the shock wave, which is retarded by the air to a greater extent than are the fragments. In effect, the fragments, which are hurled outward at high velocities, are projectiles with the capacity to inflict considerable damage to adjacent objects. Capacity for damage depends upon fragment size, velocity, and distribution.

2-157. Mechanism of Fragmentation. Flash radiographs of exploding shell. show that the usual shell steels will expand to about 1 1/2 times their original diameter before fragmentation takes place. Failure of metal occurs in shear in a direction of 45° from the normal. Photomicrographs of shell fragments indicate that some deformation is also caused by shock or impact.

2-158. Number of Fragments and the Fragment Weight Distribution for a proposed projectile design may be estimated by means of formulas proposed by Weiss,[15] Cook,[17] Mott,[1,16] or Gurney and Sarmousakis.[8] Each of these methods leaves much to be desired in terms of accuracy of prediction, but any of them will give results accurate enough for the preliminary design of projectiles that do not differ too much from current types. The work of Mott has gained the widest usage, hence it has been chosen for detailed presentation in this handbook.

2-159. Parameters Needed to Evaluate Fragmentation Effectiveness. Such ,parameters include fragment weight distribution, fragment spatial distribution, and impact velocity. In order to determine the impact velocity, it is necessary to compute from initial fragment velocity and rate of velocity loss. The latter is often expressed in terms of a remaining velocity curve. In addition to these, it is necessary to have some lethality criterion upon which to base effectiveness:. At the present time, a reliable criterion for personnel casualties is available. It is expressed as the conditional probability that a single fragment will disable the target, and is a function of the mass, area, and velocity of the fragment. Some effort has been made to obtain criteria for damage to aircraft and light armor plate; such criteria, however, are not very reliable. The subject of lethality criteria is given detailed treatment in paragraphs 2-181 through 2-185.

DETERMINATION OF FRAGMENTATION CHARACTERISTICS

2-160. Methods. The fragmentation characteristics of developmental model shells may be determined by means of fragmentation tests. Analytical methods are used to predict fragmentation characteristics during the initial stages of design.

2-161. Fragmentation Tests. Fragmentation test methods, as described in Ordnance Proof Manual 40-23 (March 1947), include:

1. Closed-pit test, for determination of number and size of fragments.
2. Panel test, for determination of fragment distribution.
3. Velocity measurement test, using velocity screen and Aberdeen chronograph.

These test methods are substantially obsolescent and have largely been supplanted. The method presently employed by the Ballistic Research Laboratories makes use of Celotex-filled recovery boxes to sample the fragment spray and Fastax high-speed cameras to determine fragment velocity. The recovery boxes are arranged around a test shell which is suspended with its axis parallel to the ground. Dural sheets are placed in front of several of the boxes and the Fastax cameras are so arranged that they photograph both the shell and the dural sheet. Thus the detonation and the flash at the sheet, when the fragment passes through, are both recorded. Elapsed time may be determined by counting the number of frames between the two events on the filmstrip.

Other systems in use are the water pit and sawdust pit. All of the later methods permit only partial recovery of the fragments but result in less fragment breakage than the earlier sand pit method.

2-162. Two-Dimensional Breakup of Shell. Mott and Linfoot proposed that fragmentation of a thin-walled shell is the result of two-dimensional, rather than three-dimensional, breakup.[18] Under this assumption, the mass distribution of fragments may be described by the equation

$$N(m) = Ae^{-\left(\frac{m}{\mu}\right)^{1/2}} \qquad (7)$$

where N(m) is the total number of fragments of mass greater than m, μ is related to the average fragment mass (in mass units), and A is a constant. If it is assumed that two-dimensional breakup holds down to the finest fragment, then

$$N(m) = \frac{M}{2\mu} e^{-\left(\frac{m}{\mu}\right)^{1/2}} \qquad (8)$$

where M (in the same units as μ) is the total mass of the shell and 2μ is the arithmetic average fragment mass. Noting that $M/2\mu$ represents the total number of fragments N_0, equation (8) may also be written

$$N(m) = N_0 e^{-\left(\frac{m}{\mu}\right)^{1/2}} \qquad (9)$$

2-163. Three-Dimensional Breakup of Shell.[18] For extremely thick-walled shell, the wall thickness will have less effect on the size of the fragment, and three-dimensional breakup, rather than two-dimensional breakup, will be the rule. The mass distribution of fragments for this case will, according to Mott and Linfoot, be described by

$$N(m) = A'e^{-\left(\frac{m}{\mu'}\right)^{1/3}} \qquad (10)$$

where A' is a constant, μ' (in mass units) is related to the average fragment mass, and N(m) is the number of fragments of mass greater than m (in the same units as μ). For fragmentation shell, equation (9) is more representative of the conditions found than equation (10).

2-164. Significance of Quantity μ. The quantity μ in the Mott equation is, for any given shell, dependent upon the characteristics of the explosive and of the metal case. It is, assuming the Mott equation to hold, a measure of the fragmentation efficiency of a projectile. To quote Gurney and Sarmousakis:[8]

"The significance of the quantity μ ...may be made clearer by stating that...the number of fragments greater than μ grams...is equal to the number of fragments having masses lying between $\mu/11$ and μ grams. Thus if, for instance, μ = 5.5 grams, we have the result that the number of fragments with masses lying between 1/2 gram and 5 1/2 grams is equal to the number with mass greater than 5 1/2 grams; (this is assuming that [the Mott equation] continues to hold down to fragments as small as 1/2 gram). We can say further, that, if [the Mott equation] were to hold for all the fragments, then the number of fragments greater than μ would comprise 37% of the total."

The value of P, in addition to being dependent upon the characteristics of the explosive and the metal, is also dependent upon the physical dimensions of the shell. To account for this variability, scaling formulas have been proposed by Mott,[1] and by Gurney and Sarmousakis.[8]

2-165. Reliability of the Mott Equation. Gurney and Sarmousakis make the following statement in regard to the reproducibility of data:[8]

"A serious difficulty in analyzing the data is the nonuniform behavior of the projectiles. There is usually a considerable variation in the numbers of fragments produced by different members of a lot of shell....Thus...one must not expect more than a rough agreement between the meager experimental data and the semi-theoretical formulae."

Chi-square tests of some foreign ammunition have shown no significant difference in round-to-round fragmentation results. American data tend to contradict this. It is felt, however, that up to this time the amount of American data examined has not been sufficient to permit conclusions to be drawn.

Solem, Shapiro, and Singleton, at the U. S. Naval Ordnance Laboratory, fired a series of experimental steel shell filled with explosives of different characteristics.[6] This was done in order to obtain values of the parameter μ, as well as to serve several other purposes. The plots of the log of the cumulative number of fragments versus the square root of the fragment mass that were obtained provide an interesting comment on the Mott equations. Several representative plots are shown in figure 2-68. Since the Mott equation predicts that this plot should be a straight line, straight lines have been drawn in by the method of maximum likelihood, and from these lines the values of μ have been determined. It can readily be seen, however, that the experimantal points in every case form a curve of increasing negative slope rather than a straight line. Assuming that the experiment was accurate, this would seem to indicate a fundamental defect in the Mott relationship.

2-166. Mott Scaling Formula. The following formula, relating the value of μ to the inside diameter d_i and the thickness t, has been proposed by Mott:[1]

$$\mu^{1/2} = Bt^{5/6} d_i^{1/3} \left(1 + \frac{t}{d_i}\right) \qquad (11)$$

where B is a constant depending on the explosive and the physical characteristics of the metal of the casing. For small values of C/M, the charge- to metal-mass ratio, this formula agrees well with that proposed by Gurney and Sarmousakis (see the following paragraph).

2-167. Gurney-Sarmousakis Scaling Formula. The following formula has been proposed by Gurney and Sarmousakis:[8]

$$\mu^{1/2} = D \frac{t(d_i + t)^{3/2}}{d_i} \sqrt{1 + 1/2\left(\frac{C}{M}\right)} \qquad (12)$$

where t and d_i are the thickness and inside diameter (in the same units).

2-168. Effect of Explosives on μ. Solem, Shapiro, and Singleton, of the U. S. Naval Ordnance Laboratory, conducted a series of firings to compare the fragmentation efficiency of agroup of explosives.[6] The test shell were cylinders made of AISI 1045 cold-drawn, seamless-steel tubing, stress relief annealed, with a hardness of about 100 Rockwell B. The nominal dimensions of the shell were 2.0 in. in I. D., and 0.25 in. in thickness. Exact dimensions are given in table 2-31. The pressed explosives were formed into 2-in. diameter pellets of 1-in. height at a pressure of 16,000 psi.

Table 2-31 lists the explosives tested, the dimensions of the cylinders, the values of the C/M (charge- to metal-mass) ratio, and the computed values of $\mu^{1/2}$. The Mott and Gurney-Sarmousakis scaling constants are included to permit calculation of $\mu^{1/2}$ for other shell sizes.

Note

Calculations will be applicable only to shell made from a metal with fragmentation characteristics similar to those of AISI 1045. See the following paragraph for information on the effect of metal characteristics on the value of μ.

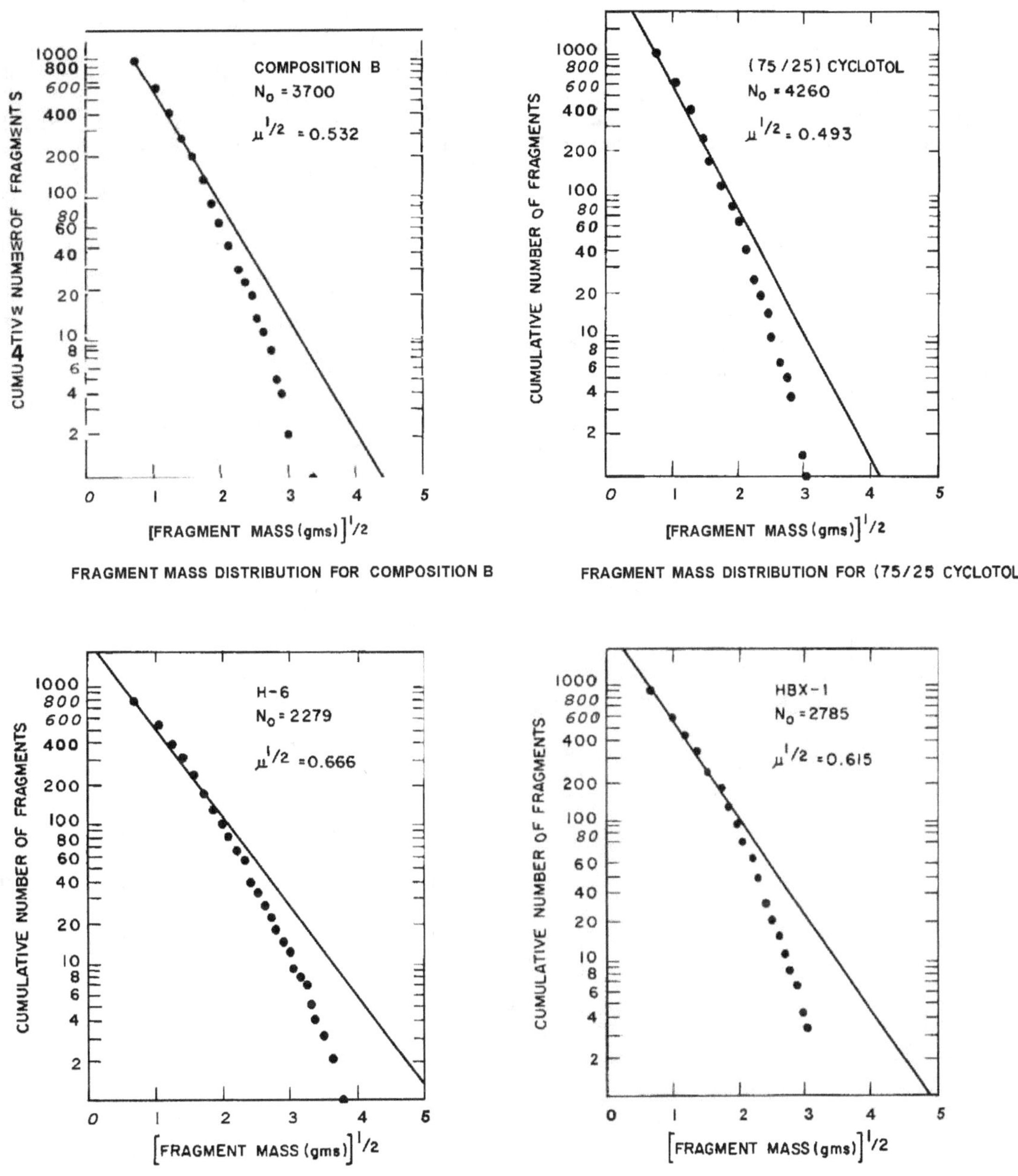

Figure 2-68. Fragment mass distributions

2-169. Effect of Properties of Metal on μ. The Ballistics Research Laboratories have under way, at this time, a program of basic research into the fragmentation characteristics of a wide variety of materials, including steels and non-ferrous castings. Reports thus far issued are listed in the bibliography following paragraph 2-207.

The results of the program indicate that there is some correlation among the density, tensile strength, and static reduction of area of the metal and the mean fragment mass.[22] Semi-empirical relations for the mean fragment mass of ring-type experimental shell have been obtained in terms of these quantities, the physical dimensions of the shell, and the initial fragment velocity. These results are, however, applicable to shell design only on a qualitative basis.

2-170. Fragmentation Characteristics of Ductile Cast Iron.[10] Experimental ring and cylinder type shell, made up from three grades of ductile cast iron — nodular ferritic, nodular pearlitic, cupola malleable — showed on detonation the following fragmentation characteristics:

a. The three grades of ductile cast iron showed little or no difference in fragmentation, notwithstanding the substantial difference in strength and ductility.

b. The fragments from the ring-type shell were coarser than those from the cylinder-type shell, which is a reversal of the behavior of steel shell under the same conditions of firing.

c. The ductility of these three grades when subjected to high rates of strain — as measured by the reduction in cross sectional area of the ring-type, and wall thickness of the cylinder type after detonation — was practically identical

Table 2-31

Mott and Gurnev-Sarmousakis scaling constants

Explosive	t (in.)	d_i (in.)	C/M	$\mu^{\frac{1}{2}}$	Mott scaling constant (B)	Gurney-Sarmousakis scaling constant (D)
			Cast explosives			
Baratol	0.254	2.000	0.562	1.237	2.73	2.55
Comp B	0.253	1.999	0.377	0.532	1.18	2.14
Cyclotol (75/25)	0.253	1.999	0.380	0.471	1.05	1.01
H-6	0.254	1.999	0.395	0.666	1.47	1.34
HBX-1	0.255	1.999	0.384	0.615	1.36	1.30
HBX-3	0.255	1.999	0.403	0.781	1.72	1.65
Pentolite (50/50)	0.254	1.999	0.366	0.596	1.32	1.27
PTX-1	0.254	1.999	0.367	0.534	1.18	1.14
PTX-2	0.254	1.999	0.373	0.546	1.21	1.17
TNT	0.254	2.000	0.355	0.751	1.66	1.61
			Pressed explosives			
BTNEN/Wax (90/10)	0.251	2.009	0.379	0.427	0.95	0.92
BTNEU/Wax (90/10)	0.251	2.012	0.367	0.507	1.13	1.10
Comp A-3	0.252	2.012	0.367	0.474	1.17	1.13
MOX-2B	0.248	2.008	0.461	1.289	2.91	2.79
Pentolite (50/50)	0.252	2.011	0.363	0.638	1.41	1.27
RDX/Wax (95/5)	0.253	2.010	0.370	0.509	1.13	1.09
RDX/Wax (85/15)	0.251	2.014	0.350	0.566	1.26	1.23
Tetryl	0.254	2.011	0.371	0.660	1.45	1.41
TNT	0.253	2.012	0.348	0.972	2.15	2.10

and showed no relationship to the static ductility.

d. The half-weight — the particular weight of a fragment which divides the individual fragments into two groups, each containing half the total weight of fragments — appears to be of the order of 1/4 that of steel shell stock for the ring-type and 1/16 for the cylinder-type.

e. Although the ductile cast irons used inthe test were centrifugally cast, the physical properties, after heat treatment, of the tubes from which the shell were fabricated are comparable to those obtained from sand casting. It is therefore reasonable to assume that the results of this test are representative of what would be obtained from conventional sand castings given the proper heat treatment.

f. The use of ductile cast irons in ammunition seems to be attractive in those cases where steel gives too coarse a fragmentation, and where strength requirements are comparable with the properties of the ductile cast irons. Making mortar shell from ductile cast iron is one application that shows promise.

A later report, BRL Technical Note No. 894, gives a comparison of the wounding effectiveness of a series of cast casings. In this study, the general rank of families of casing material in order of increasing effectiveness against personnel is gray cast iron; cast, and forged steel; and the ductile and malleable cast irons, which have about equal rank.

2-171. Application of Metal Fragmentation Characteristics Data to Design of Shell. If the properties of metal, as shown by one of the BRL reports on fragmentation characteristics (paragraph 2-170), seem promising, the results of the report usually can be used to predict, by means of the Mott equation, the probable mass distribution for a proposed projectile design.

In order to apply the Mott equation, one must be certain that it is applicable. That is, the experimental data must be based upon natural fragmentation rather than controlled fragmentation (for example, data on ring-type shell are not usable), and the plot of the square root of fragment mass versus the log of the cumulative number of fragments must be a straight line.

If the above conditions are met, the value of μ for the proposed design can be computed as follows.

1. Determine the value of the constant in the Mott or Gurney-Sarmousakis scaling formula (paragraphs 2-166 and 2-167) by substituting into the scaling formula the necessary physical dimensions and C/M ratio (Gurney-Sarmousakis only), as well as the value of $\mu^{\frac{1}{2}}$ (square root of 1/2 the average fragment mass) for the experimental shell.

2. Substitute the physical measurements and C/M ratio (if required) for the proposed design into the scaling formula and, using the computed constant, determine the new value of μ.

3. Using the obtained value of μ, compute the mass distribution from the Mott equation (paragraph 2-166).

Note

In the event that the explosive to be used in the proposed design is different from that used in the experimental shell, it will be necessary to correct the value of μ to apply to the chosen explosive. See paragraph 2-168.

2-172. Prediction of Initial Fragment Velocity. The initial velocity of fragments is quite accurately predicted by the Gurney formulas:[19]

for cylinders

$$V_o = \sqrt{2E}\sqrt{\frac{C/M}{1 + 0.5(C/M)}} \qquad (13)$$

for spheres

$$V_o = \sqrt{2E}\sqrt{\frac{C/M}{1 + 0.6(C/M)}} \qquad (14)$$

where

V_o = initial fragment velocity (fps)
$\sqrt{2E}$ = a constant (fps) for each type of explosive
C = weight of explosive charge (units the same as M)
M = weight of fragmenting metal (units the same as C).

2-173. Gurney Constant. Table 2-32 gives the value of $\sqrt{2E}$ for most of the commonly used high explosives. For information on the exact composition of these explosives, see table 2-8.

2-174. Graphical Calculation of V_o. The graphs in figures 2-69a and 2-69b simplify the calculation for V_o in terms of the outside diameter

(d_o) and thickness (t) of the shell, the ratio of the density of the explosive to that of the metal case (ρ_c/ρ_m), and the Gurney constant ($\sqrt{2E}$) of the explosive. Knowing the value of t/d_o, figure 2-69a solves for $\frac{C/M}{\rho_c/\rho_m}$. C/M may be found by multiplying by ρ_c/ρ_m. This value of C/M may be used to find $V_o/\sqrt{2E}$ from figure 2-69b.

Table 2-32

Explosive	Gurney Constant, $\sqrt{2E}$ (fps)
Composition C-3	8,800
Composition B	8,800
Torpex 2	8,800
Composition H-6	8,400
Pentolite	8,400
Minol 2	8,300
HBX	8,100
TNT	7,600
Tritonal	7,600
Picratol	7,600
Baratol	6,800

Multiplying by the appropriate value of $\sqrt{2E}$ from table 2-32 will give V_o.

2-175. Remaining Velocity of Fragments. The remaining velocity (V) of a fragment, at a distance x from the point of burst, is expressed by

$$V = V_o e^{-\frac{\bar{C}_D \bar{A} \rho}{m} x}$$

where

x = distance from point of burst in ft
V = velocity of fragment at x feet from point of burst in fps
V_o = initial velocity of fragment in fps
$\bar{A}$ = average presented area of fragment in ft2
m = mass of fragment in lb
ρ = density of air in lb per ft^3
$\bar{C}_D$ = average drag coefficient (dimensionless)

2-176. Relation Between Mass and Presented Area of a Fragment. For any homologous class of regularly shaped fragments, the mass and average presented area are related by the equation

$$M = K(\bar{A})^{3/2}$$

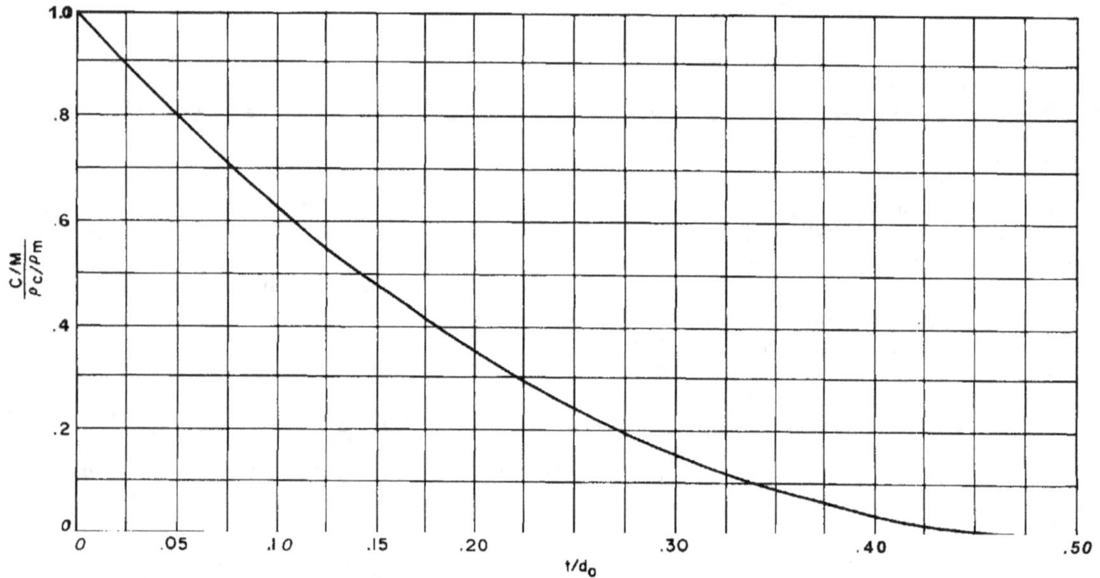

Figure 2-69a. Graph for computation of V_o

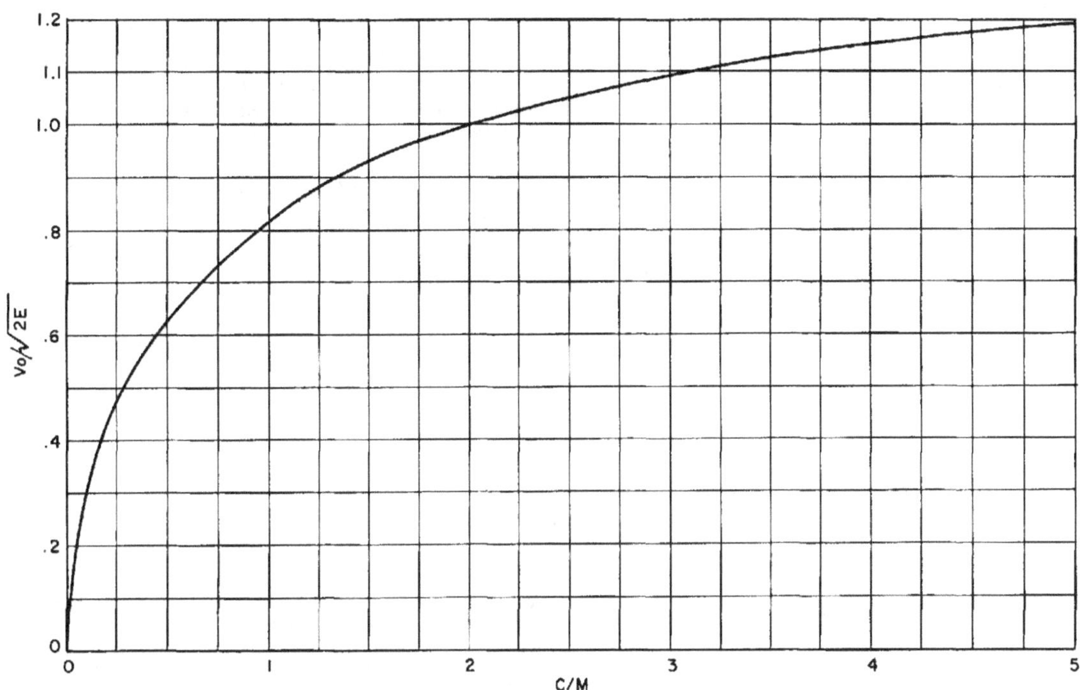

Figure 2-69b. Graph for computation of V_o

where K is a constant for the class. It has been found experimentally that the value of K is roughly constant for the fragments projected for a particular shell. Values of K are given in BRL Reports 501, 536, and M915 for a variety of shell and bomb fragments.

2-177. Measurement of the Presented Area of a Fragment. The method in use at present involves the measurement of the presented area of the fragment for each of 16 positions corresponding to the orientations of 10 of the 20 faces of an icosahedron plus 6 orientations corresponding to the 12 vertices of the icosahedron — the remaining 10 faces and 6 vertices are symmetrical to these — and the use of the arithmetical average of these values for $\bar{A}$.[28] The instrument used to obtain the presented area is known as an icosahedrongage.

2-178. Discussion of Fragmentation Patterns. When a projectile or warhead bursts, fragments are projected in many directions. If the projectile were spherical, and stationary when detonated, the density of fragments would be substantially constant, regardless of the direction. If the projectile were entirely cylindrical, the greatest density of fragments would be close to the equatorial plane, with practically all fragments contained in a narrow sidespray of the order of 20° width. For an ordinary artillery projectile, the curve of distribution with angle is peaked, and resembles the "normal error curve." An example is shown in figure 2-70.

Projectiles and warheads almost always have circular symmetry about their longitudinal axis. Hence, we may describe the distribution of fragment mass and velocity as functions of the angle θ measured from the nose of the shell. Let ρ (8) be the fragment density in fragments per unit solid angle. Then N_o, the total number of fragments of the given shell is given by

$$N_o = 2\pi \int_0^{\pi} \rho(\theta) \sin\theta d\theta$$

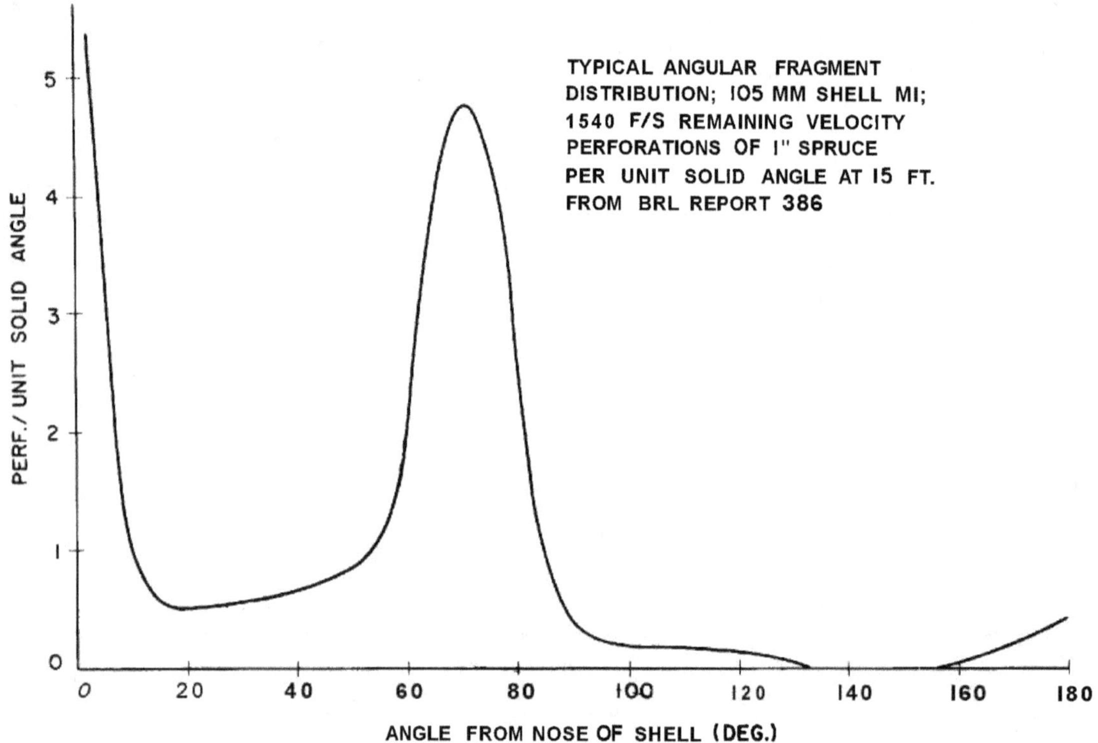

Figure 2-70. Typical angular fragment distribution

2-179. Fragment Emission. If a small target of vulnerable area A_V is exposed to the spray of a distance r from the burst point, the expected number of strikes E on the vulnerable area of the target is

$$E = \frac{\rho(\theta) A_V}{r^2}$$

and the probability of at least one strike on the vulnerable area is

$$\rho(r, \theta) = 1 - e^{-\frac{\rho(\theta) A_V}{r^2}}$$

or

$$\rho(r, \theta) = 1 - e^{-E}$$

Here, it is assumed that the fragments travel in straight lines (usually an acceptable simplification).

When the shell bursts in flight, each fragment has added to its velocity the forward velocity of the projectile at the instant of burst. This throws the principal fragment spray forward, and also increases the density of fragments in the forward hemisphere, at the expense of the rear.

2-180. Influence of Surface Contour on Fragment Distribution. The most important factor in determining the spatial distribution of the fragments is the shape of the metal casing. The shape of an artillery shell is limited by the requirements of exterior ballistics and manufacturing. For discussion of the effect of contour, see references 2, 15, and 39.

LETHALITY

2-181. Lethality Criterion. To evaluate the effectiveness of fragmenting antipersonnel weapons, a quantitative casualty criterion is necessary. One criterion of wounding power is the 58 foot-pound rule, which states that missiles that hit with less than 58 foot-pounds of kinetic energy do not kill, and that those that hit with more than **58** foot-pounds do kill. However, this criterion was never intended to be more than a rule of thumb. In **1944,** Gurney suggested that mV3 was a more suitable criterion than the kinetic energy of the ability of projectiles to wound human targets (m is the fragment mass, and V is the fragmentvelocity).

2-182. Experiments to Determine Penetration Data. The volume of a temporary cavity is a structural, rather than a functional, concept. It may be regarded as a measure, simple and physical in nature, of structural damage, but it is not real incapacitation. Because of the unreliability of the cavity studies for the evaluation of casualty criteria, McMillen and Gregg[54] concentrated on wounds which they consideredto be "fatal" or "severe." These could be caused, they assumed, by the projectiles' reaching certain vulnerable regions inside the body after penetrating protective layers of skin, soft tissue, and bone, provided that the projectiles reached the vulnerable regions with velocities in excess of **7,500** cm per sec. A constant thickness of skin was adopted, and the thicknesses of soft tissues and bones were determined from anatomical charts depicting cross sections of the human body at approximately one-inch intervals from head to foot. The velocities necessary to penetrate the various thicknesses of skin, soft tissue, and bone were based on experiments with steel balls, and in some cases involved living animals.

Although the work of McMillen and Gregg was an improvement, their results cannot be extended reliably to actual practice, because weapons do not have spherical projectiles. However, the same basic experimental data of penetration through skin, tissue, and bone are still used. The requirement that all projectiles emerge from the protecting layers with more than **7,500** cm per sec of velocity, regardless of mass, is no longer used. The new requirement is that all projectiles emergefrom the protecting layer with 2.5×10^7 ergs of kinetic energy in order that they may cause comparable damage, regardless of their mass, when they enter the vulnerable regions. The vulnerable regions are defined by McMillen and Gregg in their report.

2-183. Types of Incapacitation. There arethree types of incapacitation considered. The first is type K, which is incapacitation within five seconds. Type A is incapacitation within five minutes. The degree of incapacitation considered is that adequate to prevent an enemy either from firing a gun or offering resistance of any sort. Type B incapacitation is fatal or severe wounding, in an indefinite periodof time.

Type K incapacitation was considered in the design of new type hand grenades, since the infantryman throwing the grenade should be very close to his targets, and would be subject to retaliatory action if incapacitation were not instantaneous (within five seconds) or complete. The experiments were conducted in **1951** at the Army Biophysics Laboratory in Edgewood, Maryland. The statistical data were interpreted by T. E. Sterne, and outlined in reference **24.**

2-184. Probability of Immediate Incapacitation. From the experiments, Sterne derivedtheprobability P_{hk} that a random hit on a man by a single fragment would cause his complete incapacitation within five seconds. Sterne's criterion is reproduced, in altered units, in figure **2-71.** The abscissa is the quantity $(MV/A)10^{-6}$, where M is the mass of the fragment in grains, V is its striking velocity in feet per second, and A is its projected area in square inches, averaged over all directions of projection.

The vulnerability of a man to five-second incapacitation comes almost entirely from severe injury to the spinal cord above the second or third thoracic vertebra, or to parts of the brain. Five-second incapacitation can also be caused by multiple severe wounds to other anatomical regions. This latter type of incapacitation is not considered, however, as it is only at very close distances that a sufficient number of severe wounds are at all likely, and at such distances, five-second incapacitation by separate hits is probable. The probability of multiple severe wounding decreases with increasing distance like a large negative power of the distance, and hence far more rapidly than the probability of five-second incapacitation from single wounds, which decreases only somewhat more rapidly than the inverse square of the distance.

2-185. Status of Wound Ballistics. All of the information on lethality criteria and wound

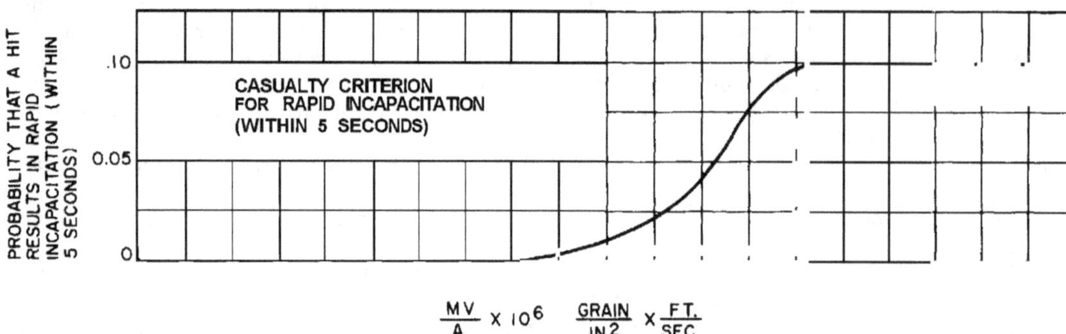

Figure 2-71. Casualty criterion for rapid incapacitation (within 5 seconds)

ballistics is still of a provisional nature. The bone penetration data that have been employed have introduced considerable uncertainty, as they were penetration data rather than data on the loss of velocity by perforating missiles. Other deficiencies are caused by the scaling of the bones, and the method of suspension and support. Then, a difficulty arises when the results with spheres are extended to random fragments from shell or grenades. Also, a critical evaluation has not yet been made of the amount of energy required on the emergence of a fragment from the protective layers in order that the fragment may cause fatal or severe wounds when it hits the vulnerable regions underlying the layers.

LETHAL AREA COMPUTATION

2-186. Introduction. The task of lethal area evaluation of antipersonnel fragmentation weapons is obviously important, as these weapons (that is, mortars, howitzers, guns, artillery rockets, and so on) accounted for the overwhelming majority of casualties in the Korean and Indochinese conflicts. It should furthermore be noted that the percentage of antipersonnel fragmentation rounds employed by armor during World War II ranged from an average value of **16** percent to a maximum value of **24** percent, depending on the theater. Let us now formulate the problem in broad outline, leaving the details until later. Evaluating the lethal or deadly area of a projectile means arriving at a figure of merit which permits one to predict how many casualties a shell will produce upon detonation under specified conditions.

2-187. Lethality. The basic concept used in arriving at a suitable lethality index is that of the "expected number of occurrences of a rare event," that is, the expected number of incapacitations of personnel targets. Consider a shell detonated above the ground (figure **2-72**) at point S (x, O, z).

The expected number I of incapacitations is then given by

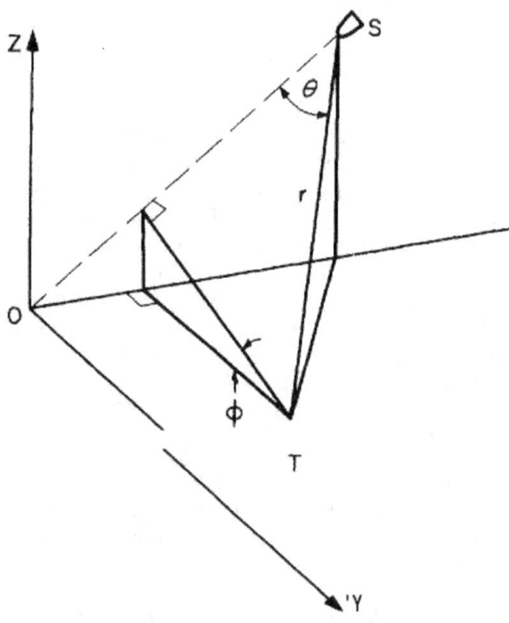

Figure 2-72. Shell detonated above the ground

$$I = \int_A\int \tau\ (x, y)\ P\ (x, y)\ dx \cdot dy \qquad (15)$$

where τ is the density of the targets and P is the probability of a target's being incapacitated within the area dx . dy. Assuming constant target density, the above expression becomes

$$I = \tau \int_A\int\ P\ dx \cdot dy \qquad (16)$$

I/τ, which has the dimensions of area, is then defined as the lethal area L and is seen to be the number which upon multiplication by the target density yields the expected number of casualties. Since not every hit by a fragment causes incapacitation, it is desirable to decompose the probability P into a function of three probabilities: the probability of the target being exposed, the probability of the target being hit if exposed, and the probability of the target being incapacitated if hit. Thus equation (16) may be written

$$L = \int_A\int P_{i/h/e}\ dx \cdot dy \qquad (17)$$

It now remains to insert the appropriate input data into the function $P_{i/h/e}$. The intrinsic parameters are the following:

1. Shell Descent Angle
2. Shell Residual Velocity
3. Shell External Geometry
4. Shell Internal Geometry
5. Shell Casing Composition
6. Shell Filler Composition
7. Target Presented Area, which depends on (see figures 2-73 through 2-77)
   a. target attitude
   b. cover, natural or artificial
   c. fragment aspect angle
8. Target Incapacitation Criterion.

Static fragmentation input data of the kind depicted in figures 2-78 through 2-80 are used invariably.

The incapacitation criterion in use at the present time is given in BRL TN 370.

Upon insertion of the input data the expression for the lethal area becomes

$$L = \int_A\int G(\theta, \phi) \cdot \left\{1 - \exp\left[T_1(\theta, \phi) \cdot \int_0^{\infty} S_1(\theta) \cdot S_2(m, \theta) \cdot T_2(m, \theta, \phi)\ dm\right]\right\} \cdot J(\theta, \phi) d\theta \cdot d\phi \qquad (18)$$

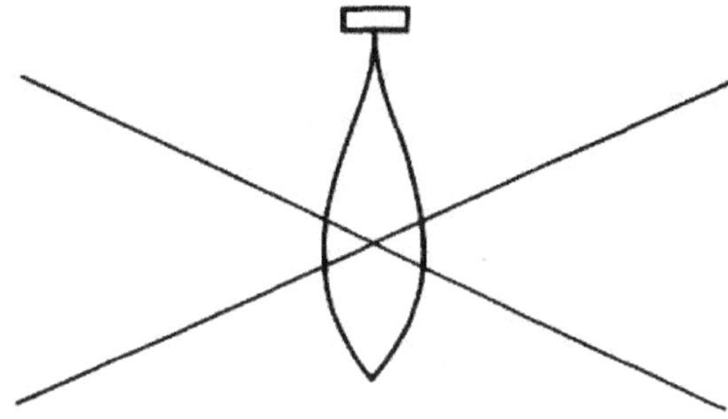

Figure 2-73. Static detonation

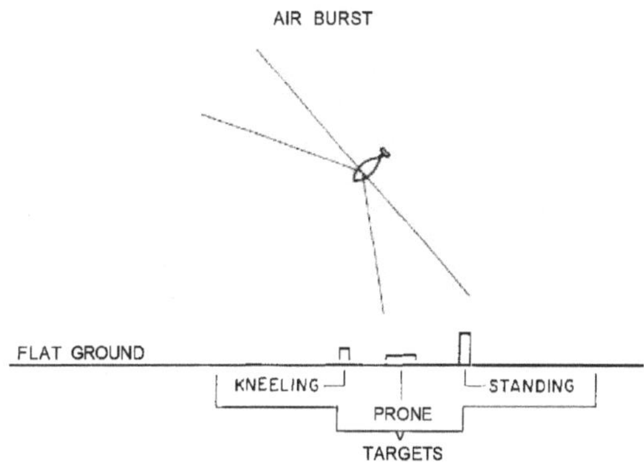

Figure 2-74. Air burst — flatground

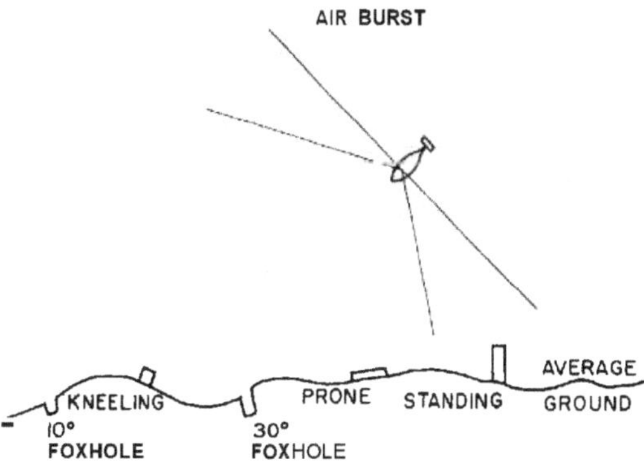

Figure 2-75. Air burst — average ground

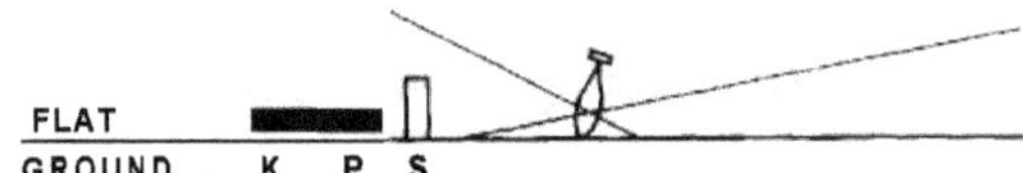

Figure 2-76. Ground burst —flat ground

with

$G(\theta, \phi)$	indicative of geometric properties of the ground
$T_1(\theta, \phi)$	indicative of geometric properties of target
$S_1(\theta)$	indicative of physico-chemical properties of shell
$S_2(m, \theta)$	indicative of physico-chemical properties of shell
$T_2(m, \theta, \phi)$	indicative of physiological properties of target
$J(\theta, \phi)$	indicative of transformation of coordinate system.

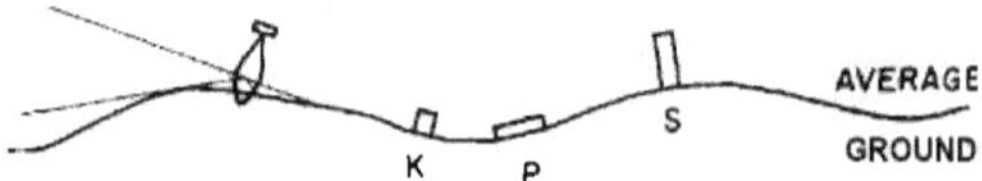

Figure 2-77. Ground burst — average ground

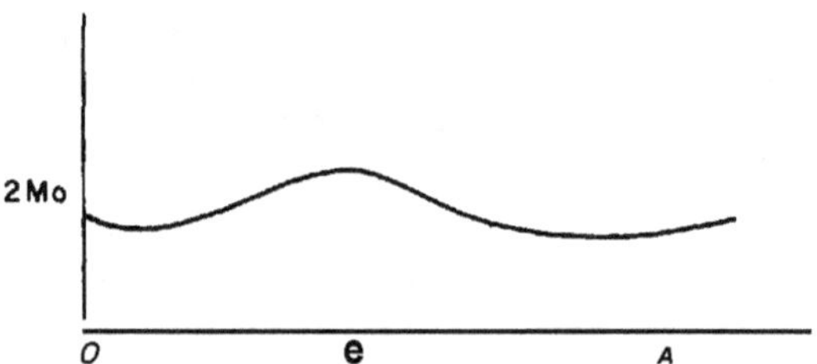

Figure 2-79. Average fragment mass

The complicated nature of equation (18) renders numerical integration via high-speed computing machinery inescapable.

2-188. Simplifications. The lethality determination method discussed so far is unquestionably the most comprehensive and accurate one but is — it goes without saying — no more accurate than the fragmentation data that enter into it. Certain reasonable simplifications that may be effected in the computation procedure will now be discussed, the principal one of which is the assumption that all fragments depart from the shell within a side spray of specified angular width, say of the order of 20°, and that the angular position of the centerline of this side spray may be fixed by inspection of the fragmentation test data. It turns out that in this case the double integral of equation (16) will reduce to the single integral

$$L = 2\Delta \sin\theta \int_{\phi} P_c\,(r, \theta)\, r^2 \sec\psi \,\, d\phi \qquad (19)$$

This integral is clearly attackable by standard numerical methods. The lethal areas and single-

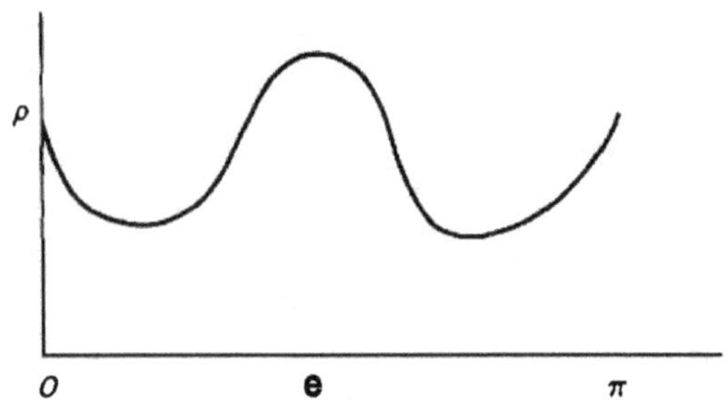

Figure 2-78. Fragment density

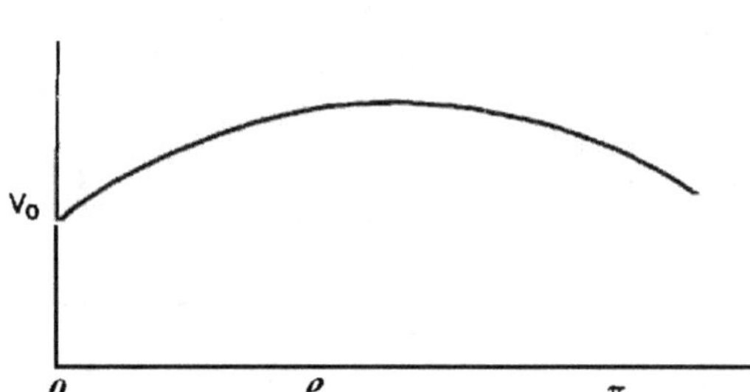

Figure 2-80. Initial velocity

shot probability are, in fact, now obtainable within three to four minutes via high-speed computing mach nery procedures. Moreover, the ground-burst lethal area under these conditions is then also easily obtainable by a procedure not involving numerical integration at all.

2-189. Lethality of Hypothetical Shell. So far, lethality determination has been based on fragmentation test data. It remains to discuss the procedure applicable to those cases where such data do not exist, such as the case where one wishes to determine in what direction to move when a choice of projectile designs is considered in the course of the investigation of a series of hypothetical shell. Here one may consider a series of shell which differ, say, in caliber, external geometry, filler fraction, filler composition, and wall thickness. In order to obtain the lethal area and — for that matter the single shot probability and ground-burst lethal area as well — it seems reasonable to assume again that all fragments depart within a side spray of specified angular width, as the principal object of this assessment is a comparison of the members of a series of hypothetical shell to one another rather than to existing projectiles.

One then proceeds to assume that these fragments are distributed with respect to mass according to equation

$$(m/m_o)^{1/K} + \ln(n/n_o) = 0 \qquad (20)$$

where

- n is the number of fragments of mass greater than m
- n_o is the total number of fragments
- m_o is a constant, characteristic of the shell
- K is a constant, indicative of break-up dimensionality.

Equation (20) was devised by Professor N. F. Mott. It is known to occur in particle size analysis of crunching processes. Making the assumption of two-dimensional fragment break-up, the constants n_o and m_o in equation (20) are related to the total mass of the fragmenting metal by the following equation:

$$W = n_o \cdot 2 \cdot m_o \qquad (21)$$

where W is the weight of the fragmentingmetal. It follows, therefore, that once either n_o or m_o becomes fixed, the other constant is determined. The two principal avenues of attack are as follows: one may calculate n_o by the following equation

$$n_o = c \cdot l \cdot d_o \, , (t/d_o)^{-2/3} \qquad (22)$$

where
- l = length in calibers
- d_o = outside diameter
- t = wall thickness.

This equation is due to K. C. Cook. An alternative is to compute m_o from

$$m_o = ct^{5/3} d_i^{2/3} (1 + t/d_i)^2 \qquad (23)$$

where d_i is the inside diameter. Here m_o depends on the shell caliber and wall thickness, with the constant being a function of filler and metal compositions.

The fragment density is then assumed to be uniform over the entire side spray and the initial fragment velocity will also be determined theoretically, namely by the following equation suggested by Gurney:

$$v_o = c \, [\alpha/(1 + \alpha/2)]^{1/2} \qquad (24)$$

Here α is the ratio of explosive charge and mass of fragmenting metal.

2-190. Weapon Effectiveness. The lethality indices obtained via equation (18) as function of Z (see figure 2-81) — with the values of all the other parameters temporarily held constant — are the cornerstone upon which the assessment of antipersonnel fragmentation weapons rests. So far, there has been secured an index significant only if the projectile arrives at its scheduled destination. However, a satisfactory measure of effectiveness cannot be obtained unless one accounts for the various types of dispersion involved. These problems are again handled more efficiently, by several orders of magnitude, with the use of high-speed computing machinery.

While the lethal area constitutes the appropriate figure of merit in the case of large targets, such as men uniformly and randomly distributed over a large sector, a different situation exists when fire against a small target, such as a

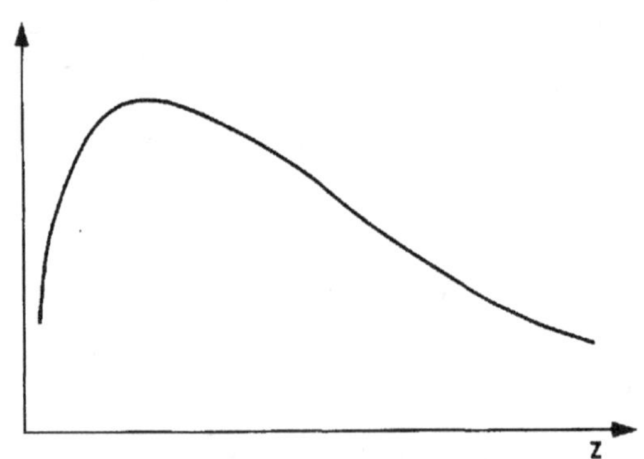

Figure 2-81. L versus Z

gun emplacement, is required. Here one wishes to compute the single-shot probability, the determination of which requires knowledge of weapon accuracy, that is, of range dispersion and of deflection dispersion.

Another type of dispersion, namely fuze dispersion, needs to be introduced in order to obtain what might be called the expected lethal area. This is the air-burst lethal area reduced to some new value by virtue of the fact that, firstly, a certain fraction of the fuzes will fail to function altogether, and secondly, that when the fuzes do function the burst heights will be distributed in some manner about the mean burst height. It is observed that the optimal air-burst height is not the one read from figure 2-81 but a value somewhat larger — by the very nature of figure 2-81, which applies qualitively to practically all shell — which is obtained by combining the L versus Z curve with the aforementioned fuze data.

A further most important use is made of the lethality index in the analysis of area fire effectiveness. For example, to find the answer to the question of how many shell of specified type need be fired into a rectangle of dimensions a x b in order to cause a prescribed fraction of incapacitations, one wishes to determine the probability that a target is incapacitated by at least one fragment within n rounds, that is to say, evaluate

$$P_n = \int_B\int \tau(x, y)\ [1 - \int_A\int \phi(x^*, y^*) \cdot P(x - x^*, y - y^*)dx^*dy^*]^n\ dx\ dy \qquad (25)$$

where $P(x - x^*, y - y^*)$ is the probability of the target $T(x, y)$ being incapacitated by a burst at $S(x^*, y^*, z)$ and the distribution function ϕ is indicative of aiming and dispersion; then, with simplifying assumptions, one may approximate

$$\int_A\int P(x - x^*, y - y^*)\ dx^*dy^* \doteq L \qquad (26)$$

Effective treatment of the above expressions makes the use of high speed computing machinery again imperative.

2-191. Weapon Systems Analysis. Results obtained by the methods of the two preceding sections are prerequisites for two types of further investigation, the first of which is antipersonnel weapon systems analysis. Here various criteria, related to effectiveness, cost, and so on, may be chosen to effect system optimization for typical tactical situations. The second type of investigation is combat analysis.

2-192. Combat Analysis. In contrast to the first type of investigation, which deals with a (in a certain sense) static case, one now considers the dynamic case of hypothetical engagements on specified terrain with either actual or experimental weapon systems. The methods of engagement outcome prediction will, once more, require integrating into them effectiveness data on the weapon systems chosen, and in particular, the antipersonnel weapon systems.

Here, once more, the use of high-speedcomputing machinery comes to the fore since the assessment, by stochastic methods, of combat models deemed satisfactorily realistic from the military point of view, presents problems of such complexity that attempts to solve them have, in fact, not been undertaken until recently.

CONTROLLED FRAGMENTATION

2-193. Purpose of Controlled Fragmentation Shell. In uncontrolled fragmentation, the range of masses and velocities is very great. To secure more effective fragments, it is desired to solve for the optimum mass (dependingonthe lethality criteria), and to design a shell that would emit all fragments with this mass. Refer to reference 15 for a discussion of methods for determining the optimum mass. In this way, the probability of damage is greatly increased, and the results can be estimated more correctly.

2-194. Methods of Controlling Fragmentation. The methods for controlling fragmentation are:

1. Preformed fragments (with or without matrix)
2. Notched or grooved rings
3. Notched or grooved wire
4. Notched casings
5. Multiple walls
6. Fluted liner.

Although all of these methods can be used on experimental shell with static firing, the fluted-liner method and multiple-wall method are most practical for artillery shell, because of the presence of setback forces. All the methods will be described in the paragraphs that follow.

2-195. Description of the Preformed Fragment Method. In this method the fragments are made of the desired size, and are then incorporated into the shell wall. This can be done by placing them within a plastic matrix, which forms the shell wall, or by enclosing them between two concentric, thin-walled steel shells. Neither of these methods can be utilized for artillery shell at the present time, as there is not enough strength to overcome setback.

2-196. Description of the Notched-Ring Method. In this method, notched rings are fitted over a liner, which can be plastic or thin metal. Figure 2-82 shows a typical assembly of a grooved-ring shell. It is assumed that in the fragmentation control the liner has no noteworthy fragments. The factors that have to be considered in shell of this sort are:

1. Quality of steel in rings
2. Spacing of the grooves
3. Groove depth
4. Width of rings
5. Liner
6. Length-to-diameter ratio
7. Ring finish.

A medium-carbon low-sulfur steel, heat treated to a hardness of 95RB, is thought to be desirable. However, no tests have been conducted to show the effect of carbon content, sulfur content, or hardness in the uniformity of fragment weight. The steel should be sufficiently hot-worked to break up segregated inclusions, and assure their uniform distribution.

ASSEMBLY OF GROOVED RING CONTROLLED FRAGMENTATION SHELL

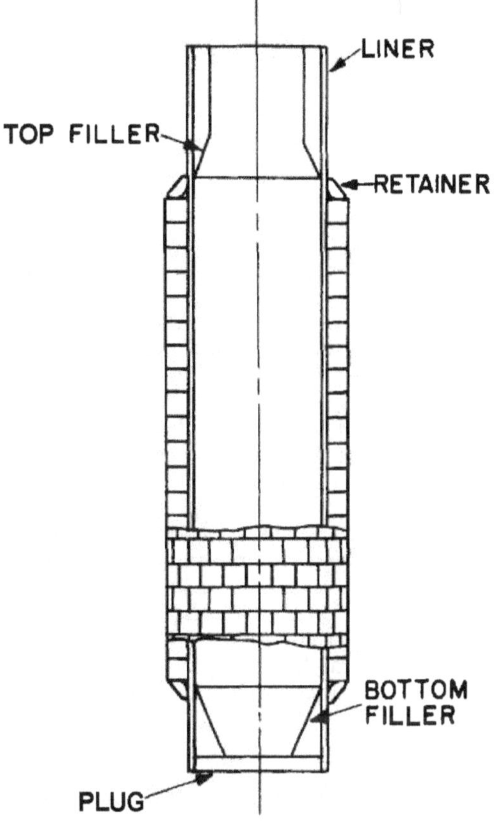

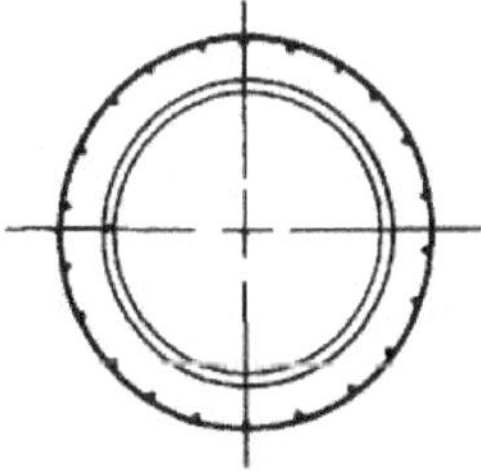

Figure 2-82. Assembly of controlled fragmentation shell

The groove spacing can be determined from the following formula.

$$G = \frac{\pi(2R - t)}{W} = \frac{\pi(2 - a)}{\theta a}$$

where

G = number of grooves per ring
R = outside radius of case, in inches
t = thickness of case, in inches
W = mean width between grooves, in inches
α = t/R (dimensionless)
θ = W/t (dimensionless).

The depth of the groove should be from 5 to 10 percent of the ring thickness. Excessive groove depth causes the fragments to break up. Preferably the grooves should have sharp bottoms, although grooves with rounded bottoms produced satisfactory fractures in some cases.

The rings should have a ground or smooth lathe finish. Sharp scratches in the surface should be avoided.

The liner should be made of a material that will produce no important fragments; it should be kept as thin as is consistent with practical manufacture and considerations of strength. A thickness of 5 percent of its radius has been found satisfactory for laminated phenolic plastic tubing.

In general, the width of the ring should be made equal to the thickness; more details are not known. The length to diameter ratio of the case should not be less than 1 1/4 to 1. The desirable length is between 2 1/2 and 5 calibers.

It has been shown that better fragmentation results can be obtained by inserting a soft porous liner between the explosive and rings. This liner prevents the fragments fromformingslivers.

For more detailed information on controlled fragmentation by grooved rings, refer to references 51, 44, and 49. The grooved-ring method has been used in static firing tests only. At the present time, this shell does not have enough strength to overcome setback forces.

2-197. Description of Notched-Wire Method. In general, this method is similar to the notched (grooved) rings, except that notched wires, spiral-wrapped about the liner or warhead, are used. Notched wires are used when the ring thickness would be too thin for economical manufacture. A detailed description of notched wires can be found in reference 46.

2-198. Description of Notched Casings. Four types of notched casings, listed below, have been tested for applicability to controlled fragmentation.

1. Cylinder, 4-in. O. D., 1/4-in. wall, with 1/8-in. holes in diamond pattern, punched and plugged; holes, 1/2 in. apart in row, rows 1/2 in. apart.
2. Cylinder, 3 1/2-in. I. D., with linearly tapered steps cut on outside, steps 1/2 in. long.
3. Cylinder, 4-in. O. D., 1/4-in. wall, with left-hand and right-hand helical grooves cut at 45" to axis, and spaced 1/2 in. apart. Groove profile V-shaped, with included angle of 60".
4. Cylinder, 4-in. O. D., 1/4-in. wall, with hexagonal pattern impressed by shearing.

Each cylinder was approximately 12 inches long, and each was provided with brass end-plates to increase the confinement of the explosion.

Both the tapered cylinder and the one with grooves gave a great percentage of very small fragments. The cylinder with punched and plugged holes gave slight indication of control, and only the hexagonal sheared pattern showed an excellent degree of control.

For more details, refer to reference 50. At the present time the results mentioned in this paragraph have not been applied further to any specific shell. Also, at present this method of control is not applicable to artillery shell, because of setback.

2-199. Multiple Walls. The multiple-wall shell are made by using close-fitted cylinders, each with thickness t/n, where t is the thickness of a one-wall shell and n is the number of walls. The multiple-wall shell do not give complete fragmentation control, however, for only the thickness of the fragments is uniform. The number of fragments is approximately n times the number of fragments of a single-wall shell, where n is the number of walls. The partial control achieved, however, is an improvement, because the average of fragment mass is reduced and the number of fragments emitted is increased. However, the increase in lethality is much less than expected. See paragraph 2-159.

2-200. Fluted Liners. With this method control of fragment mass is achieved by shaping that part of the high-explosive charge which is adjacent to the metal casing. A thin metal, paper, or plastic liner inserted within the casing forms the pattern on the explosive surface during filling. Projections on the liner are frequently long and triangular in section, form grooves in the explosive, and have been called flutes. The method is an application of the Munroe effect in multiple since, on detonation, jets corresponding to each groove cut the casing.

Successful application of the shaped charge depends on several factors. The proper depth of groove must be determined, so that the optimum depth of cut may be achieved. If the cut is too deep, excessive erosion and chipping of the edges of the fragment will result. If the groove is not deep enough, then the jets will not cut deeply enough into the steel shell, and poor fragmentation control will result.

The spacing and overall pattern of the fluted liner must be adjusted to the most suitable combination for each size and type of shell. A V-shaped groove with an apex angle of about 75° gave the best results with the British 3-inch U. P. shell. The height of the groove is usually made equal to the wall thickness of the casing. Fluted liners also have been used with the ring method of controlling fragmentation. This method gave a very good shape factor, but the efficiency of control was not too high.

AIRCRAFT DAMAGE

2-201. Aircraft Damage Evaluation. In order to interpret the results of experimental firings against aircraft the following standards have been set up:38

a. "Damage" is divided into four categories and assessed as follows:

1. "A Damage" is damage such that the aircraft will fall out of control within five minutes after damage occurs. "K Damage" denotes an aircraft that will fall out of control immediately. "KK Damage" denotes an aircraft that will disintegrate immediately in the air (damage that would render a kamikaze attack ineffective).

2. "B Damage" is damage such that the aircraft will fail to return to its base as a result of the described damage. The distance of the base will be fixed for the particular type of mission being assumed. This probability will include the five-minute period immediately after the aircraft is hit, as well as the time required to return to base after the five minutes have elapsed. Thus, the "B Damage" assessments will always be equal to or larger than "A Damage," but will never exceed 100. The sum of "A Damage" and "B Damage" may exceed 100.

3. "C Damage" is damage such that the particular attack will not be successfully completed. It is possible to have "C Damage" even though no "A Damage" or "B Damage" exists. Thus, damage to the bomb release system, to controls that would affect the prosecution of the attack, and to personnel involved in the attack would be classed as "C Damage." The assumption that the attack is 2 1/2 minutes away when the damage is incurred is an important one when evaluating "C Damage."

NOTE

The word "Kill" is sometimes used in discussions and memoranda to denote an assessment of 100 under any of the three categories of damage.

b. When assessing any category of damage, it is assumed that the crew would remain with the aircraft and attempt to complete the entire mission, even though "bailing out" would be the most likely procedure in actual warfare.

c. It will be noted that the numbers used to evaluate the various categories of damage are described as probabilities and written as percentages. However, strictly speaking, this is not correct, since the damage suffered by the aircraft (or engine) either will or will not cause it to go out of control, assuming a set of standard conditions under which the aircraft (or engine) is operating. The assessments between 0 and 100 therefore represent the uncertainty of the assessor as to whether the damage would result in a kill, or not. But if assessments are not biased, the expected value of many such assessments on various parts of the aircraft would be the correct value "in the long run" that one should arrive at for the vulnerability of the aircraft. The parameters of the error distribution can be estimated if one has a large number of cases in which the same damage has been assessed by different assessors.

2-202. Types of Damage Assessed. To facilitate the determination of the vulnerability of different components of the aircraft, damage is assessed under three general types: "Engines," "Structures," and "Fuel System."

1. "Engines" will be that portion of a tractor-type reciprocating power plant, accessories, controls, engine mount and fairing forward of the firewall; and that portion of a jet unit from the air-intake to the tailpipe, inclusive. Damage to oil cells/tanks, oil coolers/lines, propeller anti-icing equipment, and engine control linkages will be assessed under "Engines," even though the damage may occur in the "Structures" area. Damage to that portion of the "Fuel System" in the engine area will be assessed under "Engines," and also under "Fuel System."

2. "Structures" will be all parts of the aircraft except the engine(s) and controls, engine accessories and controls, and the "Fuel System"; and will include the aircraft structure, surface control cables, armor, armament, landing gear and actuating linkage, pyrotechnics, oxygen equipment, electronic equipment, and dummy personnel. On aircraft where the landing gear has been extended, all projectiles and fragments that pass through that part of the aircraft where the wheels and actuating linkage would have been shall be assumed to have struck that component. An approximation shall be given for probable damage (based on velocity of the missile and extent of damage actually done). Naturally, if a missile has been assumed to have punctured a tire, or otherwise damaged a landing gear that was not retracted, no description or assessment for the damage that actually occurred can be given.

3. "Fuel System" will be all fuel cells/tanks, selector valves, transfer and booster pumps, strainers, hoses, lines, and fittings. Damage to any portion of the "Fuel System" that would cause damage to an engine shall be assessed under "Fuel System," and also under "Engines."

2-203. Aircraft Vulnerability. Aircraft may be kept from accomplishing their mission by damage to any of the following:

1. Fuel
2. Engines
3. Structures
4. Personnel.

The probability of obtaining a kill on an aircraft as a whole is the product of the probability of obtaining a kill on any of the components. Firings have been conducted in an effort to establish lethality criteria for each of these. (See references 37, 14, 13, 9, and 5.) These firings are still under way at the time of writing.

2-204. Aircraft Vulnerability Studies. Knowledge of the vulnerability of aircraft to various types of shell would obviously be of great value to the shell designer. Of still more value would be lethality criteria by means of which this vulnerability could be computed. Studies of the vulnerability of specific aircraft to specific shell have been carried out.[37] The results of these firings can logically be applied to other aircraft of similar construction, but there is no way to apply them to shell which differ from those that were tested in the original firings. Controlled fragmentation firings have been conducted against B-25 aircraft in order to obtain data from which criteria for the lethality of fragments from airburst shell could be determined. (See references 5, 9, 13, and 14.) Curves for the probabilities of damage to engines and to fuel were developed. These curves are of the form $P = 1 - (a + bx)^{-1}$ where $x = (MV^2)^{1/C}$; P is the probability of A damage; a, b, and c are constants; M is fragment mass; and V is fragment velocity. Using these curves, the damage probabilities for two service shell were computed. The results were compared to results obtained from actual firings of these shell against B-25 aircraft. The results indicated that data from controlled fragmentation firings can be used to determine the lethality of service shell. The data thus far obtained have limited applicability, since they were conducted under the following conditions:

1. B-25 aircraft only
2. Sea level conditions only
3. Fuel and engine damage only
4. Damage functions obtained for "forward," "fore above," and "aft above" directions only.

Investigations along these lines are still under way, and useful results may be expected.

2-205. Fuel Damage. A steel fragment, when it pierces the skin or other structural components of an aircraft, usually will cause a flash hot enough to ignite fuel. When the fragment penetrates a fuel cell at a point below the fuel level, it may cause one or more spurts of gasoline to issue from the tear in the tank. Depending upon the self-sealing qualities of the tank and the size of the rip made by the fragment, the flow of fuel may be continued after the initial spurts. If the following conditions are satisfied, a fire may be started and sustained.

a. Conditions for starting: fuel fire.

1. Fragment must cause flash when penetrating metal structure.

2. Fragment must cause spurt when fuel cell is penetrated.

3. The flash and spurt must meet, that is, the distance between the structure and the fuel cell must not be excessive.

4. The spurt must not be "heavy" enough to smother the flash.

5. Sufficient air for combustion must be present.

b. Conditions for sustaining fuel fire.

1. A sufficient air flow must take place.

2. The air must be dense enough to support combustion (at extremely high altitudes a fire cannot be sustained).

3. The rip in the fuel cell must be large enough to prevent self-sealing.

4. The fire must be internal; external fires will be blown out by the slipstream.

2-206. Fuel Tank Vulnerability.

a. Self-sealing tanks. At sea level, self-sealing tanks are, in general, vulnerable (that is, fires may be caused) either to fragments with a minimum mass of 100 grains or a minimum velocity of 3,000 fps. At the minimum mass, the fragment velocity must be well above the minimum in order to cause a fire. The converse of this is also true.

b. Nonsealing tanks. At sea level, the minimum fragment mass required to cause a fuel fire is limited only by the ability to penetrate both structure and fuel tank—probably less than 30 grains. The minimum fragment velocity is 2,000 fps.

2-207. Defenses Against Fuel Fires. The following passive defenses may be taken against fires, and should be considered in design of ammunition.

a. The use of self-sealing fuel cells.

b. Using plastic foam or other inert material to fill the space between structure and cell.

c. Filling the space between structure and cell with an inert gas.

REFERENCES AND BIBLIOGRAPHY

1. Mott, Fragmentation of H.E. Shells; A Theoretical Formula for the Distribution of Weights of Fragments, AC 3642 (British), March 1943.

2. Weiss, H. K., Optimum Angular Fragment Distributions for Air-Ground Warheads, Ballistic Research Laboratory, Aberdeen Proving Ground, Report No. 829.

3. Shaw, A Measurement of the Drag Coefficient of High Velocity Fragments, Ballistic Research Laboratory, Aberdeen Proving Ground, Report No. 744, October 1950.

4. Weymouth, The Effect of Metallurgical Properties of Steel Upon Fragmentation Characteristics of Shell, Ballistic Research Laboratory, Aberdeen Proving Ground, Memorandum Report No. 585, April 1952.

5. A Critical Comparison of Damage Probabilities to Aircraft Components Predicted from Controlled Fragmentation Shell Data to Those Obtained from 75-MM & 105-MM Air-Burst Shell Experimental Firings, Project Thor, The Johns Hopkins Institute for Cooperative Research, Report No. 2, August 1949.

6. Solem, Shapiro, and Singleton, Explosives Comparison for Fragmentation Effectiveness, Naval Ordnance Laboratory, Report NAVORD 2933, August 1953.

7. Benson, The Basic Method of Calculating the Lethal Area of a Warhead, Picatinny Arsenal Report, June 1954.

8. Gurney and Sarmousakis, The Mass Distribution of Fragments from Bombs, Shell, and Grenades, Ballistic Research Laboratory, Aberdeen Proving Ground, Report No. 448, February 1944.

9. An Analysis of Fragment Damage to the B-25 Fuel System, Project Thor, The Johns Hopkins Institute for Cooperative Research, Ballistic Research Laboratory, Aberdeen Proving Ground, Report No. 836, January 1953.

10. Weymouth, Fragmentation Characteristics of Three Grades of Ductile Cast Iron, Ballistic Research Laboratory, Aberdeen Proving Ground, Technical Note No. 600, June 1952.

11. Benson, The Influence of the Surface Contour of an Exploding Body on Fragment Distribution, Picatinny Arsenal Report, February 1953.

12. Fragmentation Tests, Ordnance Proof Manual 40-23, March 1947.

13. Second Report on Damage to B-25 Aircraft Components by Fragments of Controlled Mass and Velocity, Project Thor, The Johns Hopkins Institute for Cooperative Research, Report No. 8, April 1953.

14. Damage to Aircraft Components by Fragments of Known Mass and Velocity from, Controlled Fragmentation Shell, Project Thor, The Johns Hopkins Institute for Cooperative Research, Report No. 1, June 1949.

15. Weiss, H. K., Methods for Computing the Effectiveness of Fragmentation Weapons Against Targets on the Ground, Ballistic Research Laboratory, Aberdeen Proving Ground, Report No. 800, January 1952.

16. Mott, AC 3348 (British), January 1943.

REFERENCES AND BIBLIOGRAPHY (cont)

17. Cook, K. S., Report No. 771, Appendix I, Ballistic Research Laboratory, Aberdeen Proving Ground.

18. Mott and Linfoot, AC 3348 (British).

19. Gurney, The Initial Velocities of Fragments from Bombs, Shell, and Grenades, Ballistic Research Laboratory, Aberdeen Proving Ground, Report No. 405, September 1943.

20. Sterne, T. E., Provisional Values of the Vulnerability of Personnel to Fragments, Ballistic Research Laboratory, Aberdeen Proving Ground, Report No. 758, May 1951.

21. Weiss, H. K., Methods for Computing the Effectiveness of Area Weapons, Ballistic Research Laboratory, Aberdeen Proving Ground, Report No. 879, September 1953.

22. Famiglietti, Michael, Fragmentation of Ring Type Cylindrical Shell Made of Various Metals, Ballistic Research Laboratory, Aberdeen Proving Ground, Memorandum Report No. 597, March 1952.

23. Dunn, D. J., Jr., and T. E. Sterne, Hand Grenades for Rapid Incapacitation, Ballistic Research Laboratory, Aberdeen Proving Ground, Report No. 806, April 1952.

24. Sterne, T. E., Provisional Criteria for Incapacitation by Fragments, Ballistic Research Laboratory, Aberdeen Proving Ground, Technical Note No. 556, November 1951.

25. Birkhoff and Lewy, Optimum Height of Burst for 105-MM Shell, Ballistic Research Laboratory, Aberdeen Proving Ground, Memorandum Report No. 178, June 1943.

26. Share, Probability of Damage to Guns or Personnel Due to Fragmentation of Artillery Shells, Ballistic Research Laboratory, Aberdeen Proving Ground, Report No. 636, April 1947.

27. Weiss, H. K., Justification of an Exponential Fall-Off Law for Number of Effective Fragments, Ballistic Research Laboratory, Aberdeen Proving Ground, Report No. 697, February 1949.

28. Simpson and Bushkovitch, Fragment Contour Projector and the Presentation Areas of Bomb and Shell Fragments, Ballistic Research Laboratory, Aberdeen Proving Ground, Report No. 501, November 1944.

29. Braun, Charters, and Thomas, Retardation of Fragments, Ballistic Research Laboratory, Aberdeen Proving Ground, Report No. 425, November 1943.

30. Charters and Thomas, The Aerodynamic Performance of Small Spheres from Subsonic to High Supersonic Velocities, Ballistic Research Laboratory, Aberdeen Proving Ground, Report No. 514, May 1945.

31. Sterne, T. E., The Fragment Velocity of a Spherical Shell Containing an Inert Core, Ballistic Research Laboratory, Aberdeen Proving Ground, Report No. 753, March 1951.

32. Fano, Discussion of the Optimum Characteristics of Weapons for the Most Efficient Fragmentation, Ballistic Research Laboratory, Aberdeen Proving Ground, Report No. 594, January 1946.

REFERENCES AND BIBLIOGRAPHY (cont)

33. Smith, Effectiveness of Warheads for Guided Missiles Used Against Aircraft, Ballistic Research Laboratory, Aberdeen Proving Ground, Memorandum Report No. 507, March 1950.

34. Sterne, T. E., A Provisional Casualty Criterion, Ballistic Research Laboratory, Aberdeen Proving Ground, Technical Note No. 370, March 1951.

35. Sachs and Schwarzschild, Properties of Bomb Fragments, Ballistic Research Laboratory, Aberdeen Proving Ground, Report No. 347.

36. Average Area of Fragments, Ballistic Research Laboratory, Aberdeen Proving Ground, Ballistic Research Laboratory Memorandum: Tolch to Kent, 30 September 1943.

37. Weiss, Christian, and Peters, Vulnerability of Aircraft to 105-MM and 75-MM HE Shell, Ballistic Research Laboratory, Aberdeen Proving Ground, Report No. 687, March 1949.

38. Guide to Assessors — Aircraft Vulnerability Program, Ballistic Research Laboratory, Aberdeen Proving Ground.

39. Kent, R. H., The Shape of a Fragmentation Bomb to Produce Uniform Fragment Densities on the Ground, Ballistic Research Laboratory, Aberdeen Proving Ground, Report No. 762.

40. Gehrig, J. J., Lethal Areas of Some Fragmentation Weapons, Ballistic Research Laboratory, Aberdeen Proving Ground, Memorandum Report No. 717, September 1953.

41. Weiss, H. K., Description of a Lethal Area Computation Problem, Ballistic Research Laboratory, Aberdeen Proving Ground, Memorandum Report No. 723, September 1953.

42. Schaffer, M. B., Provisional Lethality Criterion for Canister Ammunition, Picatinny Arsenal Report No. 2002, March 1954.

43. Grabarek, C. L., Characteristics of Fragmentation of MX904 Warheads, Blast With Fluted Liner, Comp B, HBX, and Tritonal Loaded, Ballistic Research Laboratory, Aberdeen Proving Ground, Memorandum Report No. 700, July 1953.

44. Shaw, The Effect of "Cushions" in Controlled Fragmentation Shell, Ballistic Research Laboratory, Aberdeen Proving Ground, Report No. 732, August 1950.

45. Famiglietti, Michael, The Relative Effectiveness of Natural and Controlled Fragmentation of Shell, Mortar, HE, 105-MM, T53E1, Ballistic Research Laboratory, Aberdeen Proving Ground, Memorandum Report No. 604, March 1952.

46. Grabarek, C. L., Mass, Spatial, and Velocity Distributions of Fragments from MX-904 (T-7) Warheads, Wire Wrapped Types, Ballistic Research Laboratory, Aberdeen Proving Ground, Memorandum Report No. 550, June 1951.

47. Grabarek, C. L., Fragmentation of Bomarc Model Warheads, Ballistic Research Laboratory, Aberdeen Proving Ground, Memorandum Report No. 621, July 1952.

48. Grabarek, C. L., Fragmentation of Multi-Walled Cylindrical Warheads, Ballistic Research Laboratory, Aberdeen Proving Ground, Memorandum Report No. 633, November 1952.

49. Shaw, Principles of Controlling Fragment Masses by the Grooved Ring Method, Ballistic Research Laboratory, Aberdeen Proving Ground, Report No. 688, February 1949.

REFERENCES AND BIBLIOGRAPHY (cont)

50. Lyddane, Fragment Velocity Law — Fifth Partial Report: Tests of Experimental Controlled Fragment Warheads, Naval Proving Grounds, Report No. 603, July 1950.

51. Shaw, Development of Controlled Fragmentation Shell, Using Grooved Rings, Ballistic Research Laboratory, Aberdeen Proving Ground, Report No. 637, April 1947.

52. Dederick, L. S., and R. H. Kent, Optimum Spacing of Bombs or Shots in the Presence of Systematic Errors, Ballistic Research Laboratory, Aberdeen Proving Ground, Report No. 241.

53. Kolmorzorov, A. N., Editor, and Dr. Edwin Hewitt, Translator, Collection of Articles in the Theory of Firing, The Rand Corporation, T-14.

54. McMillen, J. H., and J. R. Gregg, The Energy, Mass, and Velocity Which Is Required of Small Missiles in Order to Produce a Casualty, Missile Casualties Report No. 12, 1945, National Research Council, Office of Scientific Research and Development.

KINETIC ENERGY AMMUNITION FOR THE DEFEAT OF ARMOR

DESCRIPTION

2-208. Types of Projectiles. Figures 2-83 and 2-84 illustrate the various types of kinetic energy ammunition used for the defeat of armor.

2-209. Armor-Piercing (AP). The simplest type of projectile is the solid shot illustrated in figure 2-83a. This type of projectile consists merely of a decrementally hardened solid piece of high-carbon through-hardening alloy with an ogival nose. Truncated-nosed shot, with and without tips, have been made experimentally but are not in tactical use. A die-cast aluminum windshield (false ogive, ballistic cap) is epoxy-resin bonded to the nose of the shot in order to obtain the necessary exterior ballistic characteristics. (Welding and crimping have also been used to bond the windshield to the projectile.) This type of projectile is known as an armor-piercing shot or monobloc shot. The nose of the projectile must have sufficient strength to withstand the stresses developed on impact and during penetration. It must be hard enough to resist deformation, yet be tough enough to prevent cracking or shattering. The body must be of sufficient strength to withstand bending stresses.

2-210. Armor-Piercing, Capped (APC). This type of projectile (figure 2-83b) was designed to prevent premature breakup of the projectile when used against face-hardened and homogeneous armor at low and intermediate obliquities. An armor-piercing cap, made of forged alloy steel decrementally hardened to give a very hard face with a tough and relatively soft core in contact with the projectile, was soldered or crimped to the nose of the projectile. This cap was intended to break up when the projectile struck the plate, thereby absorbing the initial shock of impact.

2-211. Hyper-Velocity Armor-Piercing (HVAP). In an effort to obtain higher velocities from existing artillery, the hyper-velocity armor-piercing or composite rigid projectile (figure 2-83c) was developed. This type consists of a core of an extremely hard high-density material, usually tungsten carbide (materials other

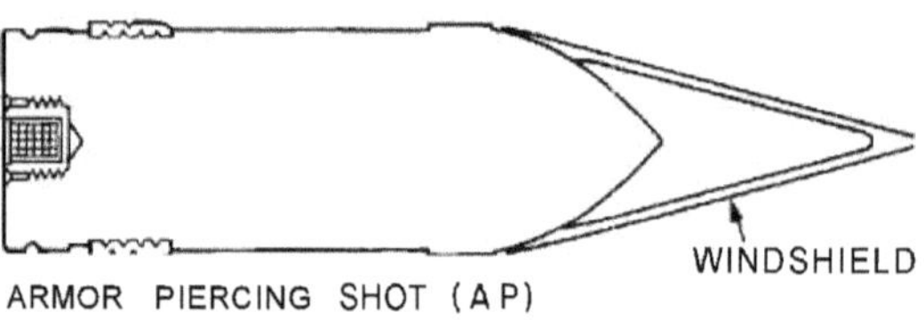

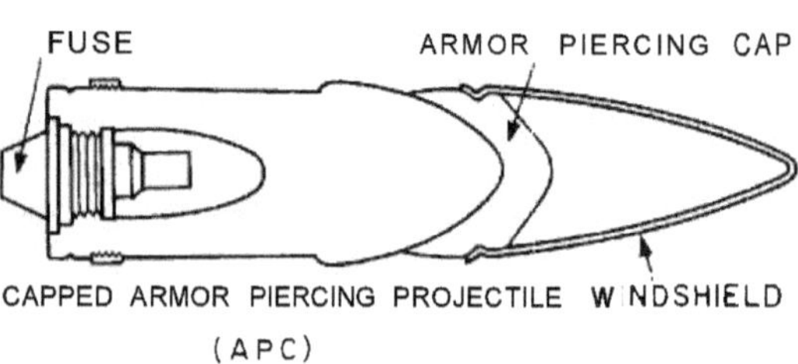

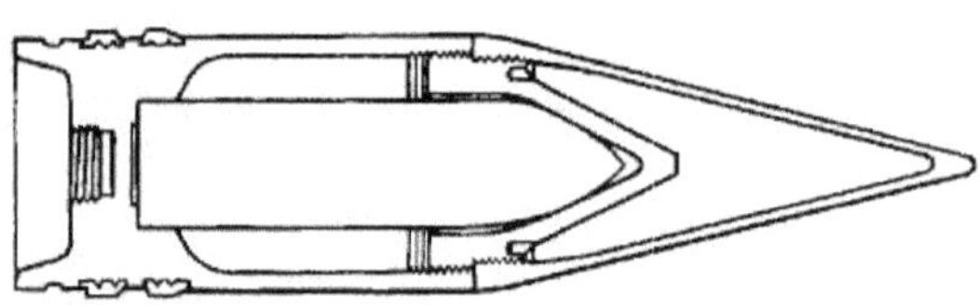

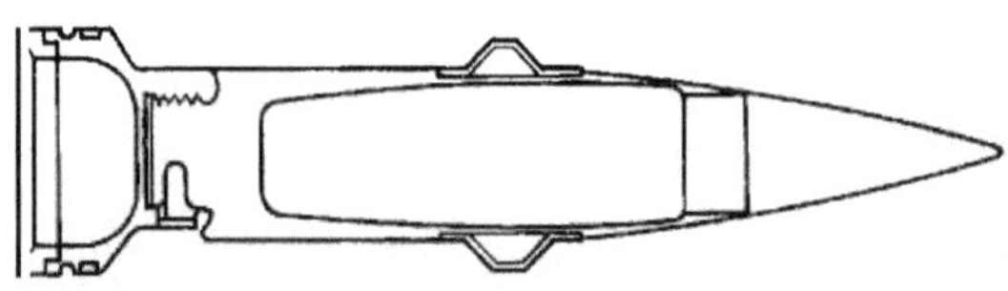

Figure 2-83. Kinetic energy projectiles (AP, APC, HVAP, squeeze-bore)

than tungsten carbide have been tested,[17] but none has been found suitable), within a lightweight carrier, usually aluminum. An armor-piercing cap similar to that used on APC ammunition may be placed on the core. The use of this lighter projectile enables velocities above 3,500 fps (hypervelocities) to be obtained without exceeding the allowable pressures of artillery designed for lower muzzle velocities and heavier projectiles. The carrier does not assist materially in penetrating the target plate, since it breaks up completely or vaporizes when it hits, leaving the core to do the damage to the target. Because of the comparatively low ratio of mass to cross-sectional area of this type of projectile, the rate of loss of velocity (slope of the remaining velocity curve) is rather high. Cost is from 7 to 10 times that of AP shot of the same caliber. Because of the development of HVAPDS shot, HVAP shot is obsolescent at present.

2-212. Skirted Projectiles. The tapered-bore gun, firing a skirted projectile (figure 2-83d), offers a means for the attainment of hypervelocity without exceeding currently existing limits on powder-gas pressures and temperatures. It produces this hypervelocity by driving a light projectile of small diameter with the same force as that applied by the powder gas to the base of a standard projectile. The combination of tapered-bore gun and skirted projectile possesses an appreciable advantage over the composite rigid arrangement in that the cross-sectional area of the projectile is diminished (with the same penetrator) and the drag coefficient is thereby reduced.

The original experimentation was concerned with the development of a gun with a tapered bore. However, the production of these was impractical, and a standard barrel gun, with a tapered adapter attached by a screw thread to the end of the muzzle, was used. This additional length of tube also resulted in an increase in muzzle velocity.

Because of the use of the tapered bore or adapter, standard HE shell cannot be fired from a gun made to use a skirted projectile, at least not without removing the adapter, a disagreeable procedure under combat conditions. As a result the skirted projectile has not received serious consideration of late.

2-213. Hypervelocity Armor-Piercing Discarding Sabot Ammunition (HVAPDS). Velocities in the 4,800 fps range have been obtained with hypervelocity armor-piercing discarding sabot ammunition. In this type of projectile (figure 2-84) a carbide core, either capped or uncapped, is placed inside a steel or light-alloy sheath to give good exterior ballistic characteristics, and this subcaliber assembly is placed inside a full-caliber carrier. This carrier is so designed that it will impart velocity and spin to the subcaliber projectile but will be discarded as it leaves the gun, thus allowing the subcaliber projectile to continue toward the target unimpeded by the carrier, or sabot. The sabot, usually made of aluminum, magnesium-zirconium alloy, or plastic, may be released from the subprojectile by a device actuated by setback, propellant-gas pressure, or centrifugal force. The actual separation of the sabot from the subprojectile is accomplished mainly by centrifugal force, air resistance, or both. The sabot, because of its poor ballistic shape and its low mass, loses velocity rapidly and leaves the subprojectile free shortly after it leaves the gun.

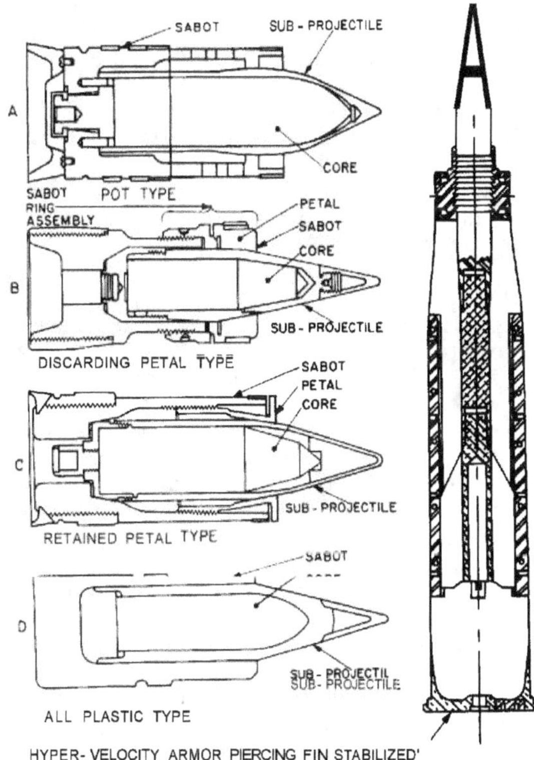

Figure 2-84. Kinetic energy projectiles (HVAPDS)

a. Method of Imparting Rotation. Basically, rotation is imparted to the subprojectile by the frictional force between the subprojectile and carrier during the acceleration in the gun. Further insurance of proper rotation may be obtained by using knurls at the rear of the subprojectile which, upon setback, engrave into the carrier; or by pins which transmit the rotation from the carrier to the subprojectile. Frequently the subprojectile is seated on a tapered surface in the carrier. This does not materially assist in imparting rotation but does serve to distribute the setback stresses more uniformly throughout the carrier. (See figure 2-84b.)

b. Method of Releasing and Discarding Carrier. Three general forms of release mechanisms have been used to ensure adequate and intact handling before firing, and proper functioning after firing. These methods are the front-release, rear-release, and sabot fragmentation mechanisms. Front-release mechanisms are classified into two types, the retained-petal and the discarding-petal types. The former acts much like a spring collet in a lathe; the individual petals, which hinge at the root, bend at the hinge under the action of centrifugal forces and release the subprojectile. (See figure 2-84c.) The discarding-petal type operates by the action of setback-induced shear forces on a narrow shear surface attaching the petals to the sabot ring assembly (figure 2-84b).

The rear-release mechanisms are ordinarily dependent upon the tracer attached to the subprojectile for retention in the sabot before firing. This tracer is usually provided with some form of lip to which the movable parts of the release mechanism are attached. During firing, some of the force available, in the form of setback, gas pressure, or centrifugal force, detaches these movable parts (quarter shoes, diaphragms, spiders, etc.), and leaves the subprojectile free in the sabot. In the example given in figure 2-84a, the tracer, fastened to the subprojectile rear sheath, butts against the fingers of the steel spider; the assembly is made tight by screwing down the tracer against the spider. During firing the spider fingers are thrown outward, both by the setback ring and by centrifugal force, leaving the subprojectile free.

In the plastic sabot design (figure 2-84d), the release of the sabot from the subprojectile, and its discard, are achieved by the fragmentation of the plastic carrier. This fragmentation is caused by the stresses resulting from centrifugal forces.

In the front-release types and rear-release types described above, actual discard of the carrier from the subprojectile is caused by air drag, which slows down the sabot faster than the subprojectile.

c. Exterior Ballistics of Sabot. From a safety viewpoint it is necessary that the lethality of the sabot or its fragments be minimized. This is done in two ways: either assuring that the sabot will break up into many small, low-energy fragments, which rapidly loose velocity, or by designing to minimize the number of secondary fragments detaching themselves from the sabot. The first is accomplished by the all-plastic sabot, and the second by the retained-petal type. The discarding-petal type is dangerous because of the high kinetic energy of the detached petals; this type is no longer being considered.

2-214. Armor-Piercing, Discarding Sabot, Fin-Stabilized Shot (HVAPFSDS).[1] In order to reach velocities in the 5,000-6,000 fps range, a discarding sabot, fin-stabilized projectile of the AP variety has been developed. This HVAP-FSDS shot consists of a solid steel shaft, approximately 7 to 10 calibers in length, with conventional ogive; the bore to subcaliber ratio is about 2.25. The projectile employs a four-segment steel discarding sabot, four-bladed aluminum tail, and aluminum., needle-nosed windshield. Launching is accomplished from a smooth-bore barrel. Because of the extremely high muzzle velocities involved, the HVAPFSDS can outperform AP, HVAP, and HVAPDS ammunition of comparable caliber. Cost is approximately double that for conventional AP shot, but only one-fifth that for an HVAPDS round. Despite its exceedingly long length, the projectile is relatively efficient. At present it is still in the developmental stage.

ARMOR PLATE FAILURE

2-215. Armor Classification. The following classifications are those most commonly used by the ordnance departments of the United States and Great Britain. In general, the common classifications refer to the hardness of plates, and distinguish between face-hardened and homogeneous types. Plates may be wrought or cast, with preference usually given the first type unless complicated shapes are required.

2-216. Face-Hardened Armor. This class includes all armor which has a hard face combined with a softer back. The U. S. Navy designates face-hardened plate as "Class A" armor, but this term is usually restricted to plate more than 1 1/4 inches thick. For light plate, the term "face-hardened bullet proof (FHBP)" is sometimes used.

Method of Production. The hard face may be produced by carburizing, in which case it is called cemented armor (British CTA, Cemented Tank Armor), or by decremental hardening, indicated by the term "noncemented armor." Armor which has been both carburized and decrementally hardened is sometimes designated as Krupp armor.

Function. The function of the hard face is to break up the attacking missile, while the softer, tougher back is designed to prevent the plate from cracking or spalling. If it performs this function, this type of armor is generally superior to other types of armor in its resistance to perforation. It is the most difficult to manufacture and consequently the most expensive.

2-217. Homogeneous Armor. As the name implies, this class of armor has the same hardness and composition throughout. This type is designated by the U. S. Navy as "Class B" armor but, as in the case of Class A armor, the term is usually applied only to plate more than 1 1/4 inches thick. Class B armor, specified by the U. S. Navy Bureau of Ships for ship structures, is known as "Special Treatment Steel (STS)." The British equivalent of Class B armor is designated "NC," meaning noncemented.

Types of Homogeneous Armor. At times a distinction is made, particularly by the British, between homogeneous hard plate and homogeneous soft plate. Homogeneous hard plate usually has a BHN (Brinell hardness number) between 400 and 475 and, like face-hardened armor, can be machined only with specialtools. As a rule it is used only for light armor and may then be called "Bullet Proof (BP)." Homogeneous soft armor commonly has a BHN between 200 and 350 and usually decreases in hardness with increase in the thickness of the plate. This type is referred to by the British as "Machinable Quality (MQ)." The U. S. Army uses hard homogeneous plate for some applications. At present it is used primarily for aircraft.

2-218. Types of Armor Plate Failure. When a nondeforming projectile perforates armor plate at normal incidence, the plate material must be removed in either of two ways: (1) by plastic flow, in which the metal is displaced axially (parallel to the direction of motion of the projectile), and radially (perpendicular to the direction of motion of the projectile), most of the material remaining in one piece and with only minor armor fragments being broken off; or (2) by driving or shearing out a comparatively undistorted piece of the plate. A plate failure which is primarily of type (1) is termed ductile, while those of type (2) are referred to as plugging.

2-219. Ductile Failure — Petalling. (See figure 2-85.) When a hard armor-piercing projectile strikes a relatively soft plate whose thickness is equal to or slightly greater than the caliber of the projectile, the first effect is a plastic deformation of the plate surface in both axial and radial directions. The material, flowing tangentially away from the projectile, gives rise to front petalling. The axial flow causes a bulge, on the front surface of the plate, surrounding the point of impact. The material in this bulge is under compression radially but under tension tangentially. If the tangential tension exceeds the tensile strength of the plate material, radial cracks develop in the bulge, the portion of the plate in contact with the projectile is bent away from the axis of the projectile, and front petalling results. If the plate is hard and not sufficiently ductile, these petals may break off, while on a more ductile plate they remain attached to the edge. As the projectile continues to penetrate the plate a bulge develops on the back surface, followed by the development of star cracks. As the projectile passes through the back of the plate, petals are formed. In perfectly ductile penetration, the petals remain attached. The "wiping off" of the petals is the first indication of a tendency to brittle failure. The energy required for ductile penetration varies approximately as d^2t, where d is projectile diameter and t is plate thickness.

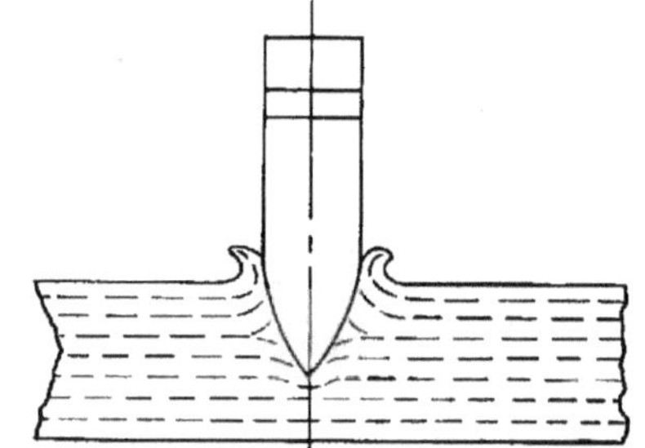

BEGINNING OF FORMATION OF FACE PETALS

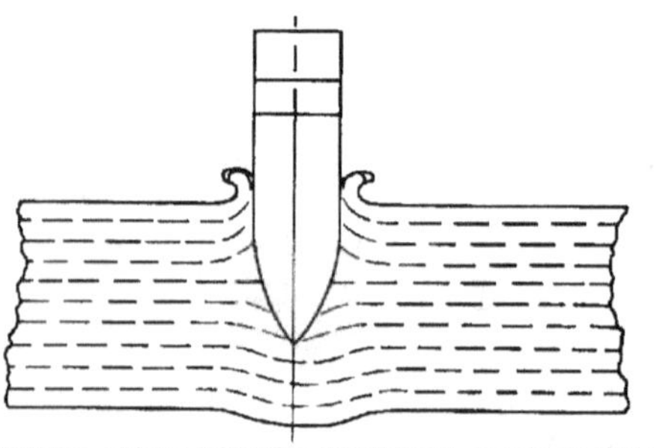

FACE PETALS FULLY FORMED. BEGINNING OF BACK BULGE

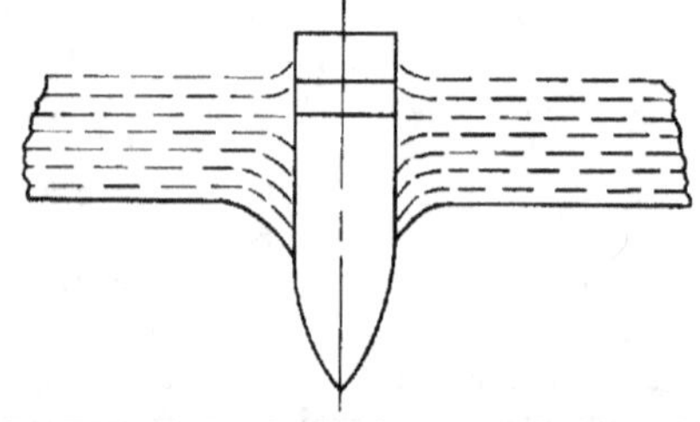

FORMATION OF BACK PETALS, COMMENCING WITH A STAR CRACK AROUND POINT OF SHOT

Figure 2-85. Attack at normal — from "Armour for Fighting Vehicles"

2-220. Spalling. In general, spalling (figure 2-86) refers to the removal of plate material from the front or the back face, leaving a hole somewhat larger than the diameter of the projectile. It may consist of any one of the following types, or a combination of them, such as (1) the detachment of petals, (2) the removal of a thin circular disk, or flake, from the back face, or (3) the ejection of a roughly conical section causing a gradual increase in the size of the hole toward the rear (usually called plugging). Spalling is apparently initiated by shearing stresses, which cause slipping between adjacent layers of the plate material.

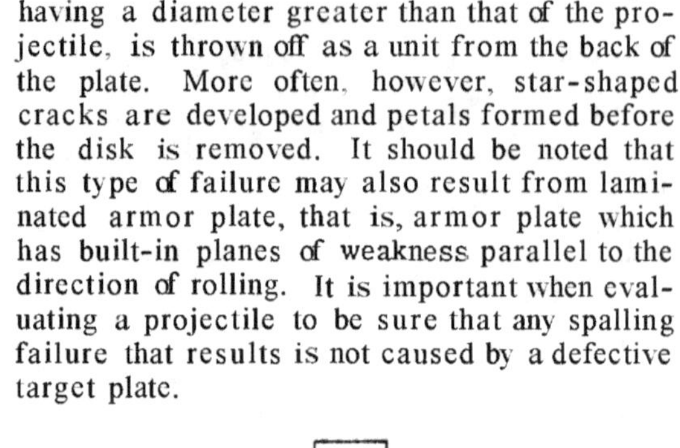

At times a complete circular disk, or button, having a diameter greater than that of the projectile, is thrown off as a unit from the back of the plate. More often, however, star-shaped cracks are developed and petals formed before the disk is removed. It should be noted that this type of failure may also result from laminated armor plate, that is, armor plate which has built-in planes of weakness parallel to the direction of rolling. It is important when evaluating a projectile to be sure that any spalling failure that results is not caused by a defective target plate.

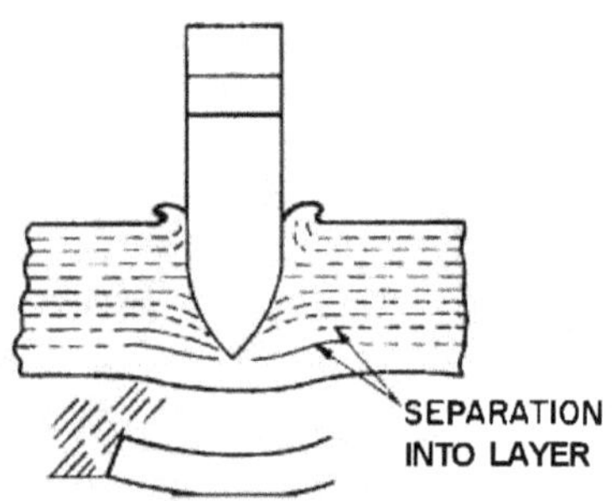

INITIAL STAGE OF SPALLING, SHOWING BEGINNING OF SEPARATION OF LAYERS

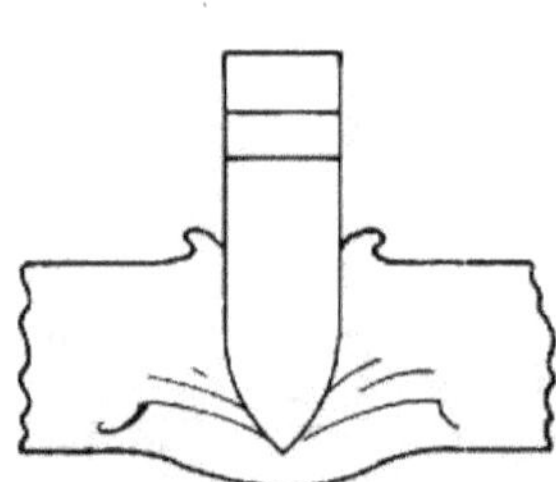

INTERMEDIATE STAGE OF SPALLING: SPALL FULLY SEPARATED FROM PLATE EXCEPT AT EXTREME EDGE

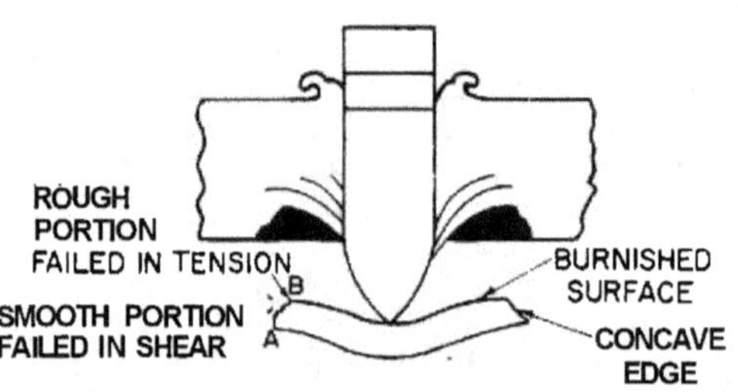

FINAL STAGE OF SPALLING: SPALLTMROWN FREE OF PLATE, INTERNAL PETALS NEARLY COMPLETED

Figure 2-86. Stages in the formation of back spalls — from "Armour for Fighting Vehicles"

2-221. Plugging. In this type of failure (figure 2-87), a plug of approximately cylindrical section is punched out of the plate ahead of the attacking missile. It is thicker than a disk and has a diameter which varies from about one-third the diameter of the missile to its full diameter. Like disking, the likelihood of plug formation increases with the hardness of the plate. As the hardness is increased, the projectile has more and more difficulty in pushing the plate material aside and there is greater tendency to push it forward. This creates shearing stresses in front of the projectile. Under the influence of these stresses, the deformation suddenly changes from one of the plastic type, over a considerable volume, to one that is confined to a very narrow region surrounding a surface of maximum shear stress. Initially, shearing is not confined to this surface, but is merely greatest at this surface. The deformation, however, takes place at such a rapid rate that little heat conduction can take place. As a result a considerable temperature rise occurs at the surface, where shear stress and the resulting strain is at a maximum. It is estimated that for a mean shearing stress of 100,000 psi, there is a temperature rise of 200°C per unit of shear strain. This increased temperature reduces the stress required for deformation and thus facilitates further deformation. Once started, the process is unstable, so that the plug merely slips out of the plate. With plate of moderate hardness there is usually some ductile penetration before the state of shear instability is reached; in fact, the first indication of such instability is usually a "wiping off" of the petals from the back face. As the hardness is increased, the unstable condition is reached at an earlier stage and less ductile penetration occurs. Even though the initial resisting force is larger for the harder plate, this may be more than offset by the shorter time required to reach the condition of unstable shear. The energy required for a shearing type of penetration is approximately proportional to dt^2, where d is projectile diameter and t is plate thickness.

2-222. t/d Ratio.[15] An important consideration in determining the type of penetration is the ratio of the armor thickness to the projectile diameter (the t/d or e/d ratio). When the t/d ratio is greater than one (armor overmatches the projectile), the penetration tends to be effected by a ductile pushing-aside mechanism, if the projectile is relatively sharp-nosed and remains intact. If it shatters, plugging may also take place. Relatively sharp-nosed shot are most effective, and the resistance of the armor increases as its hardness increases, up to a point where brittle failures set in. When the t/d ratio is less than one (armor undermatches projectile), the penetration tends to be effected by the punching or shearing out of a plug of armor in front of the shot. Relatively blunt-nosed shot are most effective under this condition of attack. Plugging is more likely to occur for very hard armor than for armor of conventional hardness. Thin plate failure, however, is not restricted to plugging, since sharp-nosed projectiles may produce the ductile type. The effect of the t/d ratio can be explained as follows. The energy required for ductile penetration (ED) is proportional to $d^3(\frac{t}{d})$ while the energy for a plugging type of penetration (E_p) is proportional to $d^3(\frac{t}{d})^2$. The ratio of E_D to E_p is $\frac{1}{t/d}$, and

$$E_p = \frac{t}{d} E_D$$

If we now assume that penetration will occur by the mode requiring the least energy, it may be seen that for $\frac{t}{d}$ less than one a plugging type

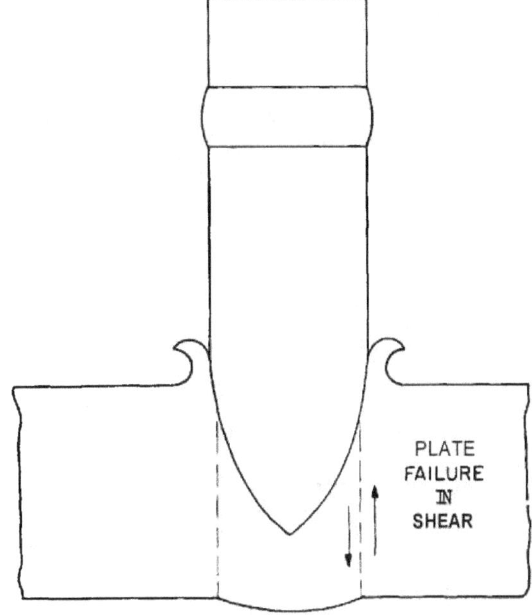

Figure 2-87. Plate failure in shear

of penetration would be expected, while for values of t/d greater than one, the penetration would tend to be ductile.

2-223. Cracking. In general, a plate fails locally at the point of impact and is permanently deformed only over a limited region. Occasionally, however, brittle fracture causes damage to the entire plate; cracks originate on the inner face opposite the point of impact and propagate to the edges of the plate, thus separating it into several large pieces. This is a result of improper heat treatment of the armor plate.

2-224. Plate Vibrations. Theories on plate vibrations claim that immediately upon impact a pressure pulse originates at the nose of the projectile and is propagated through the armor, in the form of a compression wave, with a velocity exceeding that of the projectile. Because of the multiple reflections of this wave from the faces of the plate, the region near the projectile is assumed to be in a state of steady plastic flow. As the force of the projectile on the plate is maintained, the plate tends to move forward as a whole. If it were perfectly rigid, the entire plate would move forward as awhole. Actually, the portions of the plate near the projectile will be displaced more than those further away. The result is a transverse wave which travels radially outward from the point of impact. In the region beyond the plastic flow, this wave consists mainly of an elastic distortion. Thus the motion of the projectile may be considered to take place in two ways: first, by penetration into the plate; and second, by displacement, together with adjacent plate material, due to elastic distortion. The effect of the elastic distortion is to absorb part of the kinetic energy of the projectile. This leads to an increase in the total energy required for perforation.

FAILURE TO PENETRATE

2-225. Reasons for Failure to Penetrate. Failure of AP shot to penetrate, when fired at a plate, is usually caused either by lack of sufficient hardness, insufficientkinetic energy, by ricocheting, or by shatter. If the shot has insufficient velocity it will achieve some intermediate degree of penetration and then may rebound from the plate. If the shot shatters it may or may not succeed in perforating the plate, and at the worst the sole damage may be a negligible shallow crater. A ,shattered projectile will require considerably more energy to penetrate, at low and intermediate obliquities, than one which remains intact, except against undermatching plate at high obliquity.

2-226. Shatter. The shatter of a shot usually begins with collapse of the nose and the breaking up of the shot into small pieces. If the shot shatters when the penetration is almost complete, the fragments may break through and complete the perforation. This is counted as a success, since the shower of fragments may be more damaging than a single shot. If the shot, particularly the tungsten carbide cores used in hypervelocity projectiles, can be so designed that the nose will hold together when the body shatters, the chances of successful perforation are greatly improved. The use of an armor-piercing cap reduces the peak inertial (set-forward) pressures on the nose of the shot in the early stages of penetration, and thus helps prevent the shatter of the ogive section of the projectile. The phenomenon of shatter is not yet fully understood and it produces difficult problems for the designer.

2-227. Effect of Velocity. The behavior of a given type of shot can be usefully represented by a phase diagram which relates the success or failure of the shot at a constant angle of attack with the variables striking velocity and plate thickness. Figure 2-88 shows a hypothetical phase diagram which may be used to illustrate the possible modes of behavior of a shot. It is used merely for illustration; many variations of it are possible in practice. It shows that for a given thickness, marked by the line XY, it is possible, as the velocity is increased, to pass successively through phases of (1) failure due to low velocity, (2) success with the projectile remaining intact, (3) success with shatter, (4) failure with shatter, (5) high velocity success with shatter. It may be seen that increase of striking velocity does not necessarily increase the chance of success, and that optimum values of striking velocity may exist.

2-228. Effect of Obliquity. The obliquity of a plate or the angle of attack of a projectile is defined as the angle between the tangent to the line of flight of the projectile and the normal to the plate at the point of impact. As the

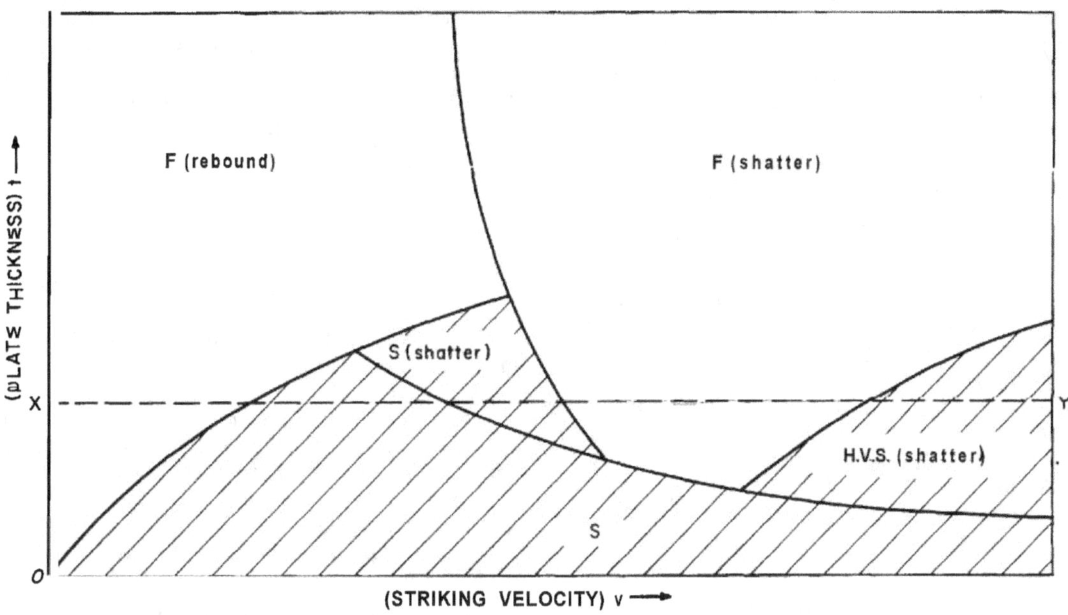

Figure 2-88. Phase diagram for representative AP shot (F = failure, S = success, H. V. S. = high-velocity success)

obliquity increases from 0" to 20" little change in the mode of penetration is noticed. Up to 20° plugs or disks and other fractures occur as for normal attack, though it is observed that the axis of the hole makes an angle with the normal of about one-half the angle of attack. At 30° marked differences in the mode of penetration are noticed. These have an important bearing on shot design. Only the normal component of the kinetic energy appears to be operative, so that the striking energy is effectively reduced by the factor cosine e. Plugs that are detached are approximately elliptical in shape with the minor axis equal to the diameter of the projectile and the major axis equal to this diameter times the secant of the angle of attack.

2-229. Bending: Stresses Due to Obliauitv. At oblique angle of attack the projectile is subjected to severe bending stresses which are of great importance to the designer. At impact the plate reaction tends to turn the shot away from the normal; as the plug begins to shear, the plate reaction weakens, and the thrust on the opposite side of the nose increases. The shot begins to rotate toward the normal, and continues to do so until the body of the projectile, now partly through the plate, impinges abruptly against the wall of the cavity. At this point the shot may fracture as a result of the tensile bending stresses.

2-230. Ricochet. At still higher angles of attack there is an increased tendency for the shot to ricochet, leaving behind it, on thick plate, a groove called a scoop. The behavior at high obliquity depends greatly upon the nose shape; shot with pointed noses are more easily deflected than blunt headed shot. A truncated ogival nose for AP shot is being considered and a double-angle nose for the tungsten carbide cores used in hypervelocity projectiles is in use.

PREDICTION OF EFFECT (PENETRATION FORMULAS)

2-231. Specific Limit Energy.** It is often convenient to discuss the perforation of armor plate in terms of the "specific limit energy," which is defined by the expression WV_1^2/d^3 where W is the weight of the projectile; d is the diameter of the projectile, that is, the

 CONFIDENTIAL

caliber; and V_1 is the ballistic limit. Experimental results may be described by formulas expressing WV_1^2/d^3 as a function of e/d and θ, where e is plate thickness, e/d is plate thickness in calibers, and θ is obliquity.

When WV_1^2/d^3 is plotted against e/d, the results for a particular combination of projectile and plate material and a particular angle θ will be found to lie along a curve or band, the width of the band representing the scatter of the results. For a different angle θ a different band will be found. The advantage of such a choice of variables is that it reduces the results obtained with all sizes of projectiles to a common basis. The fact that this procedure is possible without much error means that there is little "scale effect" in armor perforation. However, the scale effect, while small, is real and is in the direction of decreasing WV_1^2/d^3 with increasing d.

2-232. <u>The General Penetration Formula.</u> Except for minor corrections, all current perforation formulas have the following general form:

$$\frac{WV_1^2}{d^3} = \phi\left(\frac{t}{d}\,;\,\theta;\,\frac{\rho_1}{\rho_2};\,\frac{\quad}{d},\,\ldots\,\alpha_o,\,\alpha_1\,\ldots\alpha_i\right)$$

where

W	=	projectile mass
V_1	=	limit velocity
d	=	maximum projectile diameter
t	=	plate thickness
e	=	angle of incidence (obliquity)
ρ_1	=	density of plate material
ρ_2	=	density of projectile material
S_1	=	dimensions specifying projectile shape
$\alpha_o, \alpha_1, \alpha_i$	=	measures of strength of plate material
ϕ	=	general function of the quantities on which limit energy may depend.

This equation is based upon conformity to the principle of similitude, and expresses the limit velocity as a function of all the ballistic parameters necessary to fully specify the impact. The simpler equations are derived from the basic one by making assumptions as to the behavior of several of the parameters, or by limiting the conditions in order to eliminate dependence on several of the parameters.

2-233. <u>Penetration Formula by DeMarre and by Thompson.</u>[16] It has been found that most observations on armor perforation in which the projectiles are not severely deformed can be represented fairly well at obliquities near the normal by a relation of the form:

$$\frac{WV_1^2}{d^3} = R\left(\frac{t}{d}\right)^n$$

where R is determined chiefly by the strength of the plate material and n has a value between 1 and 2. If n is given the value **1.5** the DeMarre formula, in use by the Army Ordnance Department, is the result. The value n = 1 gives essentially the Thompson formula used extensively by the Navy Bureau of Ordnance. Neither of these forms fits the observations over a very wide range without changing R. In fact, for a given projectile and plate material, the behavior cannot be represented over extreme ranges of V_1 and t/d by any one set of values of R and n.

R may sometimes be expressed as a function of the t/d ratio by the following equation:

$$\log_{10} \mathbf{R} = a + b\,\frac{t}{d}$$

where a and b are constant for a given projectile at a specified obliquity.

2-234. <u>Ballistic Limit.</u> The ballistic limit is used to evaluate the performance of armor plate or armor-piercing projectiles. Ballistic limit is loosely defined as "that velocity at which a given type of projectile will perforate a given thickness and type of armor plate at a specified obliquity." In actual firings it is found (1) that it is not feasible to control projectile velocity extremely closely, and (2) that for a series of projectiles fired at an appropriate constant velocity some of the projectiles will completely penetrate (perforate) the plate and the remainder will not. This gives rise to a zone of mixed results, which may be defined as "that range of velocities in which both complete penetrations and partial penetrations are obtained."

2-235. <u>Definition of Perforation.</u> Three definitions of complete penetration are commonly used in this country. In order for a ballistic limit to have meaning, it is necessary that the definition used be specified. These definitions are:

1. Army — A complete penetration is any penetration through which at least a pinhole of light or the projectile may be seen from the rear of the plate.

2. Navy — A complete penetration is obtained when the projectile or a major portion of the projectile passes through the plate.

3. Protection — A complete penetration is obtained whenever a fragment or fragments of either the impacting projectile or the plate are ejected from the rear of the plate with sufficient energy to perforate a thin mild-steel plate (about 0.020 inch) or equivalent screen, placed (at the discretion of the proof officer) parallel to and approximately 6 inches rearward of the plate.

2-236. Critical Velocity. The British21 evaluate the performance of armor-piercing projectiles in terms of a critical velocity, which corresponds to our ballistic limit but differs in method of computation. The critical velocity is defined as that velocity at which the projectile will just pass through the plate. It is computed by firing several projectiles at velocities sufficient to perforate with some residual velocity after perforation. The critical velocity, V, the striking velocity, V_o, and the residual velocity, V_1, will be related by the following equation:

$$V^2 = V_o^2 - sV_1^2$$

where s is some empirically determined constant.

If V_o^2 is plotted against V_1^2 for two or more firing results, the points will lie on a straight line. V^2 can be found as the value of V_o^2 when V_1^2 equals zero. This method is applicable only to projectiles which remain intact during perforation.

2-237. Analysis by Statistical Method. If a curve of percent complete penetrations versus velocity is plotted, it will usually be found to follow closely a normal probability curve, thus making possible the analysis of the data by the usual statistical methods; that is, a mean value and a standard deviation can be found. Ideally, a probability curve could be plotted and the necessary data be obtained from this curve. Practically, however, this is not feasible for the following reasons: (1)in up-and-down testing the size of the step (interval) cannot be accurately controlled; (2) the number of observations that may be made is generally small in number, because of the prohibitive cost of large numbers of observations or because of the limited area of a single target plate.

2-238. Approximation of Ballistic Limit. As a result of the impracticability of making a large number of observations, it has been found necessary to develop methods of approximating the ballistic limit. Several of these methods follow. It will be noted that all of the methods presented are used to approximate the median or V-50 point (the velocity at which there is equal probability of complete penetration or of incomplete penetration). Other ballistic limits, used primarily for the evaluation of armor plate, are in use. These approach the V-0 point (maximum velocity at which no complete penetration can be expected) or the V-100 point (minimum velocity at which 100 percent perforation is obtained) rather than the V-50 point. These methods, as well as several others which are not of primary interest to the ammunition designer, are presented in "Methods of Computing Ballistic Limits From Firing Data," available from the Development and Proof Services, Aberdeen Proving Ground.

2-239. Reproducibility of Results. In conducting firing tests for the determination of ballistic limit it is just as important to use target plate of uniform characteristics as it is to use projectiles of uniform characteristics. Results will be reproducible only when characteristics of both armor and projectile are recorded and reproduced as carefully as possible. It should also be borne in mind that, due to projectile shatter, curves of percentage of complete penetrations versus velocity may show two or three peaks or even regions of discontinuity.

2-240. Computation of Ballistic Limit.[13]

Symbols and Abbreviations

BL ....Ballistic limit
PP ....Partial penetrations
CP ....Complete penetrations
V_A ....Average of all velocities in zone of mixed results
V_{HP}....Velocity of the highest PP
V_{LC}....Velocity of the lowest CP
Σ_V ....The sum of the velocities of all rounds in the zone of mixed results
N_P ....Number of PP in zone of mixed results
N_C ....Number of CP in zone of mixed results

a. 3 CP and 3 PP Within 150 fps.
   1. Firing method. Employ up-and-down firing method until 3 CP and 3 PP are obtained within a spread of 150 fps.
   2. Method of computation. Average of the velocities of the 3 lowest CP and the 3 highest PP.
   3. Uses. Most armor development programs; also projectile test programs.
   4. Number of rounds involved. 6 to 10.
   5. Comments. Permissible velocity spread may be modified to, for example, 100 fps.

b. 3 CP and 3 PP Within 100 fps, per Frankford Arsenal.
   1. Firing method. Employ up-and-down method until 3 CP and 3 PP are obtained within a spread of 100 fps. If zone of mixed results is greater than 100 fps, fire additional rounds to obtain a minimum of 3 CP and 3 PP in the mixed zone.
   2. Method of computation.
      a. If no mixed zone is obtained average the velocities of the lowest CP and the highest PP.
      b. If the zone of mixed results is less than 100 fps, average the velocities of the 3 lowest CP and the 3 highest PP.
      c. If the mixed zone is between 100 fps and 250 fps:
         (1) If the number of PP in the mixed zone exceeds the number of CP,

$$BL = V_A + \frac{(N_P - N_C)}{(N_P + N_C)} (V_{HP} - V_A)$$

         (2) If the number of CP in the mixed zone exceeds the number of PP,

$$BL = V_A - \frac{(N_C - N_P)}{(N_C + N_P)} (V_A - V_{LC})$$

   3. Uses. Tests of projectiles.
   4. Number of rounds involved. 7 to 14.

c. Method of Maximum Likelihood, per BRL Technical Note No. 151.
   1. Firing method. Employ up-and-down method.
   2. Method of computation. Method of maximum likelihood described in "On Estimating Ballistic Limit and its Precision," Ballistic Research Laboratories, Technical Note No. 151, March 1950.
   3. Uses. Suitable for all data, provided a zone of mixed results is obtained. At present, used principally as a check on other methods and where the most accurate determination possible is desired.
   4. Number of rounds involved. 5 to 20.
   5. Comments. This is the only statistically sound method available for use when a small or moderate number of rounds have been fired. Provides a standard deviation as well as a ballistic limit, though the standard deviation may not be very accurate as far as the particular plate-projectile combination is concerned. Requires 2 to 4 hours of computation to obtain ballistic limit and standard deviation, but a specially designed machine should be able to supply answers in an insignificant amount of time. When no zone of mixed results is obtained in firing, no standard deviation can be obtained and the ballistic limit can only be estimated. Ballistic limits obtained by this method differ very slightly (0 to 10 fps) from ballistic limits obtained by the 3 and 3 method.

d. Probability Curves.
   1. Firing method. Employ up-and-down method or zone method.
   2. Method of computation. Plot curve showing frequency of CP at various velocity intervals on either linear or probability graph paper. Pick off V-50 point from curve.
   3. Uses. To set up specifications for body armor materials, and for other special tests.
   4. Number of rounds involved. 150 or more.
   5. Comments. Suitable only when a large number of rounds can be fired at the same target. May also furnish reliable standard deviation. Most accurate of all methods.

e. Average of Velocities in Zone of Mixed Results, per Navy.
   1. Firing method. Employ up-and-down method.
   2. Method of computation.

$$BL = \frac{\Sigma V}{N_C \quad N_P}$$

3. Uses. Normal developmental firing by the Navy.
4. Number of rounds involved. 9 to 15.

2-241. Charts for Ballistic Limit.[4] The following method for obtaining ballistic limits, based upon empirical equations derived from firing data, is proposed by Lt. R. H. Riel of Aberdeen Proving Ground. By using the method, estimates of the V-50, protection ballistic limit for new projectiles, and for standard projectiles against untried targets, can be made within approximately ±10 percent at the lower velocity levels to within ±5 percent at the higher ranges. At the higher obliquities, ballistic limit estimations for HVAPDSFS projectiles can be made to a fair degree of accuracy.

2-242. Charts for Ballistic Limit — Design Conditions.1 Many models of the standard AP shot design, from 57-mm through 120-mm, were analyzed. All projectiles considered were from two to three calibers in length, conventional ogive, hardened to approximately Rockwell C58-62 at the nose. Ballistic limits were obtained using V-50, protection criterion against rolled, homogeneous armor. Armor plates, for the most part, were within the hardness and ductility limits prescribed by government specification. In approaching the problem the following factors were considered: plate thickness, plate obliquity, plate hardness, projectile weight, and projectile diameter. Uniform cleanliness and quality of armor plate and projectiles were assumed.

Very few models of tungsten carbide shot were available for analysis, and these varied widely in basic design. Four standard models of HVAP shot were studied, two with the tungsten carbide core incased in an all-aluminum carrier, and two in an aluminum carrier with a heavy steel base. It was observed that with shot of this latter type, lower ballistic limits could be obtained because of the additional momentum imparted to the core by the steel base. Only one HVAPDS design has been standardized to date. Other designs studied employed special features too numerous to mention. In analyzing HVAP results, only the core weight and diameter were considered, while for **HVAPDS** the entire subprojectile weight and diameter were included. In this manner, HVAP and HVAPDS data could be correlated to a fair degree of accuracy. The average tungsten carbide core, based on samples submitted by four manufacturers, contained 82-85 percent tungsten carbide, 15-17 percent cobalt and/or nickel; the remainder was free carbon and iron. The hardness range was from RockwellA83-90. With regard to the armor plate targets, the same factors were considered as for AP shot.

2-243. Charts for Ballistic Limit — Procedure for Use. In designing a new projectile, the engineer usually has predetermined a series of targets which the projectile must defeat, as well as the gun model and caliber for which it is intended. The optimum projectile weight, muzzle velocity, striking velocity at various ranges, and whether or not it will defeat the specified targets at those ranges remains to be determined. This is indeed an involved problem, but may be accomplished to a degree, in a minimum of time, by employing the following procedure (see figures 2-89 through 2-95).

1. Select the particular projectile type, and the armor plate thickness, obliquity, and hardness to be defeated.
2. Select the particular projectile weight and diameter to be investigated.
3. Select the particular gun model and operating pressure desired.
4. Select the range at which the target is to be defeated.
5. Using the information specified in step 1 above, refer to the appropriate charts "Thickness of armor defeated versus kinetic energy/diameter" (see figures 2-89 and 2-90), and read the absolute value of KE/D required to defeat the target. The units of this term are not important.
6. Using the information specified in step 2 and the value of KE/D obtained from step 5, refer to the nomograph "Kinetic energy versus striking velocity and projectile diameter versus projectile weight" (figure 2-91), and read the striking velocity required for penetration.
7. Using the information specified in steps 2 and 3, refer to the chart "Projectile weight versus approximate maximum muzzle velocity" (figure 2-92), and read the approximate maximum muzzle velocity obtainable with the chosen projectile and gun. If the selected gun does not appear on the chart, use the "Nomograph to determine gun constant for tank guns with various bore diameter, chamber volume, and projectile travel parameters" (figure 2-93) as an alternate solution. The gun constant determined

is a figure of merit directly related to a weapon's ability to launch a particular weight projectile at a certain velocity under given chamber pressure conditions. Next, use the "Nomograph to determine approximate maximum muzzle velocity for various weight and diameter projectiles fired from tank guns with various gun constants and chamber pressures using M2 or M17 propellant" (figure 2-94). This method is slightly less accurate than the chart of step 7, since it is empirical in nature. However, it has proved especially valuable in predicting the performance of design study guns and establishing a basis for future gun parameters.

8. Using the information obtained from steps 2, 4, and 7, refer to the "Nomograph to determine striking velocity at various ranges" (figure 2-95), and read the striking velocity available at the particular range considered. This value must be larger than the striking velocity, required for penetration, obtained in step 6. If such is not the case, proceed to step 9. It may be of interest in any event to determine the range at which the projectile will just defeat the target. This is easily accomplished by starting with the striking velocity (from step 6) and the muzzle velocity (step 7), and working through the nomograph in reverse.

9. Select a lighter and a heavier projectile, within practical length limitations, and repeat steps 5 through 8. If neither of these projectiles can defeat the target, then the requirement must be changed, a new gun model, or larger diameter projectile chosen.

A word of caution is offered in using the charts. The ballistic limit for any particular projectile-plate combination can never be determined precisely. A variability of 50-100 fps between projectile lots and armor plate heats is common, and to be expected. The charts were constructed from a limited number of observations. All plate thicknesses were not investigated at all obliquities and hardnesses. As additional information becomes available, the accuracy of the charts may be improved, but the method of analysis would remain the same.

2-244. Tank Damage Assessment.[12] Heretofore it has been the practice to consider a tank defeated if a penetration of any sort was achieved. Actually, in order for a tank to be completely defeated it is necessary that it be immobilized and that its guns be rendered inoperative. This may be accomplished either by damage to the propulsive system and the armament or by wounding of personnel inside the tank, which usually implies penetration. Defeat of a tank may be accomplished without penetration by damaging the suspension system and the external components of the fire control instruments. The Ballistic Research Laboratories are conducting a series of firings against tanks in order to determine the true probability of defeat of a tank by projectiles. Damage evaluations to dummy personnel and components of the tank will be used as a basis of these probabilities. It is hoped that the results of these tests will provide information that will be of aid in projectile design.

EFFECT OF VARYING ARMOR PARAMETERS

2-245. Effect of Armor Thickness on Projectile Performance. Increasing the thickness of armor increases the energy necessary for perforation. For ductile perforation the energy required is proportional to d^2t, where d is the projectile caliber and t is the thickness of the target plate. For the punching type of perforation, the energy required is proportional to dt^2. The nose geometry of a projectile determines, to a large extent, the type of perforation which takes place. The truncated nose tends to produce the punching type of perforation which is more effective against undermatching plate; however, it also increases the likelihood of projectile breakup at low obliquities.

2-246. Effect of Hardness of Plate.[24] It has been found that the optimum hardness for rolled or cast homogeneous armor plate is in the neighborhood of 280 BHN. When attacked by medium-caliber AP projectiles over a wide range of obliquities, cast armor of approximately 2 to 5 inches in thickness tends to show reduced resistance to penetration as hardness increases over 280 BHN. The optimum hardness for rolled homogeneous armor is somewhat in excess of this value, depending on the thickness involved.

2-247. Spaced Armor.[2] Refers to a structure consisting of a moderately thin plate in front of, and separated by a space from, a considerably thicker armor plate which constitutes the main armor of the vehicle under consideration. The thin front plate, called the "skirting

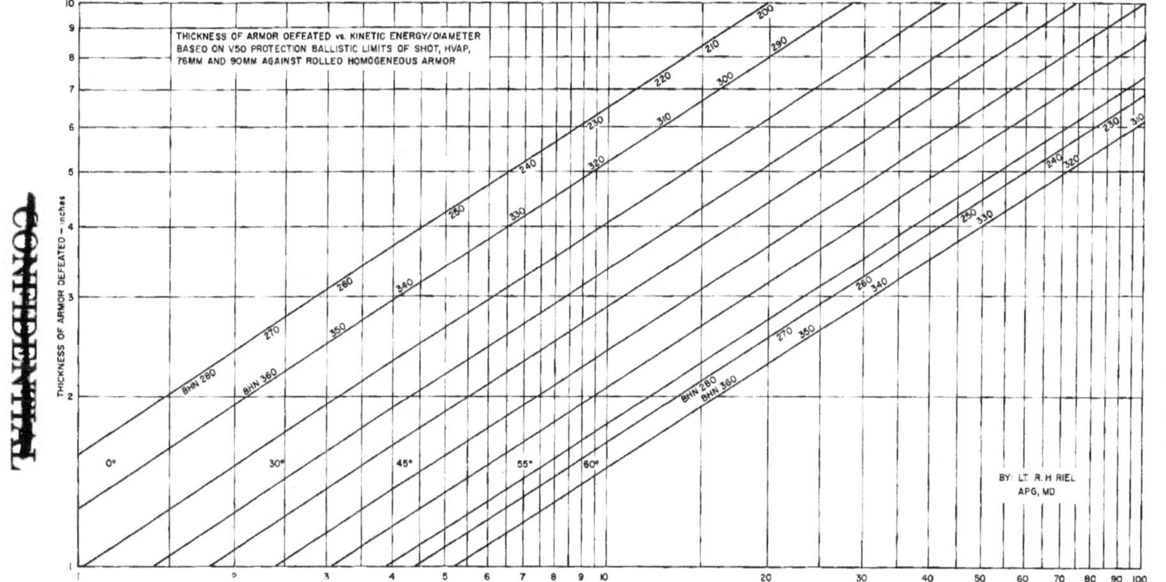

Figure 2-89. Thickness of armor defeated versus kinetic energy/diameter based on V-50 protection ballistic limits of shot, HVAP, 76-mm and 90-mm, against rolled homogeneous armor

CONFIDENTIAL

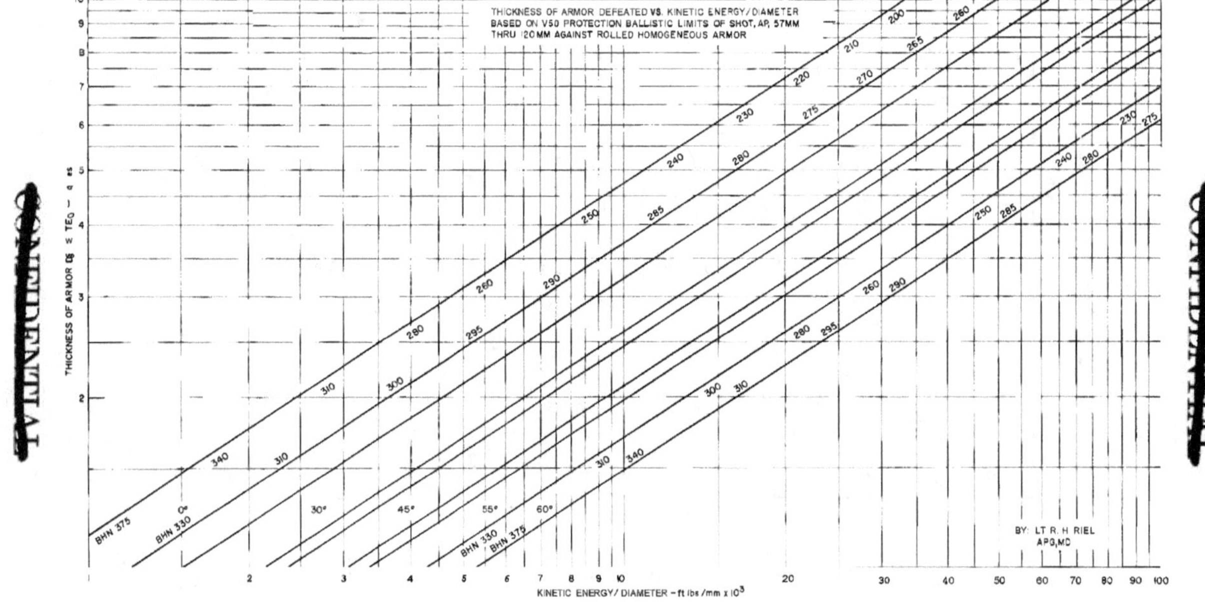

Figure 2-90. Thickness of armor defeated versus kinetic energy/diameter based on V-50 protection ballistic limits of shot, AP, 57-mm and 120-mm, against rolled homogeneous armor

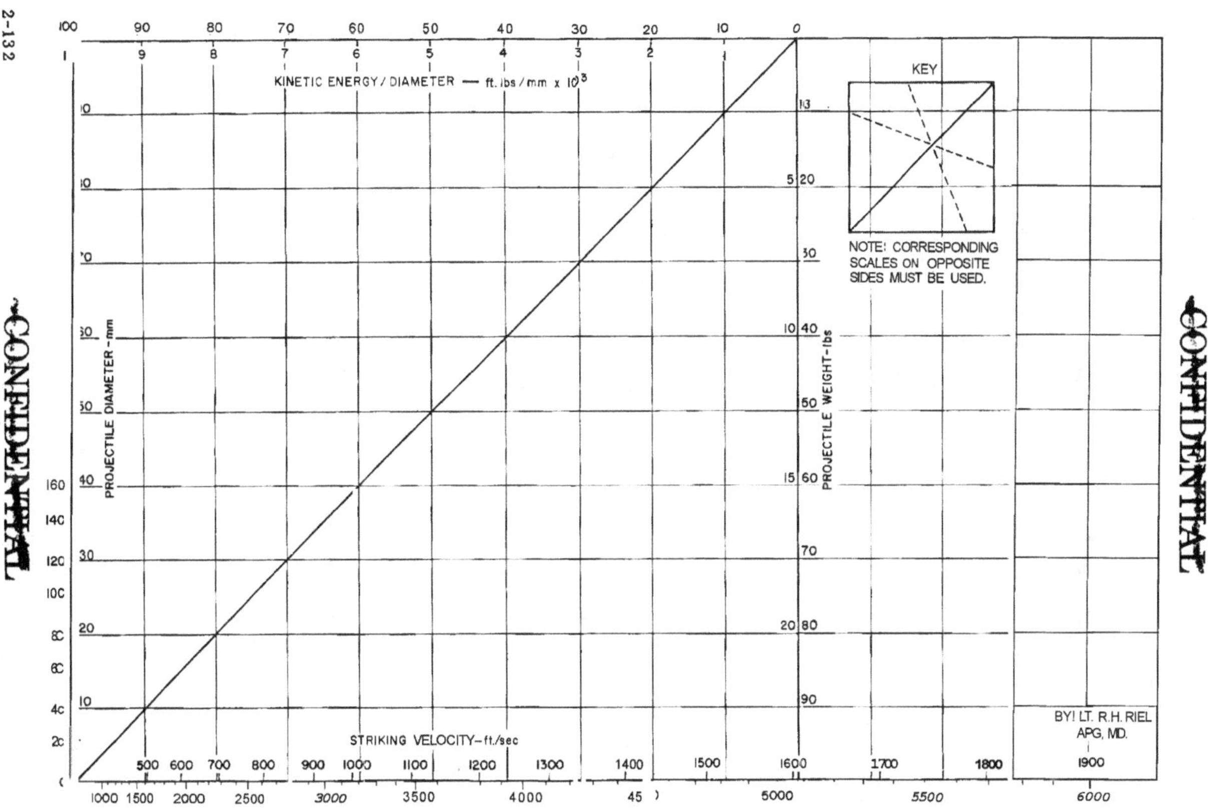

Figure 2-91. Kinetic energy versus striking velocity and projectile diameter versus projectile weight

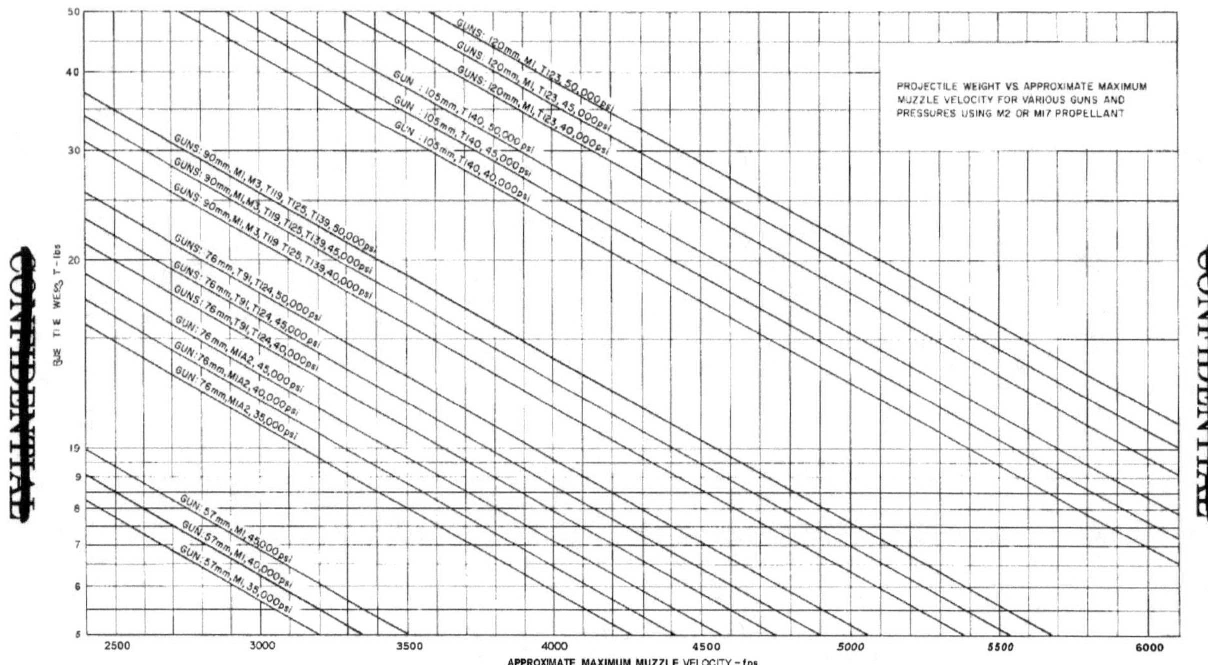

Figure 2-92. Projectile weight versus approximate maximum muzzle velocity for various guns and pressures using M2 or M17 propellant

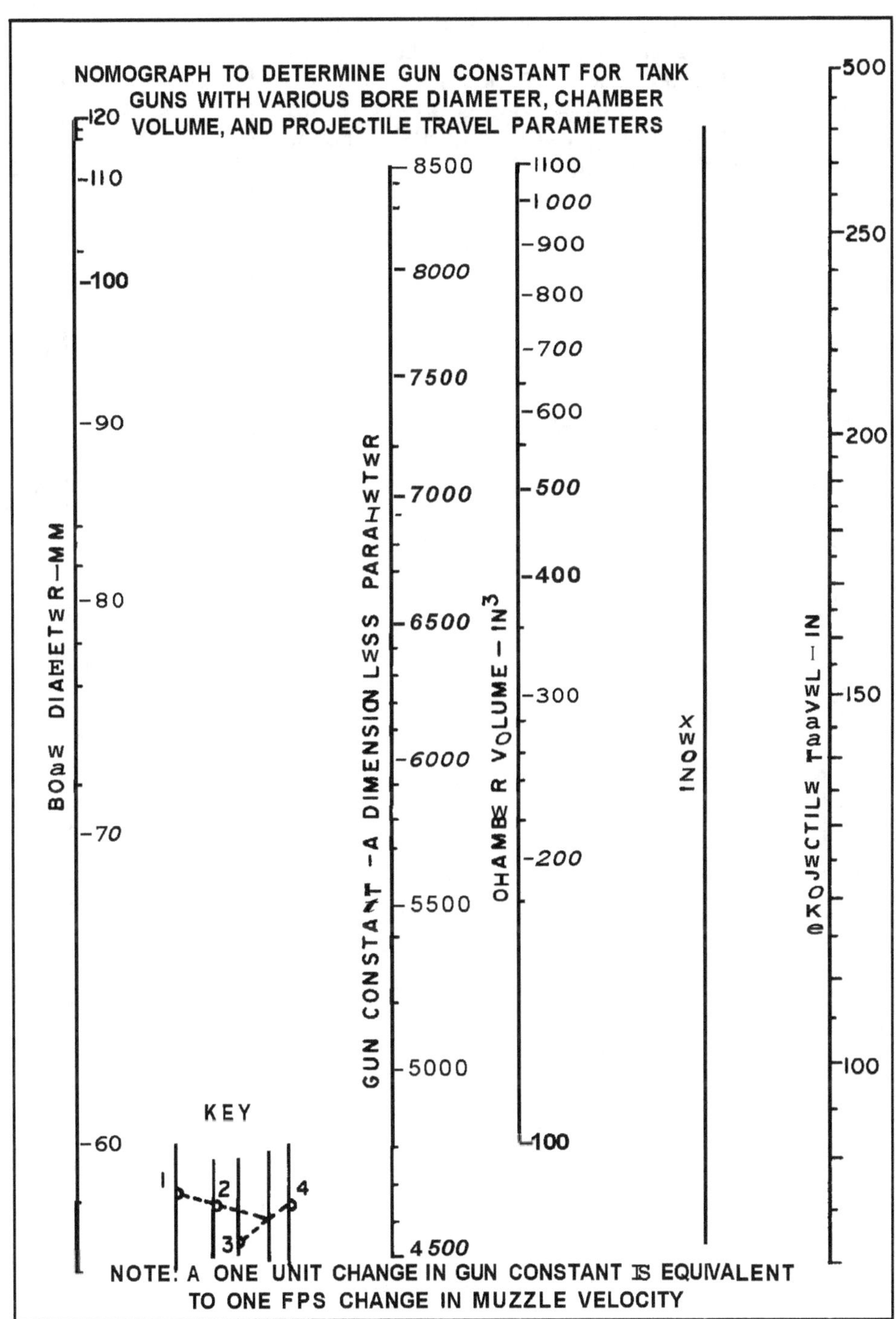

Figure 2-93. Nomograph to determine gun constant for tank guns with various bore diameter, chamber volume, and projectile travel parameters

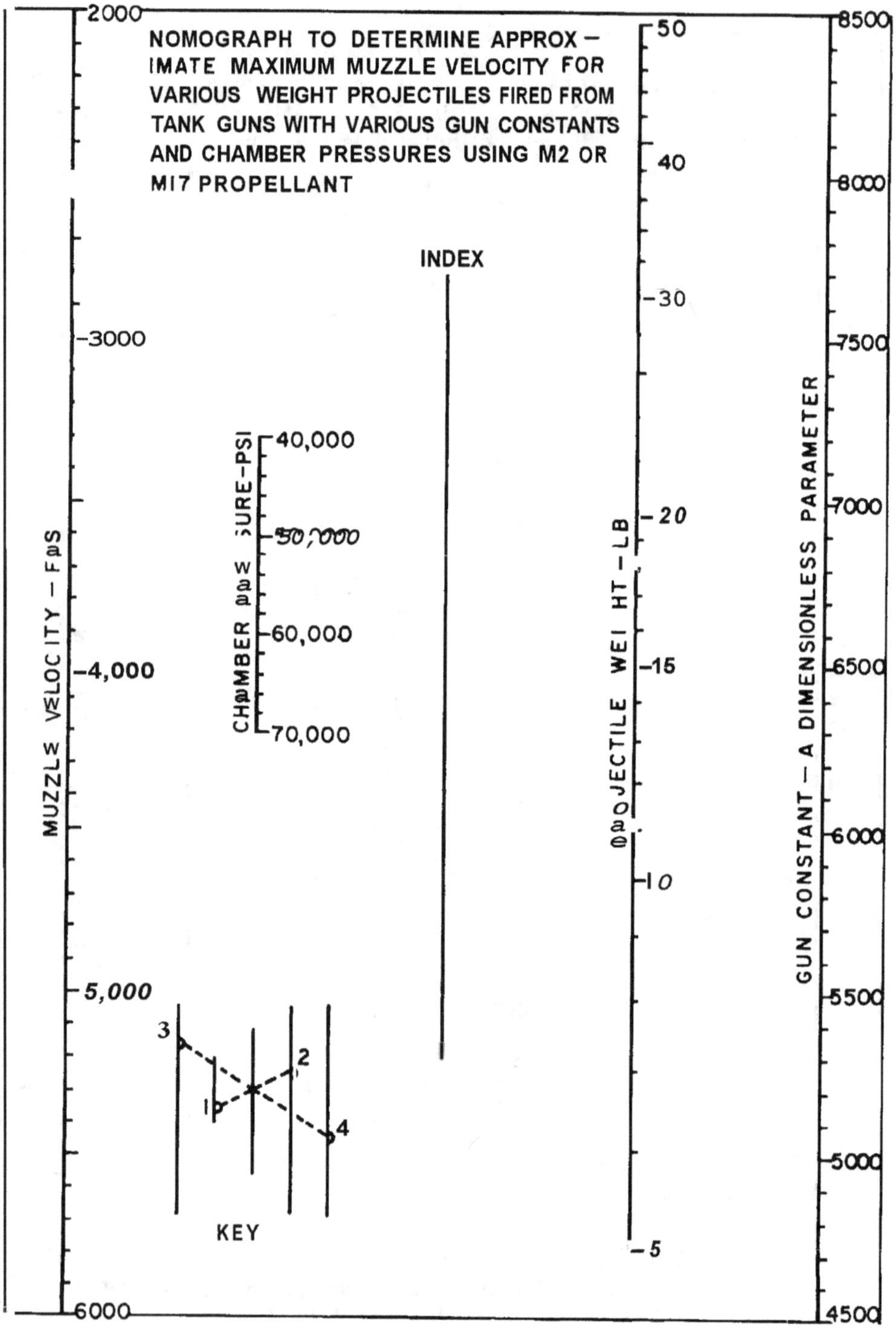

NOMOGRAPH TO DETERMINE APPROXIMATE MAXIMUM MUZZLE VELOCITY FOR VARIOUS WEIGHT PROJECTILES FIRED FROM TANK GUNS WITH VARIOUS GUN CONSTANTS AND CHAMBER PRESSURES USING M2 OR M17 PROPELLANT

MUZZLE VELOCITY – FPS: 2000, 3000, 4,000, 5,000, 6000

CHAMBER PRESSURE-PSI: 40,000, 50,000, 60,000, 70,000

INDEX

PROJECTILE WEIGHT – LB: 50, 40, 30, 20, 15, 10, 5

GUN CONSTANT – A DIMENSIONLESS PARAMETER: 8500, 8000, 7500, 7000, 6500, 6000, 5500, 5000, 4500

KEY: 1, 2, 3, 4

Figure 2-94. Nomograph to determine approximate maximum muzzle velocity for various weight projectiles fired from tank guns with various gun constants and chamber pressures using M2 or M17 propellant

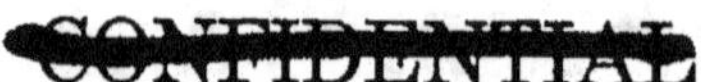

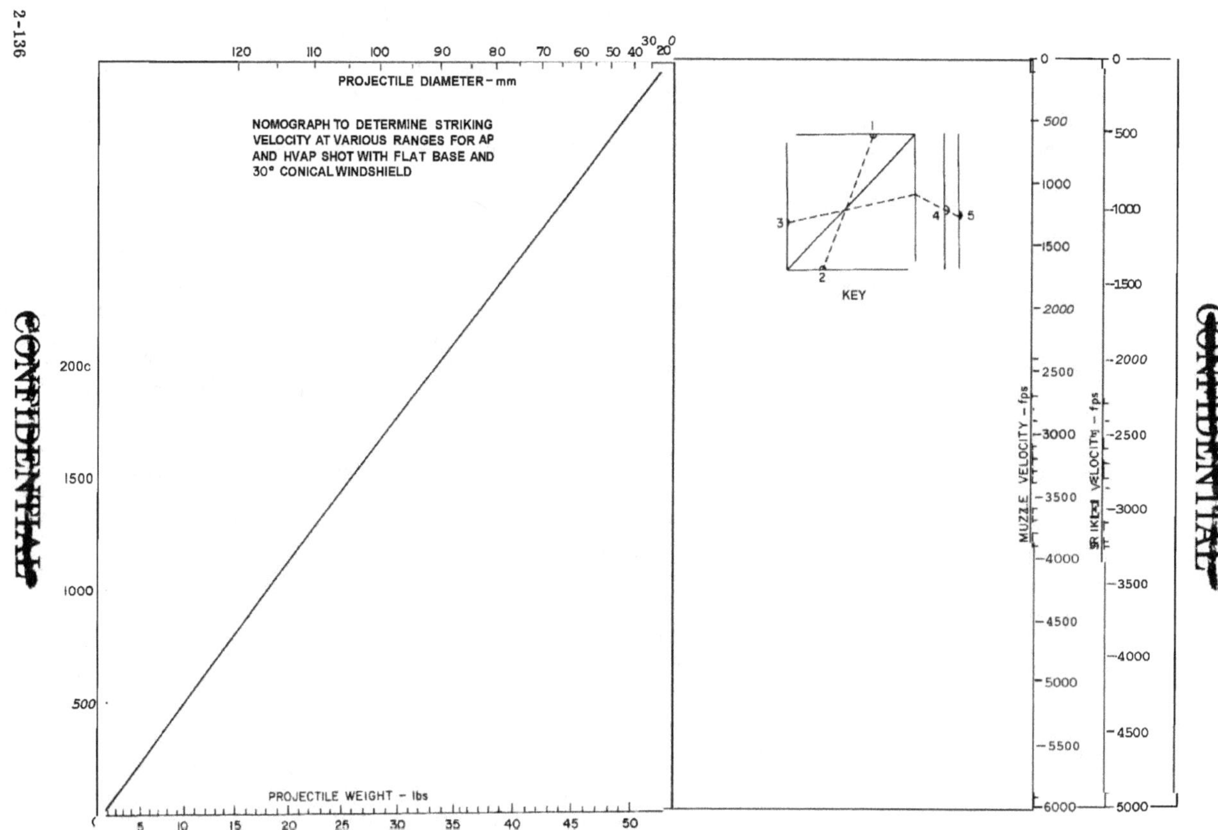

*Figure 2-95 Nomograph to determine striking velocity at various ranges for **AP** and **HVAP** shot with flat base and 30° conical windshield*

plate," faces the attack, and represents approximately 10 percent or less of the total weight of the armor. A more complex spaced armor arrangement may comprise a number of skirting plates.

2-248. The Function of the Skirting Plate[2] is not to absorb any significant proportion of the kinetic energy of attacking projectiles, but to so affect the projectiles that their performance against the main armor is drastically reduced.

The basic approach to the design of spaced armor arrangements is to have the skirting plate of the minimum thickness capable of producing the desired effect upon the projectile.

2-249. Effect of Skirting Plate.[3] The skirting plate may affect projectiles in any or all of the following ways:

a. The armor-piercing cap may beremoved, thus causing the shot to be shattered against the heavy main armor.

b. The shot may be turned or yawed so that it impacts the main armor at an increased angle.

c. The shot may be fractured upon passage through the skirting armor. The loss of the point and the dispersal of the fragments result in a marked decrease in the penetration performance.

Spaced armor is most efficient when the attacking projectile is broken up by the skirting plate. When this happens weight savings in the range of 30 to 50 percent over solid armor can be effected with no sacrifice in protection performance. When, however, projectiles are not broken, but only decapped or yawed, weight savings of 10 to 20 percent can be effected if the projectiles are yawed to impact the main armor at an increased obliquity. If the projectiles are yawed to impact the main armor at a reduced obliquity, spaced armor becomes considerably less efficient than the same weight of solid armor; becoming, in fact, even less efficient than the main armor alone.

EFFECT OF VARYING PROJECTILE PARAMETERS

2-250. Gains in Armor Penetration by Use of Subcaliber Projectiles.[16] One form of the DeMarre formula for penetration of armor plate by steel projectiles is given by the equation

$$e = d\left[\frac{WV^2 \cos^2 \theta}{\alpha d^3}\right]^{1/\beta}$$

in which e is the thickness of the plate penetrated, d is the diameter of the projectile, W the weight, θ the angle of impact relative to the normal, V the striking velocity, and α and β are constants. For nondeformable projectiles, log α = 6.15 and β = 1.43. Hence the equation reduces to:

$$e = \frac{1}{d^{1.1}}\left[\frac{WV^2 \cos^2 \theta}{\alpha}\right]^{1/1.43}$$

If the projectile energy is kept constant while its diameter is varied, the penetration improves as the diameter diminishes. This improvement continues until the impact velocity reaches the shatter value.

2-251. Improvement in Penetration With Tungsten Carbide. When tungsten carbide is used as a projectile material, α and β both diminish, and the relative penetration is increased. The constants are now, however, not independent of θ, and there is a region of θ wherein the advantage of tungsten carbide is not so great.[16] For example,[34] under conditions in which the plate is overmatching, the three types of 90-mm projectiles are rated on an equal range basis in the order of decreasing effectiveness given in table 2-33. However, the HVAPDS round is probably superior at all obliquities.

Table 2-33

Obliquity	Order of Effectiveness
0° to 45°	HVAP, APC, AP
45° to 65°	AP, HVAP, APC

2-252. Factors Limiting Penetration of Subcaliber Projectiles.[16] From the DeMarre equation it is apparent that reduction of caliber, while at the same time retaining projectile energy, improves penetration. However, in an actual gun, the powder energy is fixed, and since part of this energy is delivered as kinetic energy of the powder gas, the energy of the projectile diminishes as the caliber is reduced. Furthermore, because of less favorable ballistic conditions, the projectile strikes the plate with a further reduction in relative energy, and the advantages of subcaliber projectiles are thereby modified. In spite of this, gains in penetration can be attained by subcaliber projectiles. For a detailed explanation, and the mathematical computations, refer to section 33.2 of reference 16.

2-253. <u>British Method of Estimating the Muzzle Velocity of a Subcaliber Projectile.</u> When computing the muzzle velocity of a subcaliber (HVAPDS) projectile from the data known for the full-caliber projectile, it is necessary to take into account the energy used in accelerating the propellant gases.4 The following empirical equation, developed by the British, takes the gas acceleration into account. In effect it assumes that half the weight of the charge is being accelerated along with the projectile.

$$V_s = V_o \sqrt{\frac{M_o + M_c/2}{M_s + M_p + M_c/2}}$$

where

V_s = Muzzle velocity of subcaliber projectile

V_o = Muzzle velocity of full-caliber projectile

M_o = Weight of full-caliber projectile

M_s = Weight of subcaliber projectile

M_p = Weight of sabot

M_c = Weight of propellant charge.

2-254. <u>Comparative Effectiveness of Full-Caliber Versus Subcaliber Steel Shot.</u>4 Based upon extensive scale-model tests, the following conclusions were reached regarding the comparative effectiveness of full-caliber versus subcaliber steel shot of conventional l/d ratios.

1. The most efficient diameter for subcaliber steel projectiles employed in the penetration of rolled homogeneous armor is 7/10 of the bore diameter of the gun from which it is fired.

2. In general, there is little difference in penetrative performance against rolled homogeneous armor plate between a steel subcaliber shot of the most efficient diameter and a homologous full-caliber shot of conventional l/d ratios. The subcaliber shot exhibits slightly better penetration at the near ranges, while the full-caliber shot is superior at the longer ranges (over 2500 yds).

3. Against targets which are overmatching for the full-caliber projectile, a capped subcaliber shot of 7/10 bore diameter will exhibit terminal performance equal or superior to both capped and monobloc full-caliber shot and monobloc subcaliber shot.

4. It appears that the performance of any AP or APC projectile-gun combination (at least greater than 76 mm in size) can be improved by firing a subcaliber capped steel shot from the gun. The net result is that the armor penetration remains the same, but the subcaliber shot superiority exists because of a higher probability of registering a first-round hit on the target. (This higher probability is based entirely on the shorter travel time of the subcaliber shot.)

Based upon the high cost of present metal sabots, it would appear that the possible advantage of a subcaliber projectile would be offset by the higher cost. However, if the all-plastic sabot should prove practical, it would probably be possible to manufacture a subcaliber projectile at a low enough price to make the marginal advantages worthwhile.

2-255. <u>Effect of Nose Geometry of AP Projectiles.</u>[25] The armor penetration performance of 20-mm models of the 90-mm AP T33 (M318) projectile has been compared with that of the truncated T33 (FAP)* and the tipped truncated T33 (FAPT)* (figure 2-96) projectiles over a wide range of target conditions. Specific limit energies were calculated in order that the perforation efficiencies of the three types could be compared on an energy basis for any test regardless of nose shape and projectile weight.

For all targets for which they remained intact, the FAP projectiles were found to be superior to the FAPT and AP projectiles. For conditions where the FAP projectiles shattered, the FAPT was superior. The AP projectiles were inferior to both the FAP and FAPT except for matching or overmatching armor at high obliquity, in which case they are equal to the FAPT; and for very much overmatching plate at very low obliquity, in which case AP is superior to both FAP and FAPT. A summary of the results obtained is given in figures 2-97, 2-98, and 2-99.

* These are not official Ordnance designations, but have been used for easy reference.

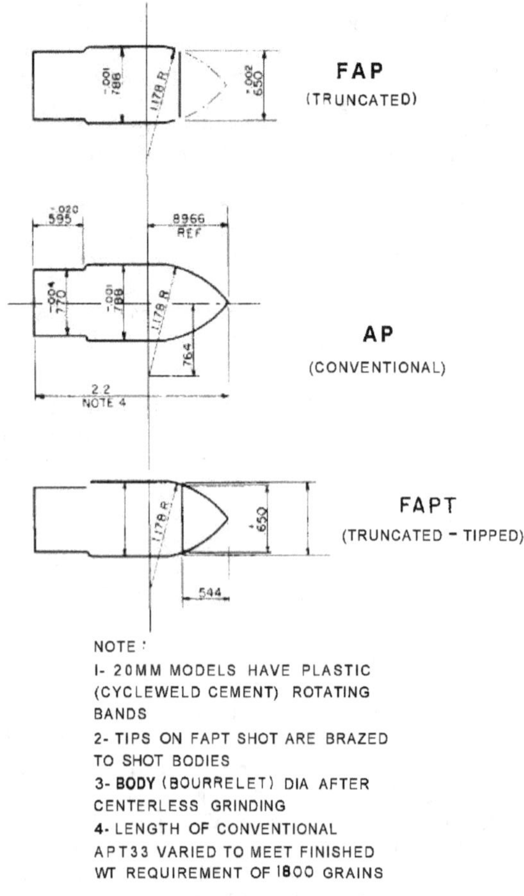

*Figure 2-96. 20-mm **AP** shot types*

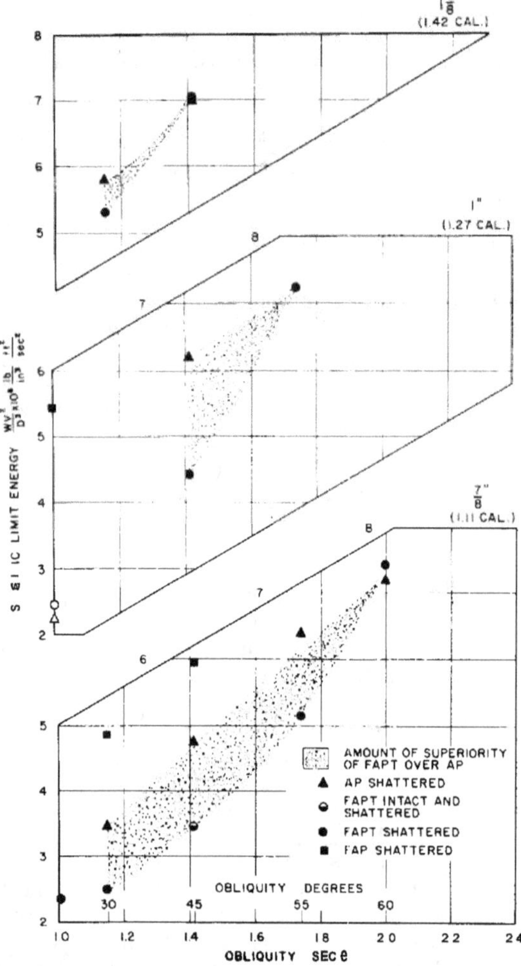

Figure 2-97. Graph of specific limit energy versus obliquity for 7/8-in., 1-in., and 1 1/8-in. plate thicknesses for conventional ogival (AP), truncated ogival (FAP), and tipped truncated ogival (FAPT) 20-mm projectiles

These results have been confirmed by limited firings of truncated 75-mm shot, truncated conical 120-mm shot, and tipped truncated 76-mm shot. These limited test firings indicate that full-caliber shot of these types can be made to show the same relative penetration performance as the 20-mm models if adequate shot hardness and ductility are provided.

2-256. Radius of Ogive.[8] Consideration of all the data obtained indicates that armor-piercing projectiles intended to defeat matching or overmatching homogeneous plate at low obliquity should have a single radius ogive of 1.25 to **1.3** calibers. This radius of ogive provides a combination of both good shatter characteristics and good penetrating efficiency.

2-257. Effect of Nose Geometry of Tungsten Carbide Cores.[19] Firings performed at Watertown Arsenal indicate that the performance of tungsten carbide cores may be improved by the use of a truncated conical nose in place of an ogival one. Figure 2-100 illustrates the .40-caliber model cores which were used in the tests. The following conclusions were arrived at.

1. Blunt truncated conical-nosed tungsten carbide cores require less energy for the penetration of rolled homogeneous armor than do the ogival nosed cores when armor under attack is less than 2.5 calibers in thickness and

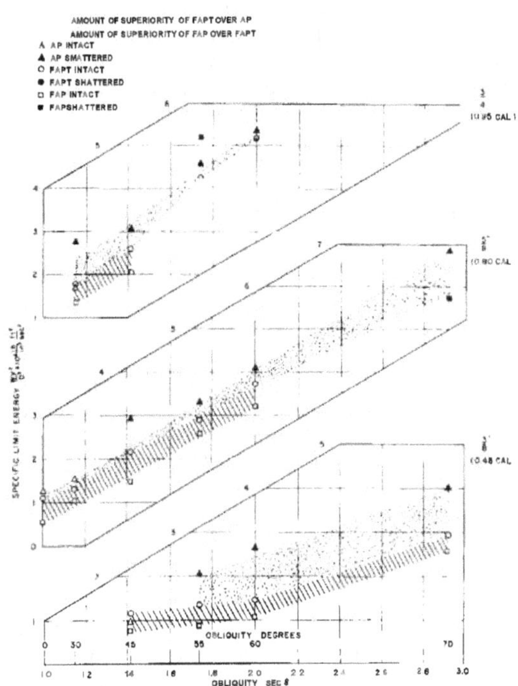

Figure 2-98. Graph of specific limit energy versus obliquity for 3/8-in., 5/8-in., and 3/4-in. plate thicknesses for conventional ogival (AP), truncated ogival (FAP), and tipped truncated ogival (FAPT) 20-mm projectiles

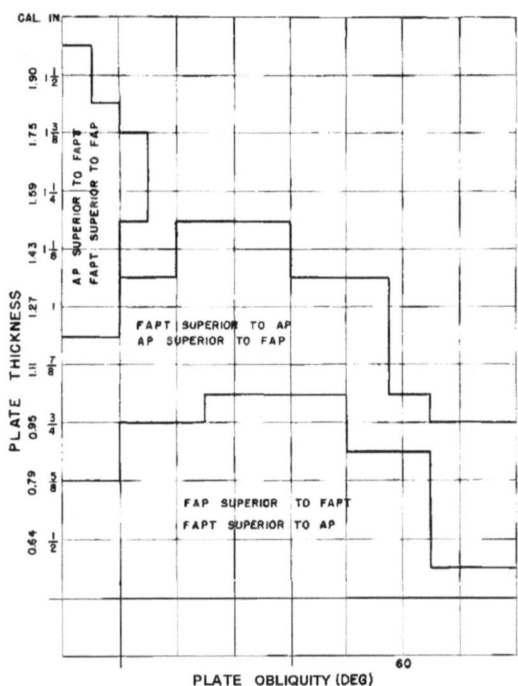

Figure 2-99. Regions of superiority for conventional ogival (AP), truncated ogival (FAP), and tipped truncated ogival (FAPT) 20-mm projectiles

is inclined at an obliquity greater than approximately 35°. At lower obliquities and against thicker targets, the standard ogival nose core penetrates with less energy.

2. The superiority of the blunt-nosed core exists because the cores suffer much less severe shatter under the attack conditions listed in subparagraph 1 above.

3. Regardless of nose geometry, a tungsten carbide core whose nose remains intact during the penetration of armor will penetrate more efficiently than one whose nose shatters.

4. A truncated conical nose remaining intact during penetration will generate a conical tip from the armor which effectively acts as a cap in reducing the stresses on the core nose.

5. No simple equation, involving the core energy as a function of core caliber, plate thickness, and obliquity, is known which can be used for accurate prediction of the terminal

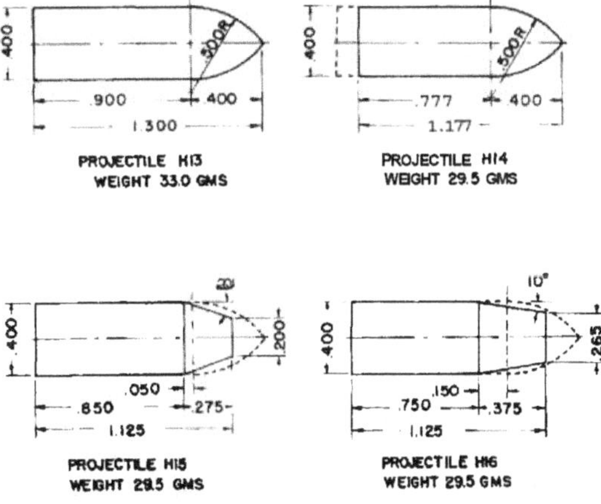

Figure 2-100. Experimental designs of scale model ogival and truncated conical-nosed tungsten carbide cores

ballistic performance of ogival-nosed tungsten carbide cores.

Figure 2-85 is a plot of ballistic limit versus obliquity, for the projectiles tested. Plots for 1.0-in., 0.77-in., and 0.50-in. rolled homogeneous armor plate are included. The results shown should not be considered as optima for the truncated nose core, further investigation is required before the optimum nose geometry can be established.

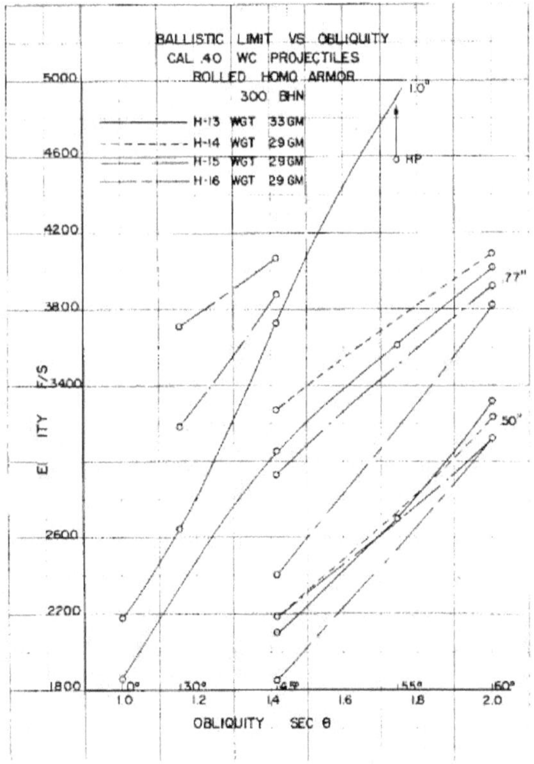

Figure 2-101. Ballistic limit versus obliquity, caliber .40 WC projectiles, rolled homogeneous armor, 300 BHN

2-258. Armor-Piercing Caps.[22] Armor-piercing caps were first introduced as a countermeasure to face-hardened armor. It was found that monobloc projectiles would deform against face-hardened armor even at low velocities and zero obliquity. The addition of the decrementally hardened cap enabled the projectile to penetrate.

By means of the armor-piercing cap, a projectile which might otherwise break-up against a specific target (either face-hardened or homogeneous) may be kept intact. The cap will lead to increased perforating ability only when the uncapped projectile deforms badly and the deformation greatly increases the energy required for perforation.

The following theories have been advanced to explain the mechanism of cap action:

1. The cap acts to break up the face of the armor. (For face-hardened armor this may be true; it is, however, difficult to support this theory for homogeneous armor which has no "face.")
2. The cap reduces the peak stresses resulting from the inertia of the plate material.
3. The cap produces a lateral pressure on the nose of the core and so decreases unbalanced compressive stresses in the region of the ogive.
4. The cap "lubricates" the core.

To assess the feasibility of using a cap, one should know the performance of the uncapped projectile over a wide range of conditions. It is necessary to determine (1) the energy required for perforation when no deformation occurs, (2) the conditions under which deformation takes place, and (3) the effect of the deformation in limiting the perforation or otherwise reducing the effectiveness of the projectile. One must then find the extent to which each of these factors is altered by the addition of the cap. All three factors are altered by changes in either projectile design or target characteristics.

If a cap succeeds in preventing projectile deformation, it will act both to retard and, in the case of oblique armor, to turn. the projectile. These effects, if not kept to a minimum, may overbalance the advantage gained by keeping the core intact.

2-259. Vulnerability Diagrams.[15] A useful way of presenting penetration data on kinetic energy projectiles is by means of vulnerability diagrams of the type shown in figure 2-102. A roughly elliptical area exists for each gun-projectile-armor combination within which penetration of the armor can be effected and beyond which the armor is invulnerable to the particular attack. The gun must enter into this consideration, since it influences the velocity and hence the kinetic energy of the shot at all ranges.

2-260. Comparative Performance of AP and APC.[15] Figures 2-87 and 2-88 show the use of vulnerability diagrams to illustrate the comparative performance of AP and APC projectiles against various thicknesses of armor sloped at different obliquities. It will be noted that, for a fixed weight of armor per unit vertical height, thinner plates sloped at higher obliquities (at least up to 53°) provide progressively more protection against APC shot. Against AP shot, however, a given weight of armor sloped at 37° obliquity provided considerably more protection than the same weight of armor sloped at 53° obliquity. A comparison between the righthand curves of figures 2-103 and 2-104 illustrates the improved effectiveness of AP shot in attacking highly sloped armor targets.

2-261. Effect of Armor-Piercing Caps on Tungsten Carbide Cores.[20] Experiments performed at Watertown Arsenal to determine the effect of steel or tungsten armor-piercing caps in preventing the shatter of tungsten carbide cores indicate that worthwhile results may be obtained. Figure 2-105 illustrates the caliber .40 model cores and caps which were tested. The following conclusions were reached as a result of the firings.

1. Either steel or tungsten caps can be employed effectively for the reduction of tungsten-carbide core nose shatter when the striking velocity is sufficiently high to cause failure of the unprotected core by shatter.

2. Either cap material, when employed in the proper design of the cap, will enhance the armor penetrating capabilities of the projectile

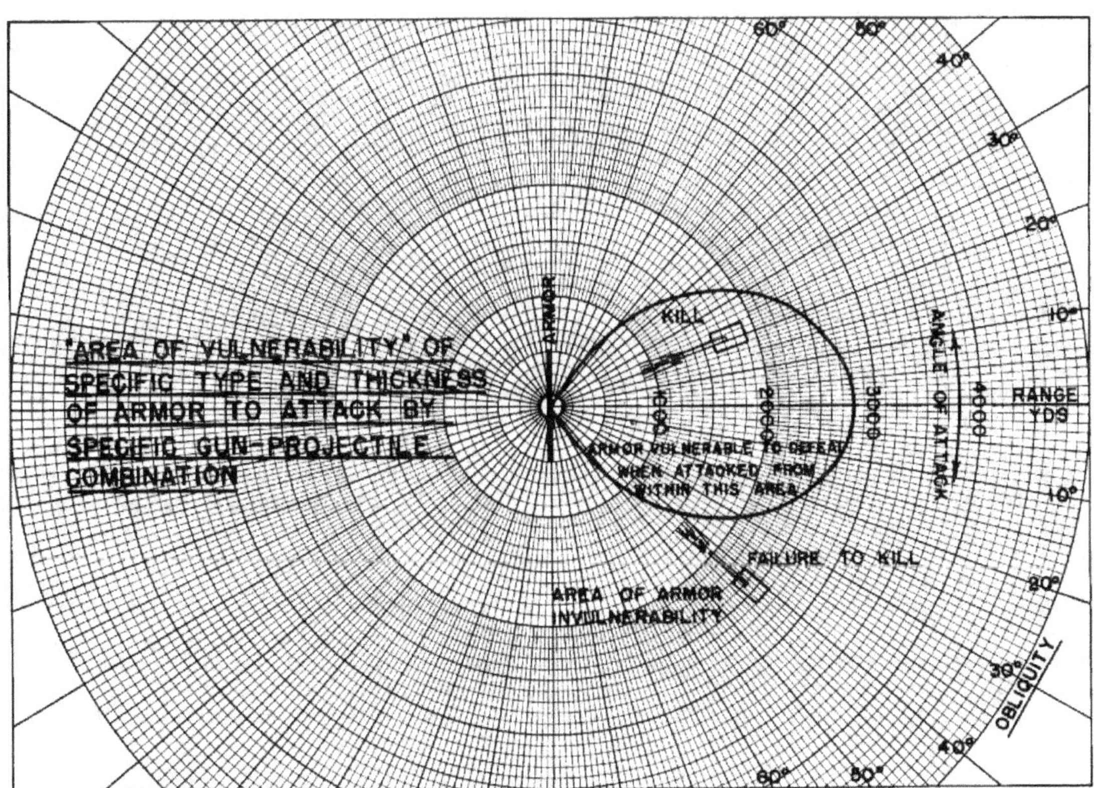

Figure 2-102. "Area of vulnerability" of specific type and thickness of armor to attack by specific gun-projectile combination

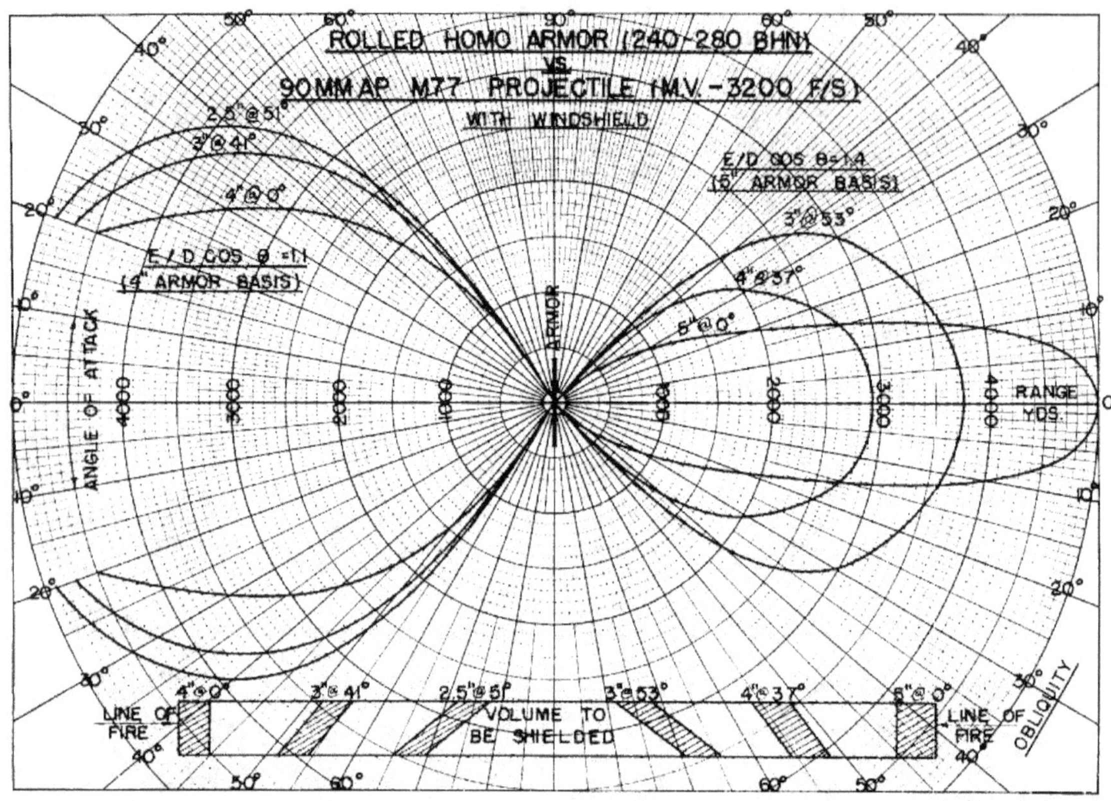

Figure 2-103. Comparison of areas of vulnerability for various combinations of armor thickness and obliquity having equal weight per unit of shielded volume when attacked by 90-mm steel projectiles

against armor less than two calibers thick at obliquities up to 55°, and possibly at higher obliquities against thicker armor.

3. There are no practical advantages to be gained by using tungsten in preference to steel as a cap material.

4. For a given cap design, there is an optimum weight of cap material which is the most effective in reducing the energy required for the perforation of a given armor target.

5. **As** the severity of the target increases (greater thickness at constant obliquity, or higher obliquity at constant thickness, or both), the weight of cap required for maximum terminal ballistic performance increases slightly. However, a cap weighing 10 percent of core weight is a good compromise. Figure 2-106 is a plot of ballistic limit versus cap weight for various thicknesses of armor at several obliquities. The two parallel dashed lines indicate the optimum cap weights.

2-262. Effect of Skirting Armor on Cap.[31]

1. Thin skirting armor can easily remove the cap of **APC** projectiles.

2. In view of the higher probability that a capped projectile will have to attack the main armor uncapped, the cardinal principle in design of a capped projectile is that the body be made to function as well as possible when unprotected by a cap.

2-263. Caps for Defeat of Spaced Armor.[32] The liability to cap fracture of **APC** projectiles by skirting armor may be reduced by increasing the tendency of the steel in the cap to flow plastically rather than to fracture brittlely. This may be done by (1) using steel of a characteristic tempered martensite structure rather than a pearlitic structure, or (2) avoiding excessive hardness. Maximum hardness is to be determined by the detailed design of

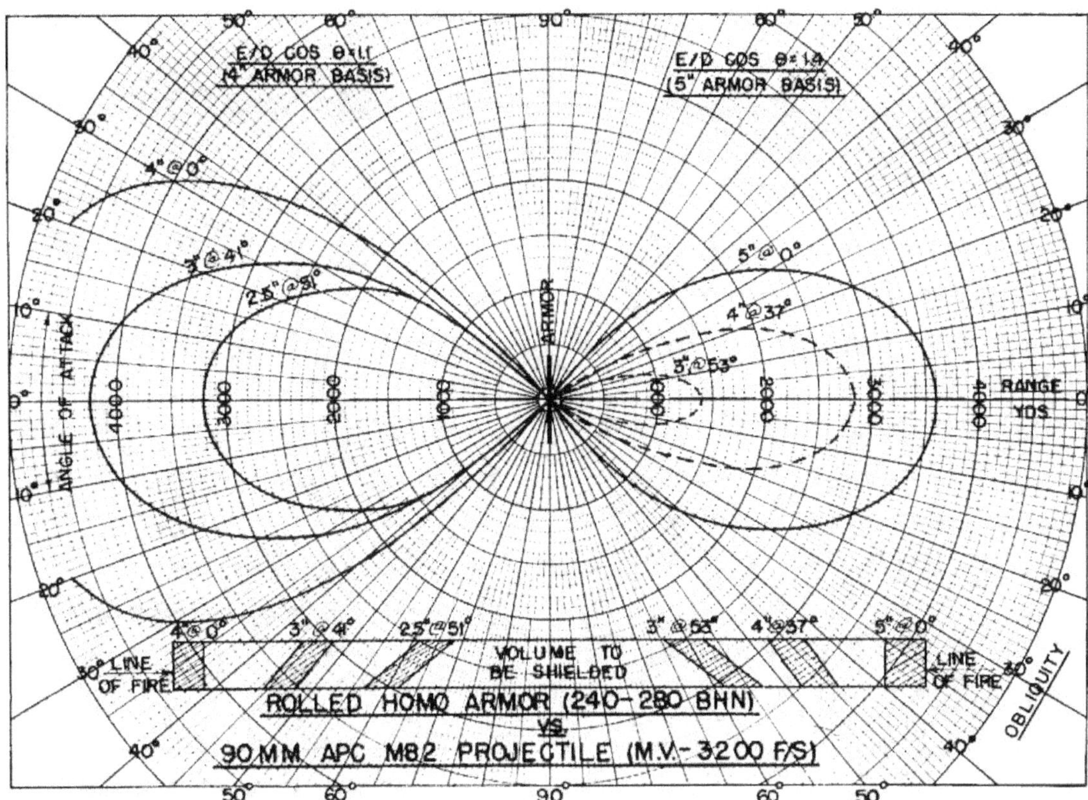

Figure 2-104. Comparisons of areas of vulnerability for various combinations of armor thickness and obliquity having equal weight per unit of shielded volume when attacked by 90-mm steel projectiles

the cap and the thickness of the decapping plate against which protection is desired.

It would not be necessary for the whole of the cap to have a low hardness level. The front face alone could be hard. This would be desirable in attacking face-hardened armor. Against homogeneous armor, experiments at Watertown have indicated that soft caps (15 R_C) have a slight superiority over hard caps with respect to protection against shatter of the projectile. No difference in the performance of the hard- and soft-capped projectiles was observed, with respect to ballistic limits; it therefore appears that AP caps need not be hardened after machining. The steel should, however, have a tempered martensitic structure, which could be given to the bar stock before machining.

2-264. Buffer Caps for Defeat of Spaced Armor.[33] By using a copper or other soft metal plug to protect armor-piercing caps, they may be protected from shatter by thin decapping plates. This plug, which may be quite small, may be placed in the windshield of the projectile in such a manner as to hold the plug against the cap when the windshield is secured to the projectile. In tests conducted at Watertown Arsenal, a 1-ounce copper plug (buffer cap) served to protect 37-mm projectiles against decapping by a 5/16-inch decapping plate. This thickness of plate corresponds to nearly 1/5 caliber. Such a buffer cap should also be useful in helping to defeat armor unprotected by decapping plates. The buffer would reduce peak inertial forces tending to cause shell breakup; this should be particularly useful against face-hardened armor.

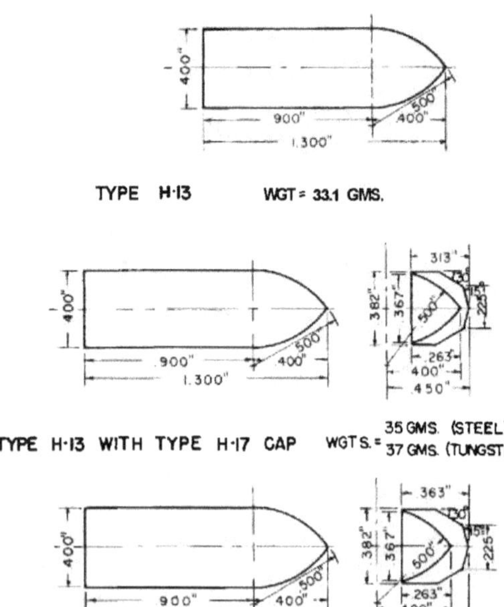

*Figure 2-105. Caliber .40 scale model of 90-mm **M304** (H-13) projectile with experimental caps employed in the study of armor penetration*

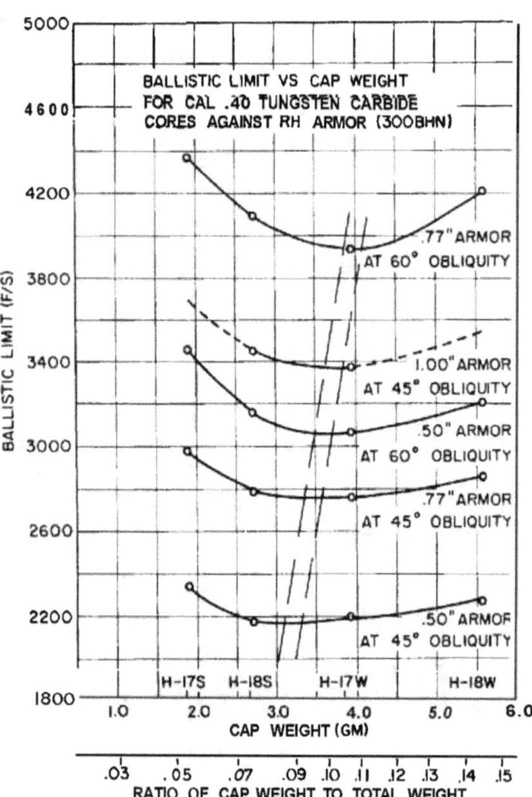

Figure 2-106. Ballistic limit uersus cap weightfor caliber .40 tungsten carbide cores against rolled homogeneous armor, 300 BHN

2-265. Comparative Performance of Kinetic Energy Shot. Against solid steel armor targets, kinetic energy shot performance may be summarized as follows.[15]

a. Monobloc steel shot are more effective than capped steel shot for the defeat of undermatching armor at all obliquities of attack and are more effective than both APC and HVAP shot for the defeat of moderately overmatching armor (up to at least 1 1/4 calibers in thickness) at all obliquities of attack above approximately 45°.

b. Capped steel shot are superior to monobloc steel shot for the defeat of greatly overmatching armor (over 1 1/4 calibers in thickness) at obliquities in the range of 20 to 45°, but both capped and monobloc shot are greatly inferior to HVAP shot in the low obliquity range against heavy armor targets.

c. HVAP and HVAPDS shot are most effective against heavy armor targets at low and moderate obliquities of attack (the 90-mm tungsten carbide cored shot can penetrate 10 to 12 inches at 0" obliquity and at short ranges), but their effectiveness is markedly degraded at obliquities above approximately 45 to 50".

The preceding statements regarding the comparative performance of AP and APC shot are well illustrated by figures 2-107 and 2-108. Figure 2-107 represents data obtained from terminal ballistic tests conducted at the Watertown Arsenal Laboratory in which .40-caliber scale models of the 90-mm AP T33 and 90-mm APC T50 shot were fired at plates from 1/2 to 2 calibers in thickness and at obliquities from 0 to 70" inclusive. The curves on figure 2-108 represent equal resistance curves; that is, all plate thicknesses and obliquities whose coordinates fall on the line designated 3000 have ballistic limits of 3000 fps. The lines furthermore represent the minimum ballistic limit for the target conditions, whether the minimum ballistic limit was obtained with **AP** or **APC** projectiles. The dashed line represents the boundary between target conditions where **AP** shot was superior and target conditions where **APC** shot was superior. It will be noted that the areas **of** superiority of the **AP** over the **APC** shot and vice versa are in accord with the previous conclusions.

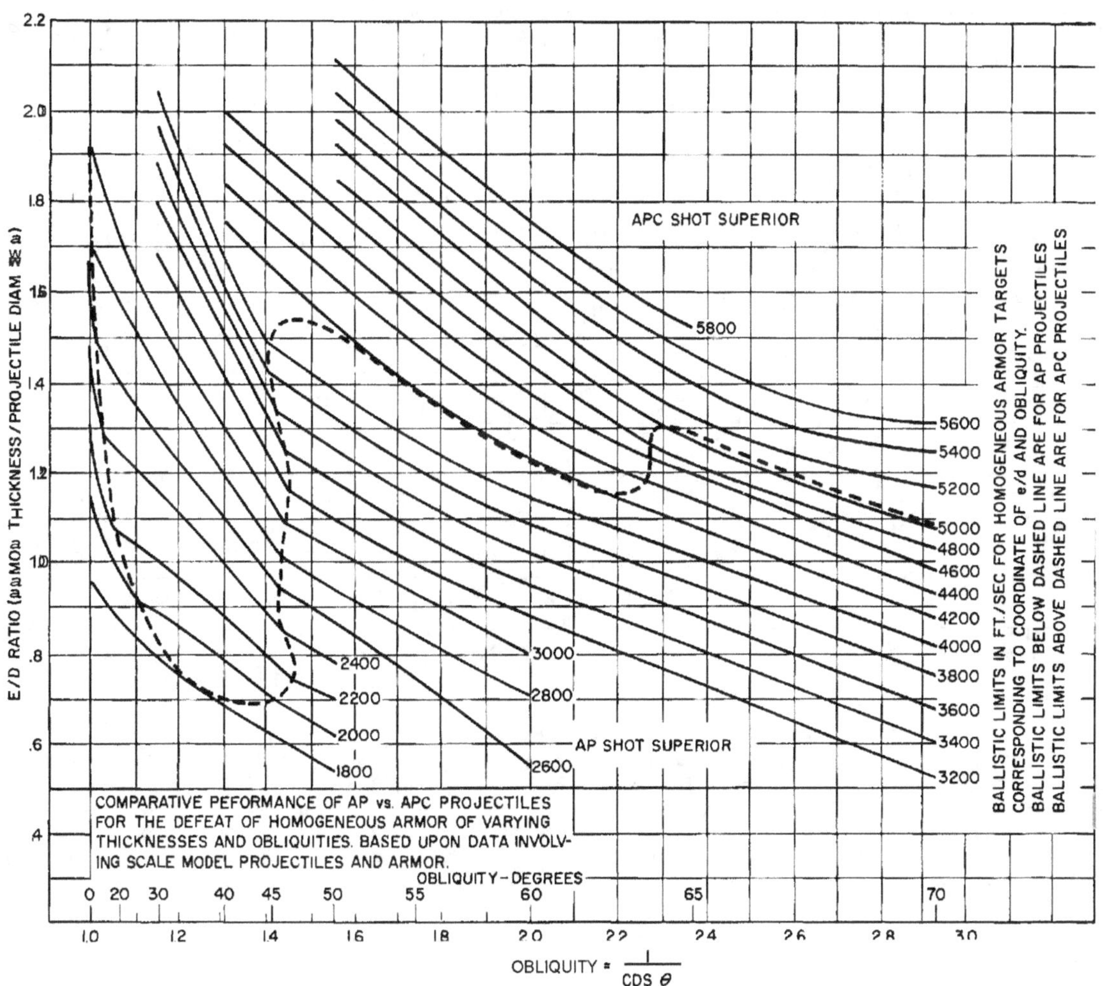

Figure 2-107. Comparative performance of AP versus APC projectiles for the defeat of homogeneous armor of varying thicknesses and obliquities, based upon data involving scale-model projectiles and armor

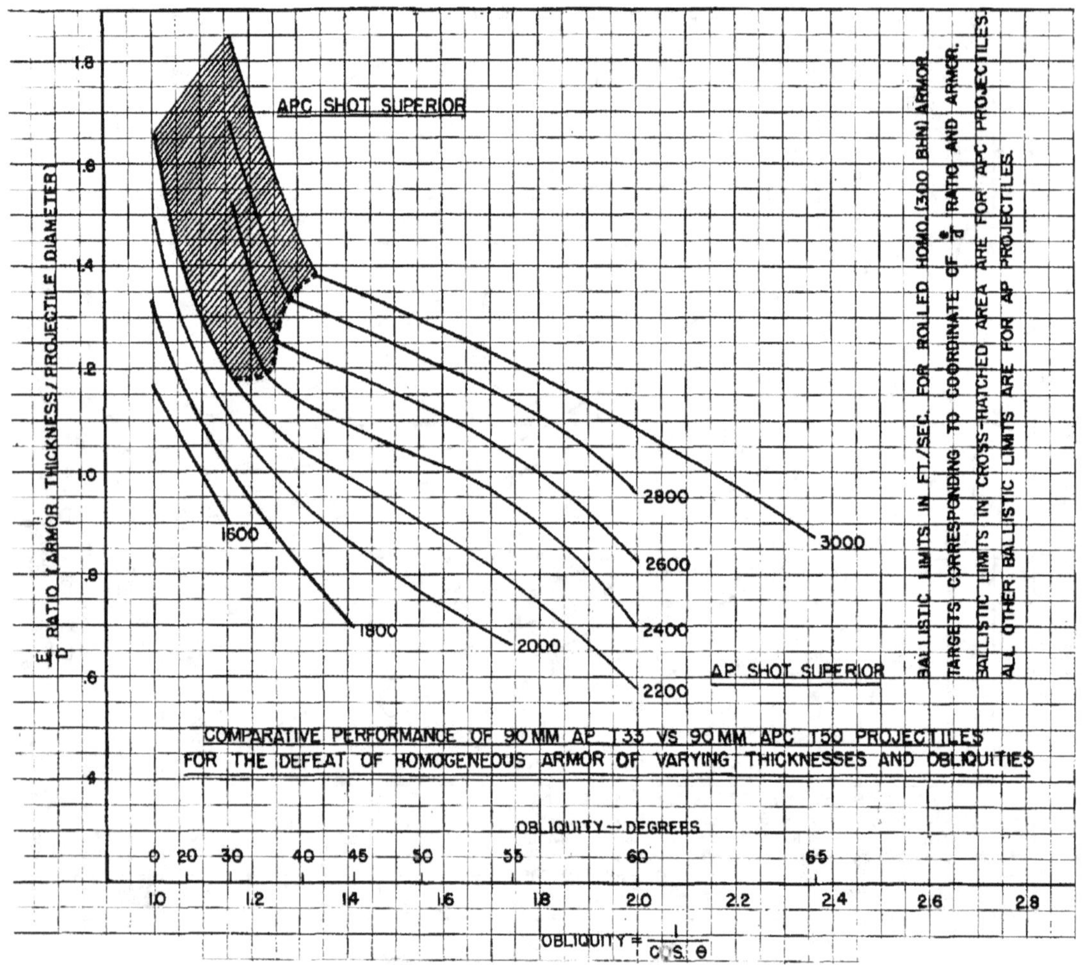

Figure 2-108. Comparative performance of 90-mm AP T33 versus 90-mm APC T50 projectiles for the defeat of homogeneous armor of varying thicknesses and obliquities

REFERENCES AND BIBLIOGRAPHY

1. Riel, R. H., An Empirical Method for Determining the Ballistic Limit of Kinetic Energy Projectiles Against Armor Plate, Development and Proof Services, Aberdeen Proving Ground.

2. Hurlich, Spaced Armor, Watertown Arsenal Laboratory, Report No. WAL 710/930-1, November 1950.

3. Wernick, Kinetic Energy Ammunition, from Report of Picatinny Arsenal Scientific Advisory Council, Second Meeting, November 1951.

4. Abbott, K. H., Full Caliber Vs. Sub-Caliber Steel Shot for the Defeat of Armor, Watertown Arsenal Laboratory, Report No. WAL 762/595, January 1952.

5. Hegge and McDonough, Discarding Projectile Carriers, Watertown Arsenal Laboratory, Report No. WAL 762/523(C), June 1948.

6. Hegge and McDonough, Discarding Projectile Carrier — Part 2, Watertown Arsenal Laboratory, Report No. WAL 762/523-1(C), November 1950.

7. Van Winkle, Principles of Projectile Design for Penetration — Fourth Partial Report, Watertown Arsenal Laboratory, Report No. WAL 762/231-4, July 1944.

8. Van Winkle, Principles of Projectile Design for Penetration — Sixth Partial Report, Watertown Arsenal Laboratory, Report No. WAL 762/231-6, October 1944.

9. Zener, Principles of Armor Protection — Third Partial Report, Watertown Arsenal Laboratory, Report No. WAL 710/607-2, June 1944.

10. Zener and Sullivan, Principles of Armor Protection — Fourth Partial Report, Watertown Arsenal Laboratory, Report No. WAL 710/607-3, June 1944.

11. Hurlich, A., Terminal Ballistics of Armor and Armor Piercing Shot, Watertown Arsenal Laboratory, Report No. WAL 710/930, March 1950.

12. Hill, Information for Tank Damage Assessors, Ballistic Research Laboratories, Aberdeen Proving Ground, Technical Note No. 263, June 1950.

13. Feroli, Methods of Computing Ballistic Limit from Firing Data, Development and Proof Services, Aberdeen Proving Ground, 4 March 1954.

14. Golub and Grubbs, On Estimating Ballistic Limit and Its Precision, Ballistic Research Laboratories, Aberdeen Proving Ground, Technical Note No. 151, March 1950.

15. Hurlich, Comparative Effectiveness of Armor Defeating Ammunition, Watertown Arsenal Laboratory, Report No. WAL 710/930-2.

16. Summary Technical Report of Division 1, NDRC, Hypervelocity Guns and the Control of Gun Erosion, Office of Scientific Research and Development, Washington, D. C., 1946.
    Ch. 9 — Terminal Ballistics
    29 — Sabot Projectiles
    30 — Tapered-Bore Guns and Skirted Projectiles
    33 — Practical Hypervelocity Guns

REFERENCES AND BIBLIOGRAPHY (cont)

17. Abbott, K. H., Principles of Projectile Design for Penetration — Tenth Partial Report — Titanium Carbide as a Substitute for WC in HVAP Cores, Watertown Arsenal Laboratory, Report No. WAL 762/231-10, January 1954.

18. Summary of Technical Report of Division 2, NDRC — Vol. 1, Effects of Impact and Explosion, Office of Scientific Research and Development, Washington, D. C., 1946.

19. Abbott, K. H., Effect of Nose Geometry on the Terminal Ballistic Performance of WC Cores, Watertown Arsenal Laboratory, Report No. WAL 762/231-8.

20. Abbott, K. H., Effect of Caps on the Terminal Ballistic Performance of WC Cores, Watertown Arsenal Laboratory, Report No. WAL 762/231-9.

21. Armaments Design Establishment Design Data Sheets, Armaments Design Establishment, Ministry of Supply, Fort Halstead, Kent, England.

22. *Curtis, C. W., The Problem of Armor Piercing Projectile Design: Its Principle Divisions and Important Phases, Frankford Arsenal, Philadelphia, Pa., Report No. R-901, February 1951.

23. *Sawyer, R. B., Mechanism of Armor Penetration, Frankford Arsenal, Philadelphia, Pa., Report No. R-902, February 1951.

24. *Curtis, C. W., Perforation Limits for Nondeforming Projectiles, Frankford Arsenal, Philadelphia, Pa., Report No. R-903, February 1951.

25. Kymer, J. R., and H. E. Fatzinger, Solid Steel AP Projectiles — Conventional, Truncated, and Tipped Truncated Ogival Types, Report No. R-1166, Frankford Arsenal, Philadelphia, Pa.

26. Gross, J. H., Control of Metallurgical Properties, Frankford Arsenal, Philadelphia, Pa., Report No. R-911, August 1952.

27. Euker, H. W., and C. W. Curtis, Design of Armor Piercing Projectiles for High Angle Attack, Frankford Arsenal, Philadelphia, Pa., Report No. R-1070, June 1952.

28. Fatzinger, H. E., Projectiles for Defeat of Sloping Armor — Ogival Contours, Frankford Arsenal, Philadelphia, Pa., Report No. R-1078, May 1952.

23. Fatzinger, H. E., Study of Penetration with 20-MM Scale Models of 90-MM Shot, Frankford Arsenal, Philadelphia, Pa., Report No. R-853, June 1948.

30. Fatzinger, H. E., and H. W. Euker, Critical Study of Shatter at the 20-MM Scale, Frankford Arsenal, Philadelphia, Pa., Report No. R-799, May 1947.

31. Zener, C., and J. Sullivan, Principles of Projectile Design for Penetration — Second Partial Report, Watertown Arsenal, Watertown, Mass., Memorandum Report No. WAL 762/231-2, 12 May 1944.

32. Zener, C., and B. Ward, Principles of Projectile Design for Penetration — Fifth Partial Report, Watertown Arsenal, Watertown, Mass., Memorandum Report No. WAL 762/231-5, 22 Sept. 1944.

33. Ward, B. C., Principles of Projectile Design for Penetration — Seventh Partial Report, Watertown Arsenal, Watertown, Mass., Experimental Report No. WAL 762/231-7.

34. Mardirosian, M. M., Evaluation of Terminal Ballistic Performance of Armor Against Standard Armor Piercing Ammunition for 57 mm, 90 mm, and 120 mm Guns, Report No. WAL 710/1060, Watertown Arsenal, Watertown, Mass., 5 July 1955.

*These reports are part of a literature survey on the problem of kinetic energy projectile design. When complete, it will comprise fifteen reports.

CANISTER AMMUNITION

2-266. Description. Canister ammunition is, in essence, similar to the ordinary shotgun shell, but is designed to be fired from rifled weapons of caliber 40-mm and greater. Figure 2-109 illustrates one of several complete-round canister designs. The projectile consists of a heavy steel base, designed to withstand the firing stresses, and a thin steel tube packed with pre-formed missiles. As the canister projectile leaves the weapon, centrifugal force causes the steel cylinder to split open and the missiles to spread out in a random pattern within a definite cone of dispersion. The velocity imparted by the propellant charge is relied upon to inflict casualties upon enemy personnel at close range.

2-267. Use of Matrix. The loading of early canister ammunition consisted of steel balls in a plaster of paris matrix. The matrix was used to prevent the balls, acted upon by setback forces during acceleration in the gun tube, from exerting excessive pressure on the bore of the weapon. It has been determined, however, that it is possible to design the steel tube so that excessive pressures are not exerted even in the absence of the matrix. Hence, in most modern practice it is eliminated (see paragraph 2-269). Firings of canister shot employing thermoplastic resins as a matrix for the missiles' indicated that such matrices can be used successfully to decrease radial dispersion; this, however, is rarely desired, except for the longer lethal ranges available with flechette-type filler.

2-268. Missiles for Canister Ammunition.[2] The missiles used at present may consist of (1) slugs —'small cylinders cut from bar stock; (2) steel balls — usually rough ball bearing stampings; or (3) flechette —stabilized fragments or darts having a pointed nose and a finned tail. Typical flechettes are illustrated in figures 2-110 and 2-111. The type shown in figure 2-110 may be fabricated on an upsetting machine similar to that used to make ordinary nails. All are designed for good exterior ballistics to permit a low range rate of velocity loss as compared to slugs or balls.. This low rate greatly improves the effective range of the round and permits the use of lighter missiles, enabling more of them to be packed into each round.

2-269. Loading of Flechettes. Loading of flechettes is done in order to obtain the greatest packing efficiency consistent with the lowest percentage of deformation from firing stresses.

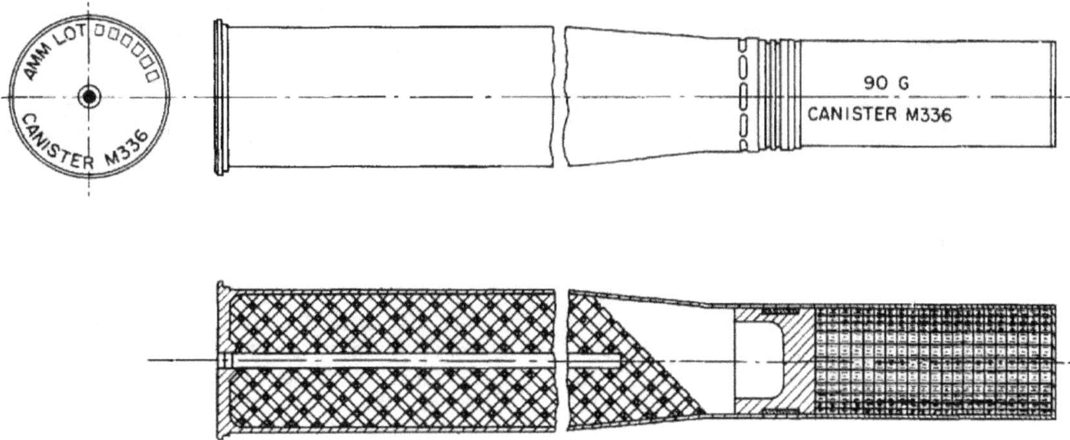

Figure 2-109. Canister complete round

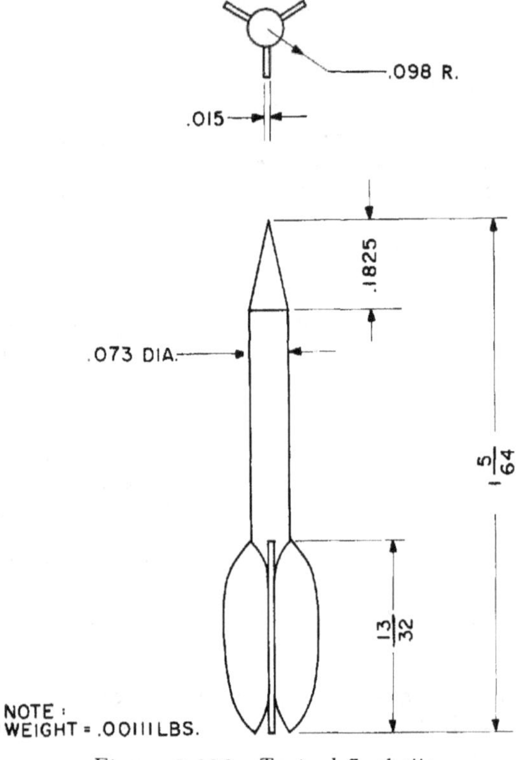

Figure 2-110. Typical flechette

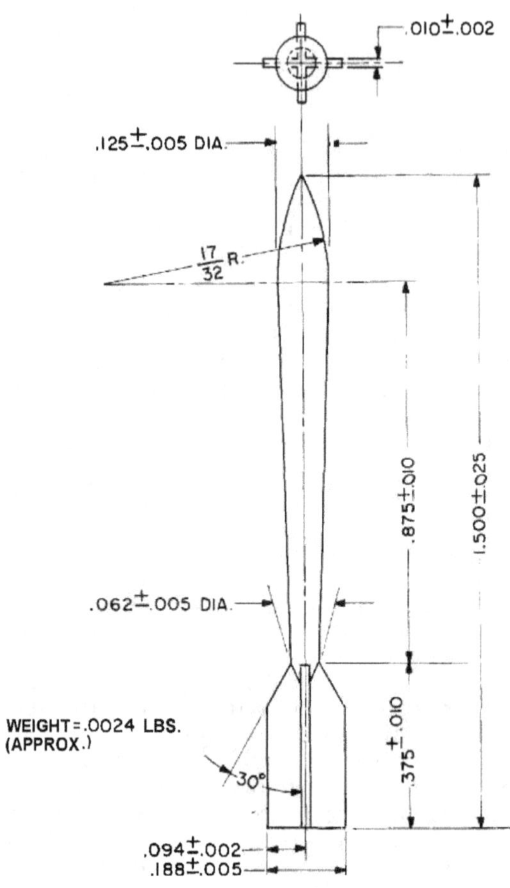

Figure 2-111. Typical flechette

They are placed in the steel tube in several rows, or layers, and are packed head-to-tail in most designs. Each row is packed into a separate container strong enough to resist the setback forces. In spite of this method of packing, flechettes still are subject to considerable damage (approximately 30 percent) upon firing. The use of a resin or derivative matrix and hardening of the flechettes are expected to be effective in reducing this. Figure 2-112 shows a typical arrangement of flechettes in the tube. It should be noted that the pack closest to the base must resist thesetback forces of all those above it; hence, it is of heavy steel construction. The packs above this are of progressively lighter construction, and those at the top may be made of molded plastic.

2-270. Assembly of Projectile. The steel tube forming the body of the projectile usually is slit in four places, 90° apart, for some part of its length, and may be further scored to ensure full opening upon leaving the weapon. Assembly of the projectile is accomplished by brazing or welding the steel cylinder to the base, loading the filler into the tube, inserting a cardboard space wad and a steel cover on top of the filling, and then soft-soldering, welding, or crimping the cover to the tube. The slits in the tube usually are filled with enamel seam sealer to prevent infiltration of moisture. The purpose of the cardboard wadding is to minimize objectionable rattling of the contents.

2-271. Weight and Weight Control. The weight of the projectile is made equal to that of the HE projectile used in the same weapon. The usual reason for this is to maintain uniform exterior ballistic characteristics and to obviate the changing of gunsight settings when switching from one type of round to another. However, for canister ammunition this practice merely serves to maintain relatively uniform weapon-recoil performance and to permit the use of a near-standard charge and components. Weight control of the complete projectile is obtained by controlling the dimension tolerances (and adjusting the number of balls when applicable) until the specified overall weight is obtained. Sufficient tolerance in all the component parts is necessary to permit this. The necessity for maintaining strict weight tolerances is not as important in canister ammunition as it is in other types; it does, however, serve to promote good workmanship and to maintain consistent muzzle velocity.

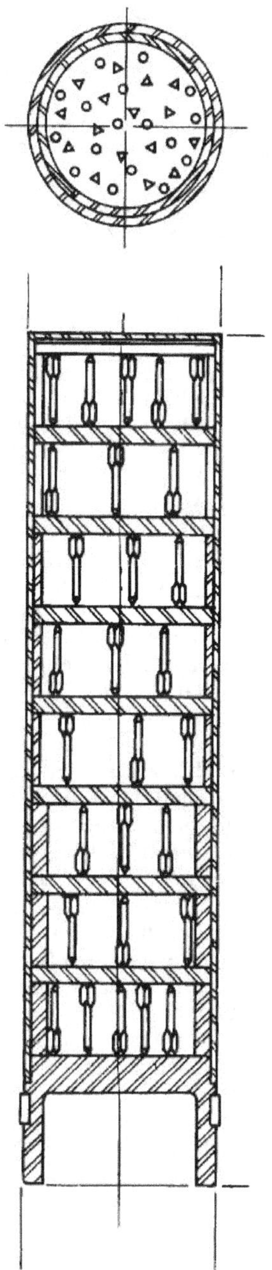

Figure 2-112. Packing of filler

2-272. Plastic Casings. The possibility of making canister projectile bodies out of molded plastic is being investigated at this time. Because of the one-piece molded construction and the elimination of the separate rotating band, this type of construction promises to be considerably cheaper than steel. It also permits the carrying of a higher weight of filler for a specified overall projectile weight. In addition, it is expected to cause less wear of the weapon.

Firings of 1.1-in. and 37-mm canister ammunition using a cord-filled thermosetting phenolic plastic as the casing material have shown promising results. The following is an extract from the Watertown Arsenal report on these firings:[1]

> "The WAL Type Discarding Carrier can be successfully adapted as the casing for canister type ammunition. Plastic carriers are sufficiently strong to withstand the acceleration forces within the gun tube, and can be made considerably lighter than steel casings. They obturate very effectively during and after engraving in the gun tube, discard quickly and uniformly beyond the muzzle of the gun, and thus promote a uniform downrange missile dispersion."

Further tests to evaluate major-caliber plastic carriers have been recommended.

2-273. Rubber Obturators. The use of rubber obturators (figure 2-113) to replace the conventional gilding-metal or steel pre-engraved rotating band has been suggested.[3] Tests now under way indicate that the following advantages may be expected from their use.

1. Elimination of the use of the gilding-metal rotating band in gun and howitzer ammunition. Gilding metal has been, at times, a strategically scarce material.
2. Less wear on the gun tube.
3. Equivalent or better obturation.
4. Elimination of the expensive pre-engraving operation on recoilless ammunition.
5. Interchangeability of ammunition between conventional and recoilless weapons of the same caliber.

6. Elimination of the need for indexing recoilless ammunit on, thus permitting a higher rate of fire.

7. Reduced sidewall pressures will permit the use of thinner base sections and permit an increased weight of payload.

Tests conducted by Picatinny Arsenal on several different types of rubber indicate that Picatinny type XP-214, with a Shore Durometer hardness of D-35, is the most suitable.

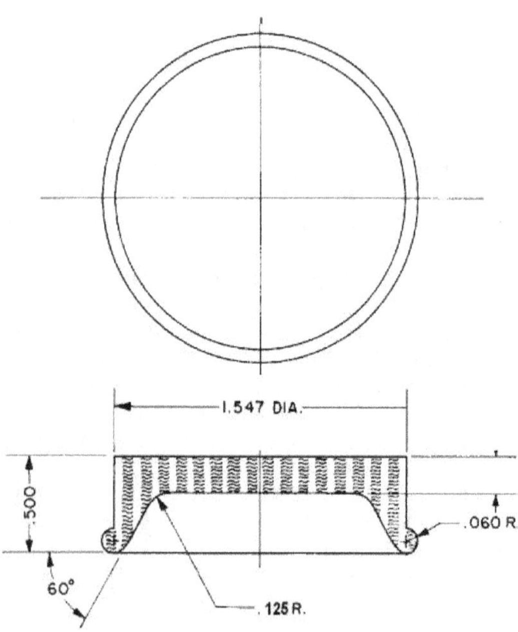

Figure 2-113. Rubber obturator

2-274. Design of Canister Ammunition.[2] The ideal canister projectile should have the following properties:

1. Greatest lethal potential in the tactical situation for which it is intended.

2. It must produce a minimum amount of damage to the bore of the gun.

3. It must remain intact while passing through the muzzle brake, if one is used.

4. It must open consistently within 50 feet of the muzzle.

5. It must have the highest muzzle velocity consistent with the interior ballistics of the weapons system.

6. Ratio of weight of filler to total weight must be as high as possible.

7. It must be cheap to manufacture.

8. It must operate within a temperature range of -65" to +125° F.

2-275. Determination of Design Parameters.

a. Muzzle Velocity. The muzzle velocity may be determined from the known weight of the projectile and the given characteristics of the weapon. (See Section 4, "Propellants and Interior Ballistics.")

b. Angle of Cone of Dispersion. The angle of the cone of dispersion may be determined from the twist of the rifling of the weapon (paragraph 2-277).

c. Optimum Pellet Size. Given the total weight of the filler, the muzzle velocity of the weapon, and the angle of the cone of dispersion, the optimum pellet size may be obtained from lethality considerations (paragraph2-278). The weight of filler, for steel canister, varies from approximately 40 to 70 percent of the total projectile weight.

2-276. Design of Base and Tube.

a. Stress Analysis. The base is designed so that the maximum combined stress on any section of the wall is not greater than the yield point of the metal. In the case of the sidewall, the tube is stressed very near to the yield point to encourage the opening of the canister as quickly as possible after it leaves the gun tube. The analysis of materials stressed to the point of failure (as is necessary in this case to achieve proper functioning) is a rather inexact and ill-defined field. Hence, it is not too surprising that a great deal of past experience and trial and error is involved in designs of this type. Of the three types of fragments used in canister projectiles (balls, slugs, and flechettes), the ball-type filler gives the most extreme sidewall stress. This type of filler is considered as a fluid; the stress formulas of the subsection "Stress in Shell," Section 4, are applicable. The tubes and bases for slug-loaded and flechette-loaded canisters are calculated in the same manner as is the ball-loaded canister, except that the additional support gained by the use of the stacked slugs, in the first case, and the separators, spacers, and matrix in the second case, are taken into account. Since the spacers and separators are subject to an acceleration of 20,000 g's, the assumption is made that if they fail, they will fail in compression. Accordingly, on these components only compressive stresses are taken into account.

b. Lengths of Slits and Scores. The lengths of slits and scores are determined from:

1. Past experience with similar canister designs.
2. Modifications made in the course of test firings. No attempt has been made to compute these lengths, since it is a simple matter to determine the optimum during the test firing.

2-277. Dispersion of Canister Shot. The angle of dispersion is dependent only on the twist (T) of the rifling, in calibers per turn.[5]

$$\text{Cone angle} = 2\left(\tan^{-1} \frac{\pi}{T}\right)$$

It may also be shown that, discounting theeffect of unequal air currents, Magnus forces, and gravity, the shot pattern will be essentially similar to the shot arrangement in the canister. Since the shot pattern within the projectile is essentially random, the fragments arrive at the target in a random pattern.

2-278. Casualty Criteria for Canister Ammunition.[4,5] An old and often misused casualty criterion for canister ammunition assumes that any fragment containing 58 ft-lb of energy is capable of incapacitating a human target. A companion, and equally misused, criterion requires that a canister round produce one perforation per 6 square feet of 1-inch yellow pine board at a prescribed range. Reference to the text on wound ballistics (paragraphs 2-185 through 2-192) will reveal that these criteria bear little relationship to the true lethal potential of a fragment. The following criteria, based upon wound ballistics and probability theory, have been proposed by Picatinny Arsenal.[4,5]

a. Criterion for Rapid-Fire Weapons. The major tactical requirement for canister information is the defeat of the human-wave, massed infantry assault. If the logistical situation is such that a large number of rounds are available, and if the cyclical rate of fire inherent in the weapon system warrants it, then a criterion that will maximize the protection to the weapon crew is proper. Such a criterion is defined as one that will maximize the cumulative probability that a randomly placed, advancing enemy will be rendered incapable of inflicting damage before the range of 100 feet from the weapon (hand grenade throwing distance) is reached. The fragment size that obtains this maximum probability is chosen.

b. Lethal Area Criteria Based Upon a Single Round. A more generally chosen criterion is that of the maximum lethal area of a single round. This is because canister ammunition, in general, is considered "special purpose" ammunition and, as such, the protection must be confined to that obtainable from only one or two rounds. (This condition is especially true in tanks, where ammunition stowage is a serious problem. Furthermore, most weapon systems have inherently low cyclical rates of fire, especially with blunt-nosed projectiles, such as canister.) An ideal criterion for the single round would be one that maximizes a range at which the enemy has a fairly low probability of surviving to do damage in the time it takes to traverse the distance from that range to the range of 100 feet. Unfortunately, however, high levels of probable incapacitation (on the order of 90 percent) are rarely achieved withcanister ammunition, even with the most efficient payloads. Hence, the next best criterion, lethal area, is utilized. The use of this criterion will maximize the most probable number of enemy incapacitations, with a single round, in the time it takes to traverse the distance from any range to the range of 100 feet. It should be noted that essentially the same criterion can be used for offensive situations, except that the time of each incapacitation is kept fixed (at 5 minutes, for example). The latter, however, is rarely used, since the major requirement is for defensive situations.

REFERENCES AND BIBLIOGRAPHY

1. McDonough, J. P., Preliminary Study of the Applications of WAL Type Discarding Carriers as Casings for Canister Shot, Watertown Arsenal Laboratory Report No. WAL 763/885, Watertown Arsenal, Watertown, Massachusetts.

2. Schaffer, M. B., Stabilized Fragments, Research and Development Lecture Series, December 1953, Picatinny Arsenal, Dover, New Jersey.

3. Monthly Progress Reports, Contract No. DA33-008-ORD-160, International Harvester Company, Refrigerator Division, Evansville, Indiana.

4. Schaffer, M. B., Provisional Lethality Criterion for Canister Ammunition, Picatinny Arsenal Report No. 2002, Picatinny Arsenal, Dover, New Jersey, March 1954.

5. Schaffer, M. B., Lethality Calculations for Canister Ammunition, Picatinny Arsenal Technical Memorandum Report No. A-1002, Picatinny Arsenal, Dover, New Jersey, June 1955.

HIGH-EXPLOSIVE PLASTIC (HEP) SHELL

2-279. Introduction. High-explosive plastic (HEP) shell are relatively new in the field of armor-defeating ammunition. Plastic explosives originally were intended for use against concrete fortifications, but now are used primarily as antitank projectiles.

In general, antitank shell are designed to neutralize enemy tanks, either by destroying them as stable, mobile gun mounts, or by killing the gun crew so that they can no longer fire their weapons. Ideally, an antitank shell will perform both these functions by igniting the tank's fuel and detonating its ammunition.

2-280. General. HEP shell are projectiles that attempt to defeat tanks without penetrating their armor. The effect of HEP shell is based on the fact that a "sufficient" quantity of explosive, of "sufficient" height for a given shape of explosive, placed in intimate contact with armor plate will when detonated result in the rupture of a portion of the opposite face of the plate. The ruptured portion is known as a spall, and is generally a rough disk. Depending on the quantity of explosive in excess of that needed to cause the rupture, the spall will attain velocities between 100 and 800 fps. The mass and velocity of the spall depend on the quality and thickness of the armor, and the mass, type, and shape of the explosive filler.

2-281. Comparison With Armor-Piercing Shot. In general, HEP shell defeat standard tank armor approximately 1.2 calibers in thickness, at angles of obliquity from 0 to 60°. When weights alone are considered, the HEP shell far surpasses the armor-piercing projectiles in destructive power.

AP shot and AP projectiles are generally expected to penetrate homogeneous armor with thickness about equal to their caliber. HVAP shot are expected to penetrate up to 3 calibers of homogeneous plate, because of the extreme velocity and the hardness and density of the core.

For armor-piercing shot, it should be noted that the thickness of the armor to be penetrated does not depend on the thickness of the armor plate alone, but also on the angle of obliquity. Thus, the thickness to be penetrated is usually greater than the actual armor thickness. Since the HEP shell shock wave is transmitted approximately normal to the surface of the armor, its spall effect can be accomplished on thicker plate than with a comparable caliber of armor-piercing shot.

2-282. Advantages and Disadvantages of HEP Shell. At the present time, not all properties and characteristics of the HEP shell ai-eknown. However, many characteristics and trends have been discovered, and these are listed below.

a. The advantages are as follows:

1. HEP shell make low-velocity weapons, such as recoilless rifles, effective antitank destroyers.

2. While the effectiveness of other antitank projectiles decreases as the angle of target obliquity increases, the effectiveness of HEP shell is not so adversely affected.

3. HEP shell are cheaper to manufacture than other types of projectiles.

4. HEP shell are light in weight, hence more desirable logistically.

5. Accuracy is comparable to, or better than, HE shell fired from the same weapon.

6. Blast and fragmentation from HEP shell provide very desirable secondary effects against primary targets (armored vehicles).

7. HEP shell are effective in neutralizing secondary targets (fortifications, weapons, emplacements, personnel, and nonarmored vehicles).

b. The disadvantages are as follows:

1. HEP shell cannot be applied to sufficiently wide ranges of velocity conditions.

2. HEP shell have a low ballistic coefficient because of their light weight and blunt head shape.

3. The plastic explosive filler of HEP shell must be press-loaded rather than cast. This method is more time consuming, and hence more expensive.

2-283. Theory of HEP Shell Performance. The theory for HEP shell phenomena is still being developed and the theoretical treatment has not yet reached the stage where engineering design criteria can be furnished. The presentation below is a purely descriptive account of the mechanism of "spalling."

Interacting Wave Front Theory. When a charge of explosive is detonated in contact with a flat steel plate, the explosive energy is transmitted into the plate normal to the surface. The shock wave produced in the steel is reflected from the rear surface of the plate as another shock wave. The shock waves meet at some line within the steel, and reinforce each other, though not simply additively, as with purely elastic stress waves. If the charge is sufficiently great (the height and shape of the explosive in contact with the plate are important parameters), the steel ruptures and a spall is driven off the rear side of the plate. This account has been overly simplified for present purposes, and it is probable that not only shock waves but a complex interaction of elastic stress waves play an inseparable role in the mechanism of spalling.

2-284. General Principles of HEP Shell. A squash charge is most effective when it is in the form of a flat cone. Since the explosive must adhere closely to the surface, it cannot be crumbly, but must have soft plastic properties like putty. Composition A-3 has proved to be best among the various fillers used, while C-4 is the second best. (Cast explosives of the TNT or Composition B type do not give HEP action.)

For the chemical composition and properties of the various explosive fillers, refer to table 2-8, "Characteristics of High Explosives." The spalling effect is produced best with explosives which have a high detonation velocity.

The most serious limitation to HEP shell is that they do not function satisfactorily outside a range of striking velocities from 1000 to 2000 fps. The maximum velocity limit exists because deflagration of the explosive filler occurs when HEP shell are fired at velocities above 2000 fps against plate with 0" obliquity. This condition makes it necessary to lower the velocity of HEP shell for guns and howitzers below the velocities of other ammunition for those guns, thereby preventing ballistic matching. The minimum-velocity limit exists because fuze functioning time and shell crush-up on the target are not properly coordinated at low velocities. At present, it is unknown whether the fuze action is too fast or too slow for the shell; either one of these is possible. The minimum-velocity requirement is a serious handicap in the development of HEP shell for recoilless weapons. There should be sufficient delay time in fuze functioning to allow for proper shell deformation. At high angles of obliquity, the delay is shorter than for low angles. Beyond that, not much is known about HEP fuzing requirements.

2-285. Effect of Spaced Armor on HEP Shell.' The British have done a considerable amount of firing of HEP shell against spaced-armor structures, and have found that this ammunition can be rendered ineffective by skirting armor, which prevents the shock wave from reaching, and being transmitted through, the main armor plate. Spaced armor (consisting of a layer of sponge rubber between the skirting plate and the main armor) has been found effective in preventing spalling by HEP or "squash-head" shell.

2-286. Accuracy and Time of Flight. The accuracy of HEP shell is generally better than HE shell fired from the same weapon. The time of flight is generally greater than that of HE shell because of the more blunt nose of HEP shell, which is required for comparable stability.

2-287. Temperature Effects. HEP shell are not affected at temperatures of -65 to +125°F. Temperature effects upon shell performance differ for the various explosive fillers.

2-288. Effect of Nose on HEP Shell Performance. Variations in nose material, nose shape, nose length, nose hardness, and nose thickness can have a marked effect on HEP shell performance, as follows.

a. Nose Material. The explosive shape at time of detonation is very important in causing a spall. Because of this, it was thought that a softer nose, like annealed copper, might be more suitable. However, tests showed that the annealed steel nose gave better results.

b. Nose Shape. The accepted nose shape is the ogival one. This has proven better than either a hemispherical or convex nose, which originally were experimented with because of the belief that the ogival nose might cause "pinching" of the explosive filler, between the folds of metal, in deforming on impact. In addition to the better explosive effect, the blunt ogive also has better ballistic characteristics.

c. Nose Length. The longer nose is the more preferable, as it provides a greater contact area upon impact. However, care must be taken in manufacture, as there is a danger of telescoping, subtracting from the contact area. More exact data on nose length is not available.

d. Nose Hardness. Although it would seem that a more brittle nose would shatter and scatter the filler, not enough tests have been conducted to state conclusively that brittle steel would be less effective than more ductile steel.

e. Nose Thickness. A thin nose gives better results than a thick nose, but it cannot be too thin or it would not withstand the explosive filler pressure. Various gages of annealed steel have been used. Data can be obtained for all the experimental models in the "Research and Development Reports" of the Chamberlain Corporation, Waterloo, Iowa.

2-289. General Conclusions on HEP Performances, Static Charges. Conclusions developed from various tests at the Aberdeen Proving Grounds are listed below.

a. If the charge weight is held constant, the weight of the spall displaced by cylindrical charges will increase as the charge diameter is increased, up to the point where the charge will have less than the minimum thickness required to displace spalls.

b. The area of a displaced spall is usually slightly greater than the area of the charge in contact with the plate.

c. Explosive charges in the shape of a conic frustum are more effective than an equal weight of explosive in cylindrical shape.

d. The most effective shape of a charge is a frustum of a right circular cone. An oblique circular cone is not as effective.

e. Tough, ductile armor is spalled less readily than higher-strength, more brittle armor. As the ductility of armor decreases, the extent of spalling and cracking of the parent metal increases. The difference in performance of armor of two degrees of toughness will be the greatest at lower temperatures. Weight and velocity of spall fragments increase with increasing brittleness of rolled homogeneous armor.

f. The spalling and cracking of rolled homogeneous armor increases as the temperature decreases.

2-290. One-Piece HEP Shell. The first experimental HEP shell were made from existing smoke shell cases. The first original design of HEP shell in the United States was of two-piece construction, and was a copy of the first British models. One-piece shell were developed in order to lower cost. They have proven to be as good as two-piece shell in regard to terminal ballistics, and also have several other advantages, listed here.

1. Production costs are less.
2. Uniformity of production is better.
3. Equipment on hand would be better utilized.
4. Higher velocities could be attained.

For details on the construction of one-piece shell, refer to Section 6.

2-291. Status of HEP Shell Development and Theory. At the present time, experiments are being conducted by Lessells' and Associates to determine engineering design criteria for HEP shell action. More work is also being done to use some of the existing knowledge, derived from experimental shell, to make standard ones. The chief reason for the lack of more conclusive information is that not enough tests have been conducted. Specific data on the types of experimental shell can be obtained from the Progress Report of the Chamberlain Corporation, on Contract DA-11-022-ORD-662.

Considerable theoretical and experimental work on HEP phenomena is being conducted at Stanford Research Institute, and a rigorous mathematical treatment of the mechanism of spalling may be found in reference 2.

REFERENCES AND BIBLIOGRAPHY

1. Hurlich, Spaced Armor, Watertown Arsenal Laboratory, Report No. WAL 710/930-1, November 1950.

2. Seminar on HEP Shell, held at Poulter Laboratories, Stanford Research Institute, 18 February 1955.

431165 O - 57 - 11

SPECIAL PURPOSE SHELL

INTRODUCTION

2-292. Function of Special Purpose Shell. The special purpose shell described in this section are designed to achieve the following terminal effects (table 2-34).

Table 2-34

Shell	Tactical function
Illuminating	Illuminating, signalling8
Colored marker	Target indicator for ground and supporting air forces
WP (white phosphorous)	Screening, spotting, anti-personnel, incendiary
Colored smoke	Control battery fire, identify troops and targets
Propaganda	Psychological warfare
Liquid-filled	Dispersion of chemical warfare agents

These shell, which are all fired from large-caliber (20-mm or larger) mortars, howitzers, or guns, are made to function by base ejection, separating burst, or explosive burst.

The base-ejection shell is normally used for guns and howitzers, but old-type mortars also use a base-ejection shell. A base plug is retained in the base of this type shell, by the use of shear pins or shear threads, until the burning of the ejection charge in the shell develops sufficient pressure to shear the pins or threads. Either a single- or double-ejection charge may be used.

The separating-burst shell is normally used on newer type mortar shell (see paragraph 2-307). It is similar to the base-ejection shell, except that a shear joint divides the shell approximately into two halves.

The burster charge of the explosive-burst shell breaks the shell and also disperses the filler. The burster charge is a high explosive of relatively low detonation rate, which may or may not be contained in a metal burster tube located axially in the shell.

2-293. General Description of Special Purpose Shell.

Illuminating Shell (see paragraphs 2-294 through 2-307) contain a parachute and flare (sometimes called an illuminant assembly), which are ejected from the shell. The flare is then suspended in the air by the parachute and lights up a desired area for the specified time of burning. Support of the flare by means other than parachutes is being considered.

Colored Marker Shell (see paragraphs 2-308 through 2-318) contain a high-explosive core surrounded by a brightly colored dye. The burster explosive breaks the shell open and provides the heat energy to vaporize the dye. When the shell explodes, the dye is dispersed to form a cloud of colored vapor.

WP Shell (see paragraphs 2-319 through 2-328) contain a filler of white phosphorous, through the center of which passes a metal tube filled with a high-explosive burster charge. The burster charge is designed to open the shell and to disseminate a low-lying cloud of WP.

Colored Smoke Shell (see paragraphs 2-329 through 2-336) contain canisters filled with a dye composition. When the canisters are ejected, they fall to the ground and the composition burns, emitting a colored smoke screen for a specified length of time. The smoke trails out along the ground increasing in height and width as it leaves the burning source.

Propaganda Shell (see paragraphs 2-337 through 2-343) contain leaflets or other propaganda materiel. The materiel is usually ejected from the shell about 150 feet above ground, for best coverage of the target area, but is sometimes ejected at ground level for pinpoint targets.

Liquid-Filled Shell (see paragraphs 2-344 through 2-349) are similar in design to the

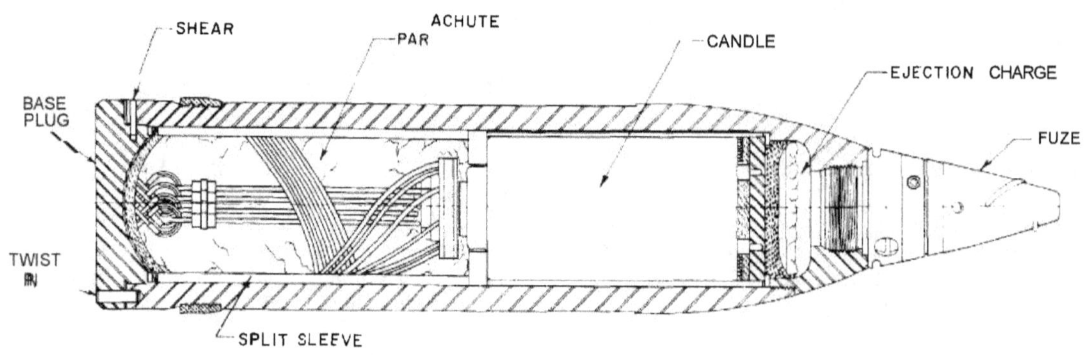

Figure 2-114. Base ejection shell — shear pin type (105-mm M314 illuminating shell)

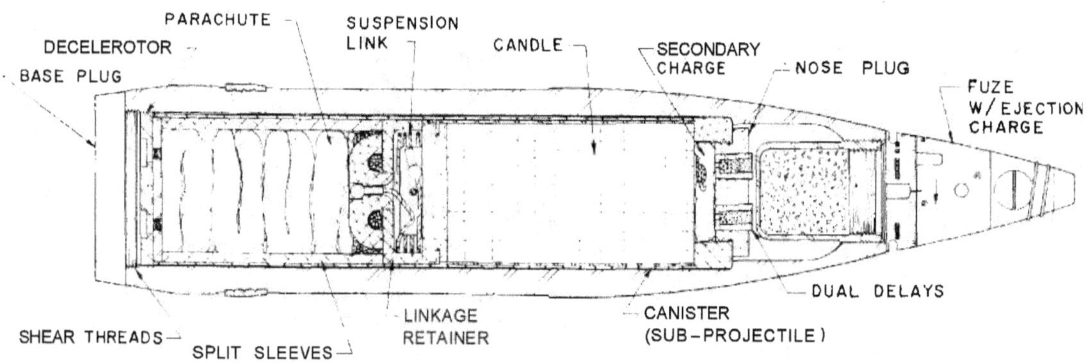

Figure 2-115. Base ejection shell — shear thread type (105-mm T107 illuminating shell)

Ejection from this type of shell is accomplished by means of an expelling charge. The base may be either of shear-pin or threaded design. If a shear pin is used, a twist pin must be used to prevent rotation of the base relative to the body. The threaded-base design permits loading in the field (multipurpose shell). The illuminating shell is fuzed for ejection at 2,000 feet; the propaganda shell is fuzed for ejection at about 300 feet; while the smoke shell is fuzea for ejection on the ground or in the air.

WP shell, except that their water-like liquid fillers require more rigid fits and tolerances to seal the shell against premature leakage of the contents. The design of these shell is controlled by the Chemical Corps.

ILLUMINATING SHELL

2-294. Introduction. Illuminating shell are essentially parachute flares specially housed in shell metal parts for launching from artillery weapons. This type of ammunition is used for the placement of battlefield illumination at the ranges covered by such weapons. Wherever tactically feasible, illumination by howitzer ammunition is preferred, but when necessary mortar-type illuminating shell are used.

2-295. Design Elements of Howitzer-Type Shell. In considering the design of illuminating shell, the design of a particular shell representative of most illuminating shell will be described here. The shell described is the T107 howitzer-type illuminating shell.

2-296. Tactical Requirements. In order of importance, illuminating shell are used for:

a. Spotlighting enemy positions while friendly troops are being deployed for attack
b. General lighting for observation and adjustment of ground fire
c. Continuous lighting to repel massed enemy attack
d. Intermittent lighting to prevent or detect infiltration.

2-297. Terminal Effects' Limitations. There are several interrelated and opposing factors affecting the design and use of illuminating shell. These include:

a. Optimum height of shell functioning for minimum light intensity over a given area
b. Relationship of candlepower to burning time
c. Relationship of candlepower to candle rotation
d. Relationship of candle size to parachute size
e. Limitation on position of center of gravity versus position of center of pressure as governed by size of flare
f. Interchangeability in field for different tactical uses.

The optimum height of illuminating shell is a basis on which candlepower requirements may be established. (See paragraph 2-366.) The length of time the flare is in the air is based on parachute design and the special requirements of application. With these requirements and a knowledge of the intensity and burning time of the candles used, an attempt may be made to design to meet the requirements. The tendency in optimum design is to slow down the speed of the payload before the parachute opens. The parachute must be moving slower than 500 fps to function properly. However, if the ejection charge is too large, the space available for payload is reduced (see paragraph 2-307). The burning time of a rotating candle (above 2,000 rpm) is only about half that of a stationary candle. A solution to this is to stop the rotation. (See paragraph 2-301.)

The setback force that occurs when ammunition is accelerated in the weapon is the chief force encountered in ammunition design. In parachute design, the destructive force exerted on a parachute varies with the square of velocity at the instant of parachute opening. This is the chief force encountered in parachute design, but it is also the chief criterion in the engineering applications of a parachute to an illuminating shell. Parachute design is discussed in paragraphs 2-365 through 2-370.

2-298. Shell Metal Parts Design. With one exception, the selection of materials for the metal parts of illuminating shell is based on three considerations: cost, weight, and strength. The use of the end item defines which of the three is the most important factor. Thus, aluminum split sleeves are used in preference to steel because the saving in weight outbalances the increase in bulk necessary for strength. In general, aluminum is to be preferred where strength is of little concern.

2-299. Base Plug. The original design of the base plug for the T107 shell was as pictured in figure 2-116a. Its principle advantage was lower cost, because of its small size. It proved inadequate in that the setback force exerted on the plug approached the shear force required to shear the pins holding the plug in place. Another failing was that, because of the tolerances involved, the split rings were not always in intimate contact with the base plate. Consequently, there was a good probability that at setback the plug might be pushed forward into the parachute, thereby fouling the chute and lessening ejection velocities by allowing propellant gases to escape. The new design (figure 2-116b) eliminates this type of failure because the plug shoulder prevents forward motion of the plug.

To design an optimum base plug, two opposing factors must be resolved:

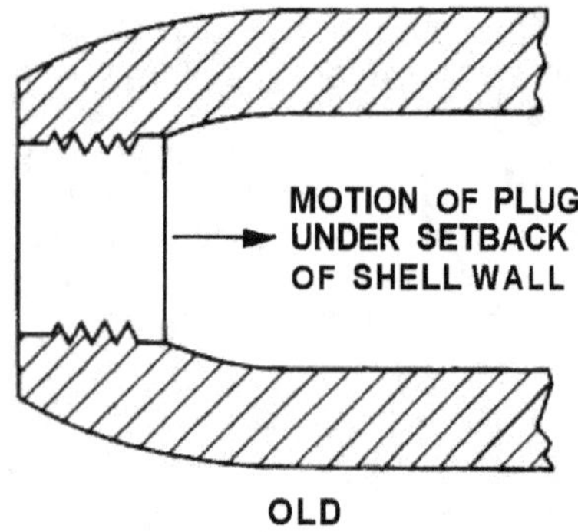

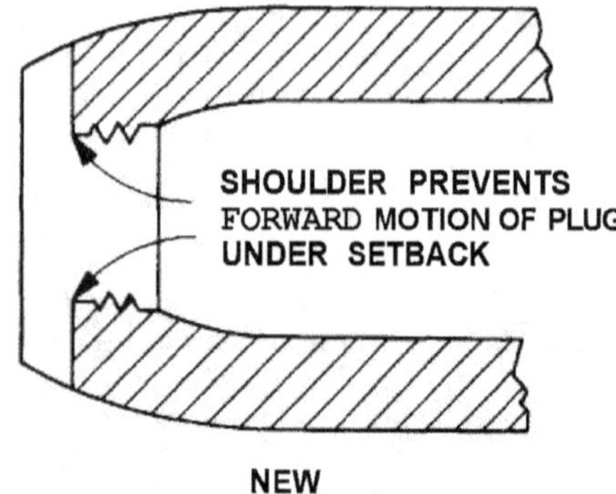

Figure 2-116. Old and new base plugs

a. The amount of pressure necessary to shear the pins or threads (whichever is used) must be less than the pressure required to cause failure in any other part of the shell.

b. If the shear resistance of the plug is low, the effective ejection pressure will be low, resulting in low ejection velocities. Low ejection velocities are undesirable in that the resultant absolute velocity of canister and parachute is too great to permit perfect parachute functioning. For proper design, some compromise must be made between extremes of pressure.

The method for designing the base plug shear threads will be illustrated by the following example, in which

S_1 = tensional hoop stress at any point on the inner surface of the propellant chamber wall

R_o = internal radius of shell at point of calculation of S_1

R_1 = external radius of shell at point of calculation of S_1

S_2 = shear stress on threads of base plug

A_1 = area of baffle plate

A_2 = root area of base plug threads.

Considering the shell to be a thick-walled vessel,

$$S_1 = \frac{R_1^2 + R_o^2}{R_1^2 - R_o^2} P_1 \qquad (27)$$

where P_1 = chamber pressure.

$$S_2 = \frac{A_1}{A2} P_1 \qquad (28)$$

Assuming that the shell parameters are fixed, then S_1 is fixed with respect to P_1, and A_2 must be varied to make S_2 a certain percentage of S_1. Assume that S_2 should be three times S_1 (a reasonably safe assumption); then

$$\frac{R_1^2 + R_o^2}{R_1^2 - R_o^2} = \frac{1}{3}\left(\frac{A_1}{A_2}\right) \qquad (29)$$

or

$$A_2 = \frac{A_1}{3}\left(\frac{R_1^2 - R_o^2}{R_1^2 + R_o^2}\right) \qquad (30)$$

A_1 is fixed by the shell size. Since the T107 base plug diameter is 3.375 inch,

$$A_1 = \frac{\pi (3.375)^2}{4} = 8.1 \text{ sq in.} \qquad (31)$$

At the rotating band, where this shell is presumed to fail,

$$R_o = 3.090 \qquad \text{and} \qquad R_1 = 3.940$$

thus

$$A_2 = \left(\frac{8.1}{3}\right)\left(\frac{3.94^2 - 3.09^2}{(3.942 + 3.09^2}\right) \qquad (32)$$

$$A_2 = 0.655 \text{ in.}^2$$

Assuming 3.375 diameter standard threads and 1/3 efficiency of threads

Major diameter = 3.375

Minor diameter = 3.3167

$$\text{Contact area} = \frac{(\text{maj. dia.}) - (\text{min. dia.})}{2 \times 3} \times n\pi D \quad (33)$$

where n = number of full threads.

Substituting in equation (5),

$$0.655 = \frac{3.375 - 3.317}{2 \times 3} \times n\pi 3.375$$

so that n = 1.56 or, roughly, 2 threads.

Thus, the base plug should be designed so that it has at least two shear threads. If it were desired to increase the ejection velocity, four shear threads could be used, giving a safety factor of about 1.5 (which is roughly a limiting value). A similar calculation could be made to determine the size and number of shear pins needed. In the actual T107 design, 2 1/2 threads are used, giving a safety factor of 2, which is sufficient.

The thickness of the base plate is determined from a consideration of the setback pressure acting on the base, tending to shear the plug. Assuming a breech pressure of 30,000 psi and an unopposed area of 7.2 square inches,

$$F = 7.2 \times 30{,}000 - 10{,}000g \times 4.5 \text{ lb (setback)} \quad (34)$$

$$F_S = 216{,}000 - 45{,}000 = 171{,}000 \text{ lb} \quad (35)$$

Shear area = $\pi D \times t = 9.4t$ sq in.

$$S_S = \frac{171{,}000}{9.4t} = \frac{18{,}200}{t} \quad (36)$$

The thickness (t) that is assumed critical is the distance from the breech edge to the shoulder, since the area of surface b is small (see figure 2-117).

Assuming the maximum allowable shear stress to be 40,000 psi, then

$$40{,}000 = \frac{18{,}200}{t}$$

or $t = 0.455$ in.

The design of the actual T107 has a t = 0.57 minimum, representing a safety factor of

$$\frac{0.57}{0.455} \doteq 1.25$$

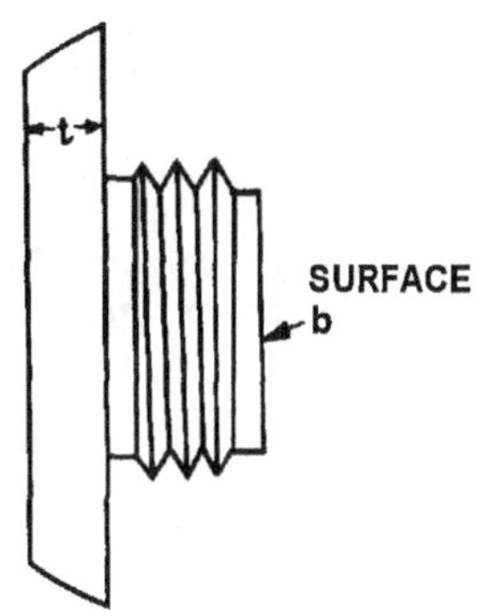

Figure 2-117. Base plug cross section

By using a hardened plug, a saving in weight and bulk could be made with a consequent increase in cost.

2-300. Accessory Parts Design. There are some accessory metal parts peculiar to illuminating shell which must be considered in design. These are: split sleeves, which protect the parachute and assist in deployment; the decelerotor, which decelerates the illuminant assembly before second ejection; and swivel, which separates the parachute from the burning flare.

Split sleeves are longitudinal, cylindrical segments concentric with the shell wall. They function as columns to transmit the ejection force to the base plug, while at the same time they support the setback weight of the candle assembly forward of the parachute. This is the critical force on which split-sleeve design is based.

The total load acting on the segments includes:

a. Weight of segments	0.27 lb	Total weight 4 lb
b. Weight of illuminant container; c. Weight of illuminant and first fire	2.3 lb	
d. Weight of shroud cleat; e. Weight of shroud plate	0.4 lb	
f. Weight of delay body; g. Weight of head	(partial) 0.4 lb*	
h. Weight of plug and steel wire	0.6 lb	

*The weight of the delay body and head is only partially supported by the split sleeves, so an estimate must be made of the support weight.

Assume a setback force of 10,000g in the shell. Then, the total force on the segments is

$$Q = 4 \times 10{,}000 = 40{,}000 \text{ lb}$$

If the segments act as columns, then Ritter's formula for columns can be applied

$$Q = \frac{S_c \cdot A}{1 + \left(\frac{l}{\rho}\right)^2 \frac{S_{cy}}{\pi^2 nE}} \qquad (37)$$

where:

Q	= external load on columns	40,000 lbs
S_c	= induced stress= S_{cy} =	35,000 critical*
A	= area of section	
S_{cy}	= compressive yield stress	35,000 psi (aluminum)
E	= modulus of elasticity	10×10^6 for aluminum
l	= length of section	3 7/8 in. is for T107
ρ	= radius of gyration	0.12 in.
n	= a constant = 2	

Substituting in equation (37),

$$Q = 40{,}000 = \frac{35{,}000A}{1 + \left(\frac{3.87}{0.12}\right)^2 \left(\frac{35{,}000}{\pi^2 \times 2 \times 10^7}\right)}$$

or $$A = \frac{40{,}000 \times 1.18}{35{,}000} = 1.35 \text{ sq in.}$$

Assuming an O. D. of 2.87 inches (figure 2-118), where

$$A = \pi Dt = 2.87\pi t =$$
$$t = 0.140 \text{ in.}$$

which represents a minimum thickness of metal.

Thus the I. D. is 2.87 - 2(0.140) = 2.59 inches. The actual I. D. of the T107 split sleeves is 2.47 inches maximum, which gives t = 0.20 inches, representing a reasonable factor of

*Critical design stress is the point at which $S_c = S_{cy}$.

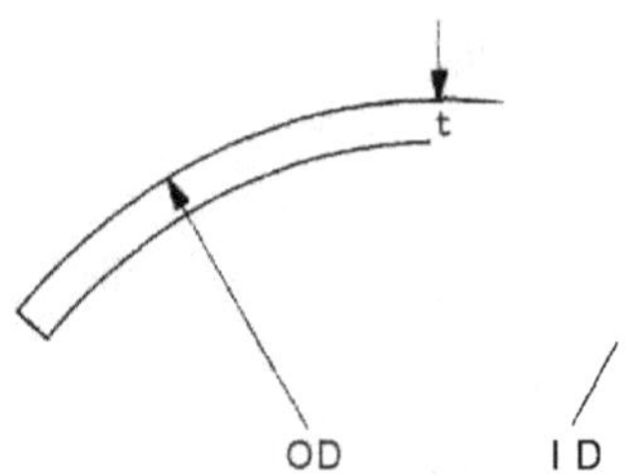

Figure 2-118. Wall thickness diagram

safety of $\frac{0.20}{0.14}$ - 1.4. There still remains the number of split sleeves to specify: the results of static compression tests on split sleeves of three materials are given below. The values given in table 2-35 represent S_c at failure.

Table 2-35

No. Sleeves	2	3	4
61ST alum-inum	37,500	33,900	37,600
4130 steel	98,900	106,400	105,800
1025 steel	73,200	77,600	81,100

It is noteworthy that no significant point of failure variation occurs with variation of the number of sleeves used; therefore, since in general it is desirable to keep the number of parts to a minimum, the best design would call for two sleeves. However, experience has shown that the use of only two sleeves causes poor canister ejection on account of binding of the sleeves; this fault is overcome in practice by specifying three sleeves, with the use of four sleeves as a second choice.

2-301. Decelerotor. To minimize parachute damage, it is desirable to incorporate some device to decelerate the canister between first and second ejection. Originally, the Navy contributed the idea of housing the parachute and candle within a subprojectile, so that it could tumble during a secondary time delay, thereby losing some of its high velocity (approximately 1,100 fps). Three factors influence the selection of a suitable deceleration device.

a. Deceleration efficiency, defined as the ratio of velocity reduction in a given time interval to the total extra volume occupied by the device in the shell.

b. Cost of item.
c. Ease and reliability of deployment.

To date, three methods have been used or proposed as deceleration devices.

a. Small deceleration parachute.

b. Decelerotor (figure 2-119), consisting of two semicircular steel plates mounted eccentrically on the rear of the canister, which swing out under centrifugal force, thereby causing an air-braking action.

c. Use of the split sleeves as a brake by attaching them to a swivel. This procedure involves complications, since it can only operate after second ejection, before parachute deployment.

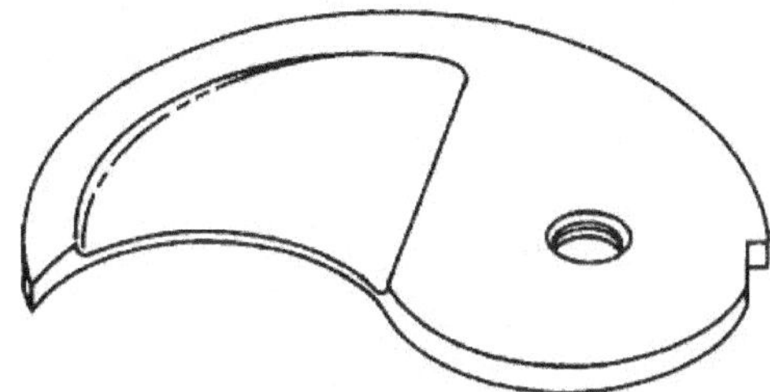

UNDERSIDE VIEW OF DECELEROTOR BLADE SHOWING SLOPED CHANNEL

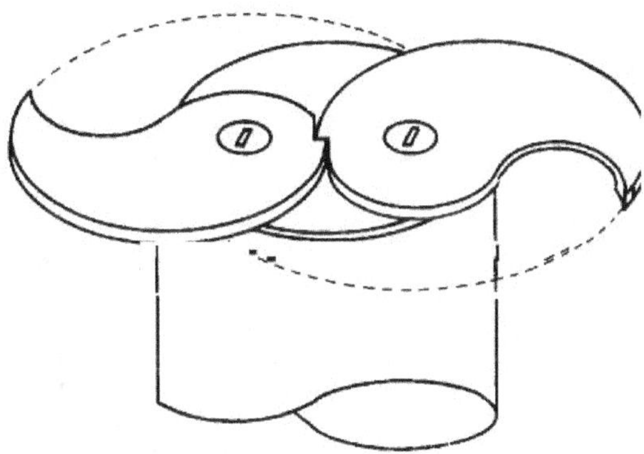

DECELEROTOR WITH BLADES FULLY EXTENDED

Figure 2-119. Decelorotor

2-302. Design of Small Deceleration Parachute. To calculate the deceleration caused by a small deceleration parachute, we may assume a four-second delay between first and second ejection. The following calculations are made for a 700-fps initial velocity; one-foot diameter parachute with drag coefficient $C_D = 1.2$; canister load of five lbs. Neglecting effects of gravity,

$$M \frac{dv}{dt} = F = -\frac{1}{2} A \rho C_D v^2 \qquad (38)$$

where all quantities are as defined in paragraph 2-368. Substituting in (38),

$$\frac{5}{32.2} \times \frac{dv}{dt} = -\frac{1}{2} \times 0.0024 \times 1^2 \frac{\pi}{4} \times 1.2 v^2$$

clearing fractions and integrating:

$$v = \frac{140}{t + 0.2}$$

which is the desired relation between time and velocity. Similar results may be obtained from curves shown in figures 2-120 through 2-124, which give various values of initial velocity v_0 versus time and parachute diameter. Since the volume and weight of the packed parachute increases as the diameter squared, it is desirable to keep the parachute as small as possible to achieve the required deceleration. From the curves, it can be seen that it is possible to reduce the velocity below 200 fps. This represents the maximum diameter parachute to be used in this instance.

The decelorotor is currently used in the T107 shell and is pictured in figure 2-119. One surface of the rotor blade is pitched so that the airscrew effect as it rotates retards the forward velocity of the canister, while a component of force acts opposite to the direction of rotation of the canister, retarding canister rotation. It is difficult to calculate the total effect of the decelerotor, since its rotational velocity varies while it is functioning. Experience has shown that more velocity reduction occurs from the use of the decelerotor than occurs due to tumbling, or by use of a flat plate of the same surface area. As a result, a parachute with a larger, lighter canopy can be used without fear of tearing.

The swivel that had been used up to the present was for the purpose of allowing free rotation of the candle while the parachute was braking. This minimized malfunctioning caused by twisting of the shroud lines. In the present T107, which has a nonswiveling attachment, few cases

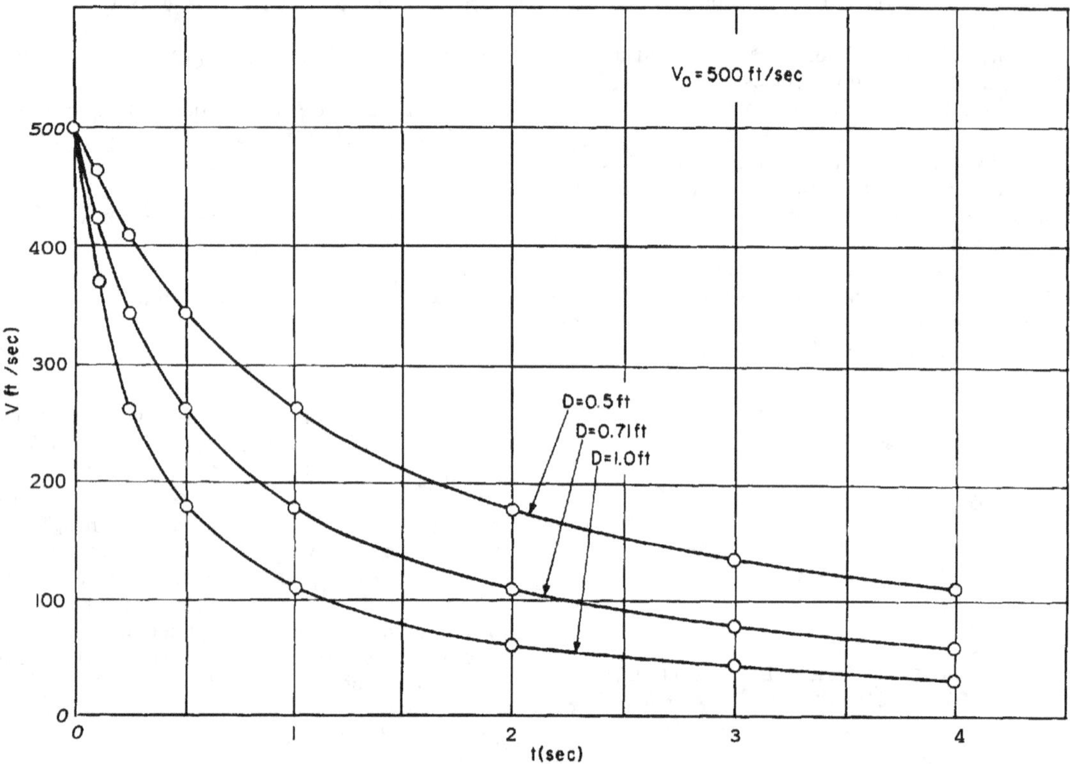

Figure 2-120. Initial velocity (v_o) as afunction of time and parachute diameter (v_o = 500 fps)

of twisted shroud lines have been found, because of (1) use of nylon shrouds instead of steel; (2) rotational deceleration imparted by the decelerotor. In addition to elimination of the swivel, another advantage is gained by reducing rotation; that is, high-speed rotation reduces burning time without any increase in candlepower. It is believed that unburned illuminant is lost by the action of centrifugal force on the flare composition, causing it to break apart.

The design of the remainder of the metal parts does not seem to be critical, since little evidence of mechanical failure has been noted in tests to date, and since their design depends only on their utility.

2-303. Design of Ejection Charge. Of major importance in attaining optimum performance of the shell is the quantity and kind of primary ejection charge used. Since it is desirable to reduce the canister velocity before secondary ejection, and since the velocity of ejection reduces the forward velocity of the canister, increase in ejection velocity is desirable.* Several requisites govern the choice of the charge material. It must

a. Possess high energy content

*Up to the point, that is, at which the velocity is so high that some propellant remains unburned, and the parachute tends to singe.

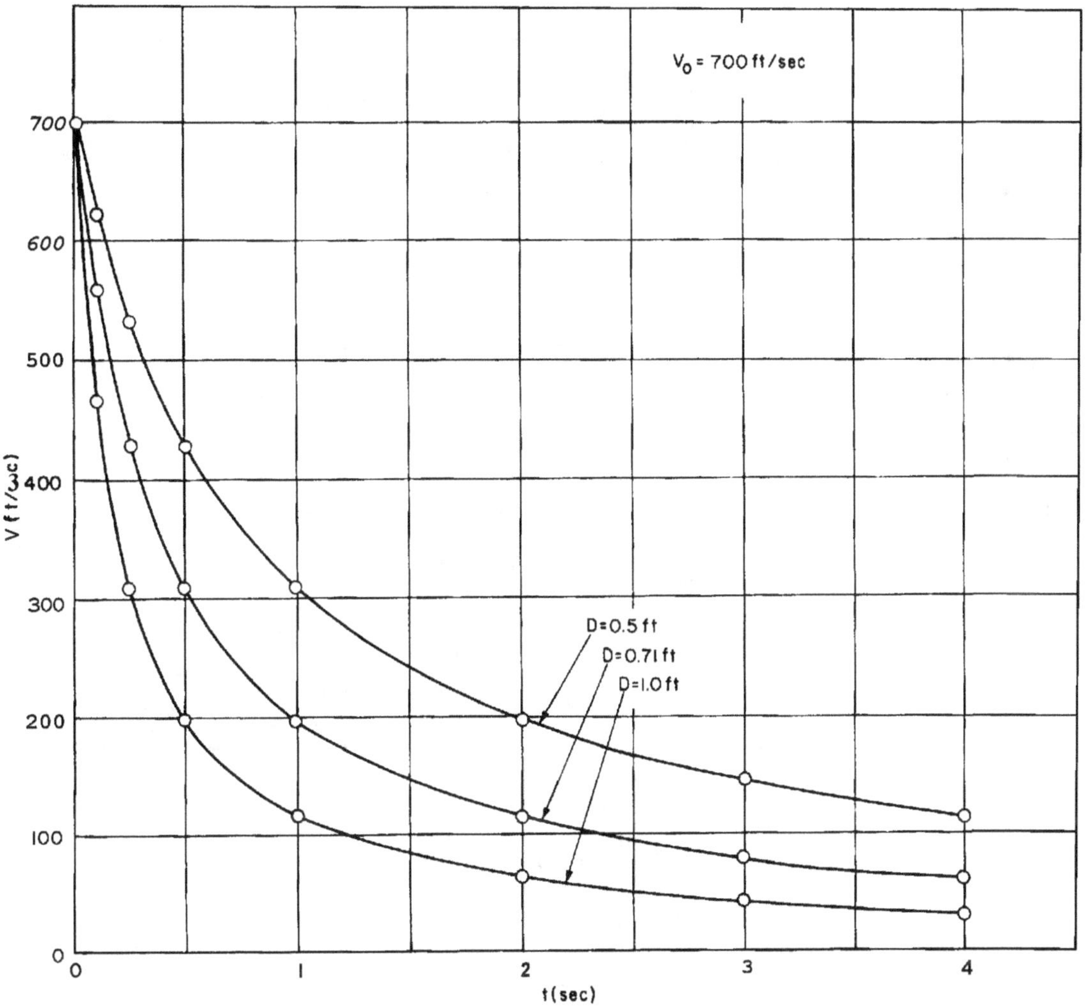

Figure 2-121. Initial velocity (v_o) as a function of time and parachute diameter (v_o = 700fps)

b. Be relatively safe
c. Be nonhygroscopic
d. Be cheap and available
e. Be relatively ash-free.

NOTE: see "Manufacture of Propellants," paragraphs **4-4** through **4-6.**

The black powder, so widely used until now, may be supplanted by a smokeless powder, such as IMR powder, which is superior in most respects. The energy content of IMR is almost double that of black powder.

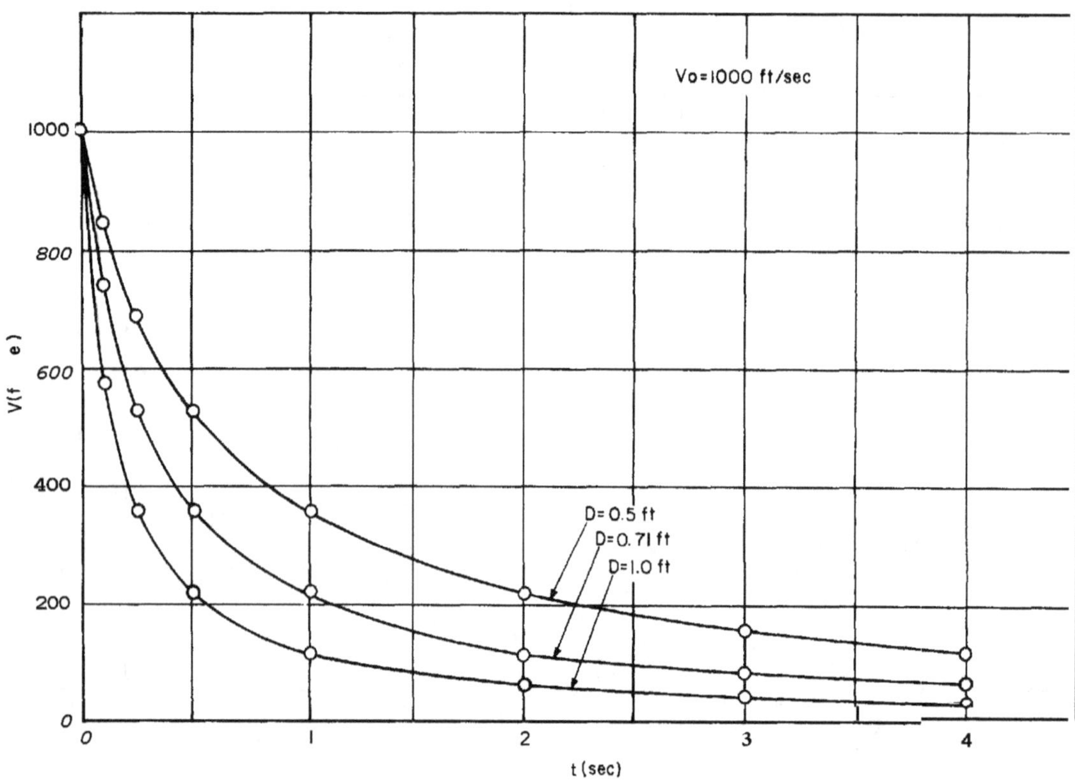

Figure 2-122. Initial velocity (v_o) as a function of time and parachute diameter (v_o = 1,000 fps)

The amount of charge used will determine the velocity reduction. It would seem that sufficient reduction of velocity could be achieved by using enough powder to eliminate the need for external decelerating devices. There is a practical limit to the amount of powder that may be used effectively. If too much is used the canister will be ejected before burning is completed, reducing the ejection pressure. Since most of the charge energy is used to decelerate the canister, rather than in shearing the base plug, a rough estimate can be made to determine the amount of velocity reduction attained using a given weight of powder and a given initial velocity. For IMR powder, the available energy is 348 ft-lb per lb x 10^3. With a 5-lb load and 1,000 fps initial velocity, the total kinetic energy available to the canister at ejection is

$$E = \frac{1}{2} M (v_1^2 - v_2^2)$$

where

v_1 = initial velocity
v_2 = final velocity.

Using 1.7 ounces of **IMR** powder,

$$E = \frac{1.7}{16} \times 348 \times 10^3 = 36{,}800 \text{ ft-lb}$$

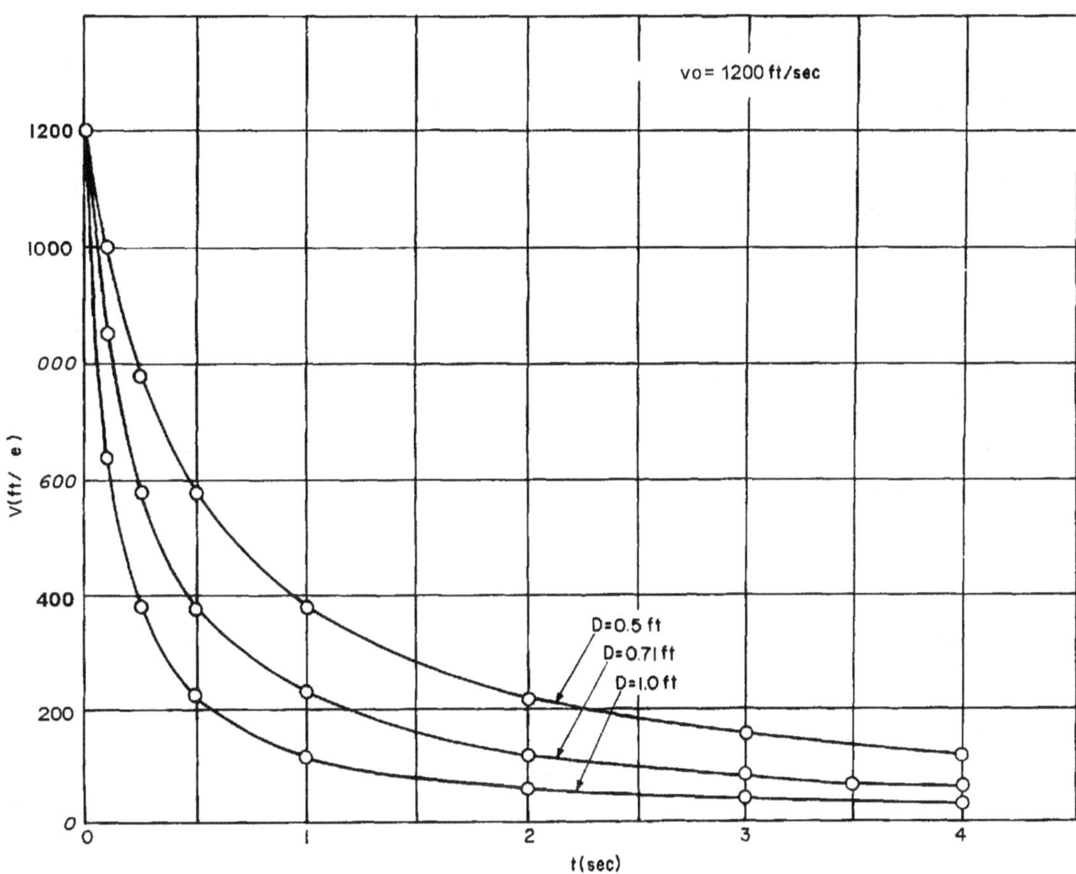

Figure 2-123. Initial velocity (v_o) as a function of time and parachute diameter (v_o = 1,200 fps)

$$36{,}800 \text{ ft-lb} = \$\left(\frac{5}{32.2}\right)(v_1^2 - v_2^2)$$

$$v_2 = 730 \text{ fps}$$

2-304. Direction of Future Designs. While the design of this shell has reached the stage where it can be frozen and put into production, the design is by no means optimum. The only essential parts of the shell are the shell body, base plug, candle, ejection charge, and suspension system. Any method of eliminating any or all of the remaining parts is desirable. Investigation is proceeding along these lines.

One major line of study is improvement of the mechanical properties of the illuminant candles to the point where no supporting cases are necessary. With the elimination of this metal part, cigarette burning, and, consequently, more light output might be achieved.

Another major field of study is the development of new types of suspension systems. One

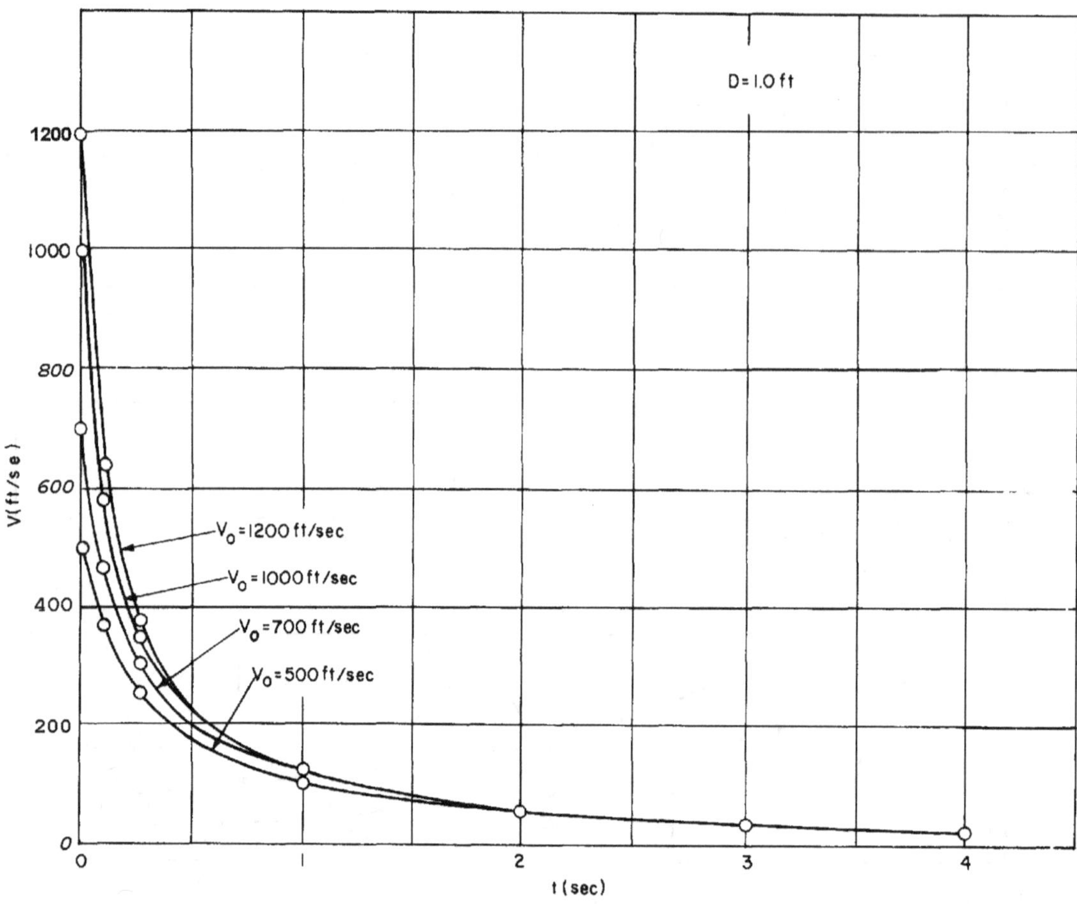

Figure 2-124. Initial velocity (v_0) as a function of time and parachute diameter (D = 1 ft)

promising system is the employment of rotating airfoil blades instead of a parachute. This design has unique advantages, including better storage properties, cheaper production cost, greater strength, more heat resistance, smaller total volume, higher deployment values, and decreased weight (because several metal parts such as split sleeves, shroud cleat, and plate and cable are eliminated). Preliminary calculations indicate that descent rates at least as low as those obtained with parachutes can be attained with a system of blades that can fit into the T107 shell.

Another line of attack is investigation of ejection powders. With ejection powders capable of providing enough thrust to slow the candle assembly to a reasonably low value, it will be possible to discard the double-ejection system in favor of the simpler single-ejection system.

2-305. Design Elements of Mortar-Type Illuminating Shell. Although howitzer-type illuminating shell are preferred, as stated in the introduction, in some situations mortar shell may be used to advantage. Design problems peculiar to mortar shell and not already treated above are covered here.

2-306. Special Design Problems of Mortar Ammunition. Figure 2-125 shows a complete mortar round ready for firing. The tail fin assembly includes the percussion primer, ignition cartridge, and propellant increments. The shell metal parts consist of a tail cone and body fastened together by shear pins. Between the face of the candle and the fuze-ejection charge is an assembly of chipboard, felt wadding, and interlaced quickmatch that protects the surface of the candle from the pressure of the ejection charge, obturates the gases from reaching the parachute, and simultaneously ignites the nongaseous first-fire composition on the surface of the candle.

Firing of the round initiates the fuze. At burst, the ejection charge pressurizes the nose cavity until the shear pins give way, so that body and fuze are projected clear of the load.

a. It can be estimated that this round is only marginally stable because the steel boom and the heavy tail cone locate the center of gravity close to the center of pressure. At one time such steel parts were the only available materials with which to meet the stresses due to chamber pressure and to prevent collapse of the tail cone or boom (which houses the ignition cartridge).

b. Because the tail cone is empty except for a bit of string and tissue paper, the payload is less than optimum. Various attempts to use this space for parachute stowage have failed, for two reasons: (1) redistribution of weight unbalanced the round; and (2) the parachute had no effective provision for vacating the tail cone. The candle and cone tumbled in tandem down to the ground.

c. At one time, a cup and compression spring were fitted into the cone, and the nose was weighted to compensate so that the round would be stable. This device was abandoned because the tumbling of the tail fouled the chute.

Figure 2-126 illustrates a complete mortar round similar to the 60-mm illuminating shell which functions the same way at burst, except that it uses a time fuze.

2-307. 4.2-Inch M335 Illuminating Shell. Figure 2-127 shows a mortar round which is complete except for the propellant increments and fuze. Development of this shell was started by the Chemical Corps during the last year of World War II and represents a conversion of a chemical shell by making a removable base

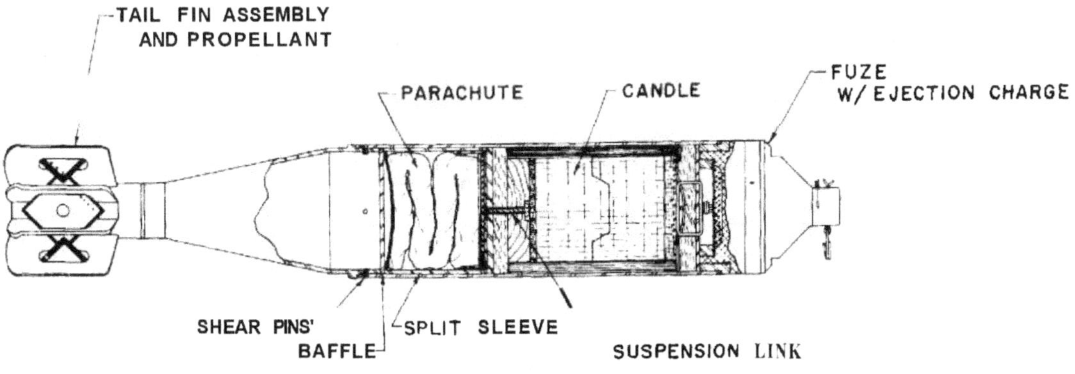

Figure 2-125. 60-mm M83 illuminating shell

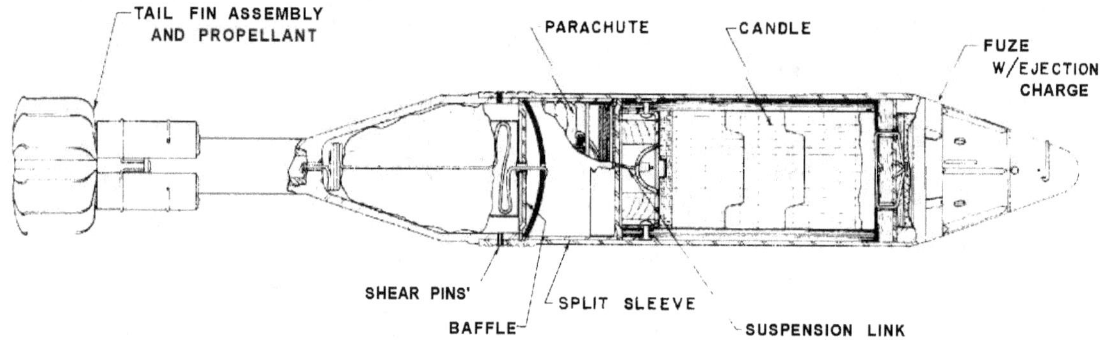

Figure 2-126. 81-mm M301 illuminating shell

plate and inserting components copied from Navy star shell. Unlike most other mortar shell, this shell is spin-stabilized. On ignition (at bottom of the mortar tube), the chamber pressure acts upon the pressure plate to force the brass obturating cup outward, where it engages the lands and grooves in the tube to impart the twist.

The functioning of this shell is the same as that of other mortar rounds. The ejection charge discards the shell body and fuze, so that the remaining components can separate in the airstream, where the parachute will open and suspend the burning candle.

In this design, the swivel attachment brought the shroud lines so close to the candle that heat and flame tended to destroy them. In an effort to solve this problem, the Navy used steel aircraft cable for shrouds. Army experience with steel cable has been unsatisfactory because (1) the heavy steel shrouds often damage the parachute; and (2) centrifugal force and setback kink the cables during firing, and upon shellburst the kinked cables tend to whip and flail and inflict damage on the unfurling parachute.

In order that the loading space in the shell might be used more efficiently, and that the

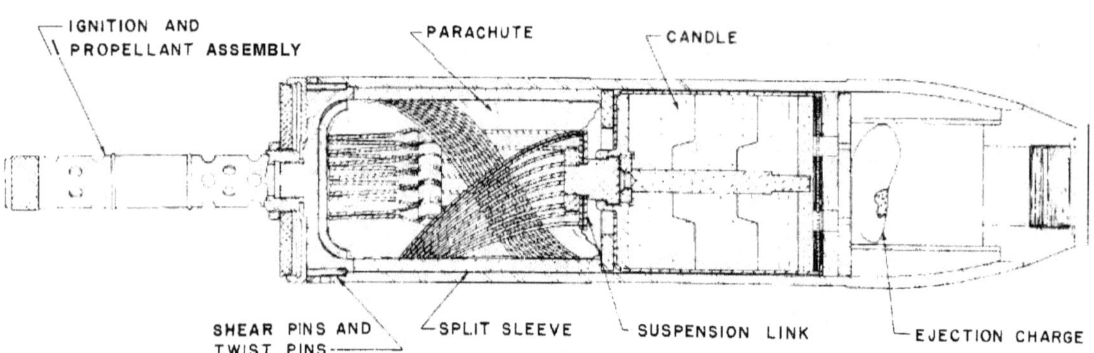

Figure 2-127, 4.2-in. M335 illuminating shell

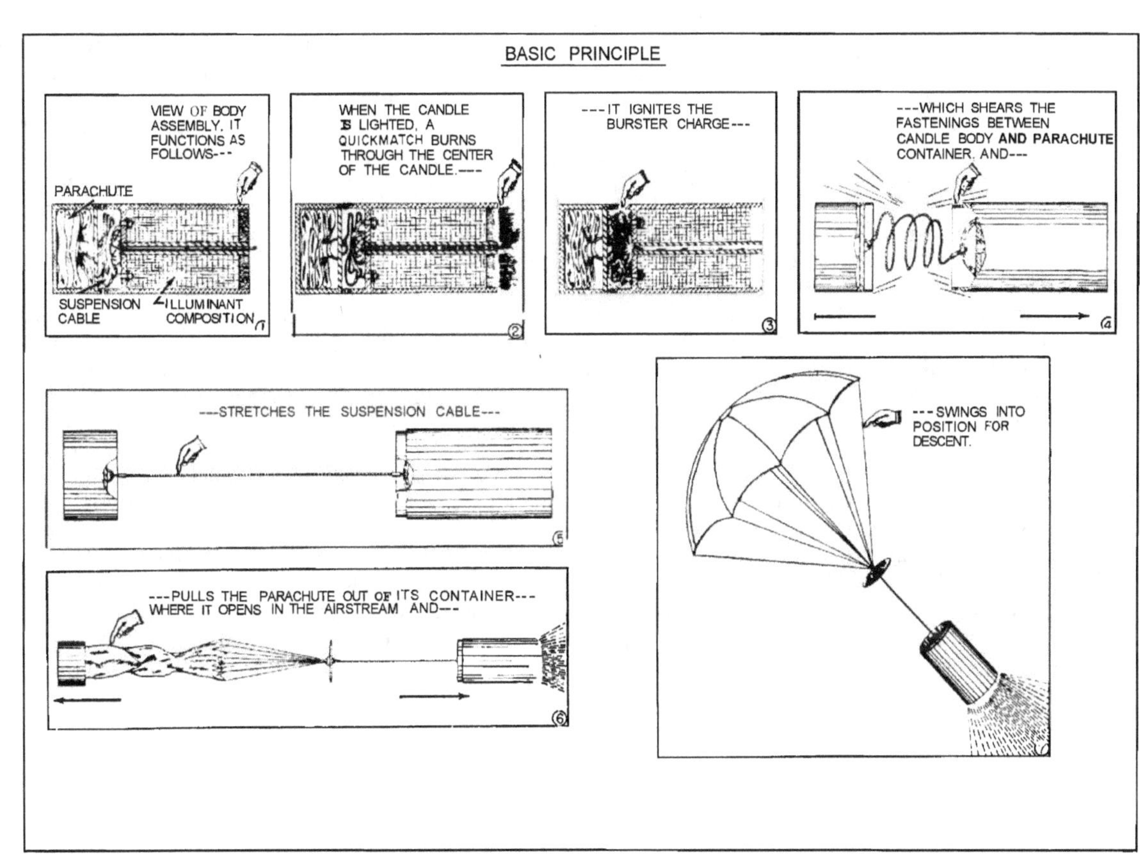

Figure 2-128. Separating burst principle

arrangement of parts might be simplified to lower the cost of manufacture, the separating burst principle was established. Features of the separating burst principle are displayed in figure 2-128. A separating charge is placed in the center of the shell to separate the two halves from each other at the shear rings. The resultant opposite motions of the two halves allow the parachute to be placed in the rear half and to be yanked out when it reaches the limit of the suspension cable. The front half of the shell is the case for loading the candle; this eliminates the double container formerly required. In addition to housing the parachute, the tail cone supports the setback weight of the candle directly; thus, split-sleeve supports are not needed.

Figure 2-129 illustrates a complete round designed to function by the separating burst principle. The shell metal parts consist of the tail cone, the body, and the fuze adapter, all of which are fastened together with shear pins. The baffle plate that protects the parachute has a connection on its other side for the suspension cable. In this design the cable is threaded through copperclad steel tubing and formed into a tight coil, so that it functions as a shock absorber. The separating charge is assembled from the rear and retained by the cable connector with sufficient clearance for gas to enter the cable cavity and press against the baffle plate to shear the pins. The body contains the illuminant composition and thus becomes the candle assembly when the shell separates.

There are still some design refinements, such as the following, necessary in this type of shell.

a. The shock absorber occupies valuable space. It still must be proved that the energy used to straighten out the tough steel coil is absorbed during the shock of parachute opening. This design should be compared to that without the shock absorber.

b. The separating charge should be accessible for inspection during long-term surveillance. A suggested solution is to omit the centrally located separating charge and to make the single nose charge perform the functions of both. The use of a slow-burning; propellant and stronger shear pins on the fuze adapter than on the tail cone may make this feasible. However, test firings of this arrangement have been limited, and the shear strength proportions required for proper control are not yet known.

c. The weight of the steel tail cone of the fin-stabilized round moves the position of the center of gravity too far to the rear. Other materials may be suitable instead of steel.

Figure 2-130 illustrates a complete mortar round being developed with an all-plastic shell body. Both the tail cone and the candle body are made of a phenolic-type plastic. Sample tail cones (molded of glass-filled phenolic) were tested for the ability to withstand a nondestructive load of 6,700 psi, equivalent to 112 percent maximum rated pressure of the mortar. It is considered that this static criterion results in an overdesigned part. But modification will have to wait for considerable ballistic experience in the absence of any realistic dynamic test method.

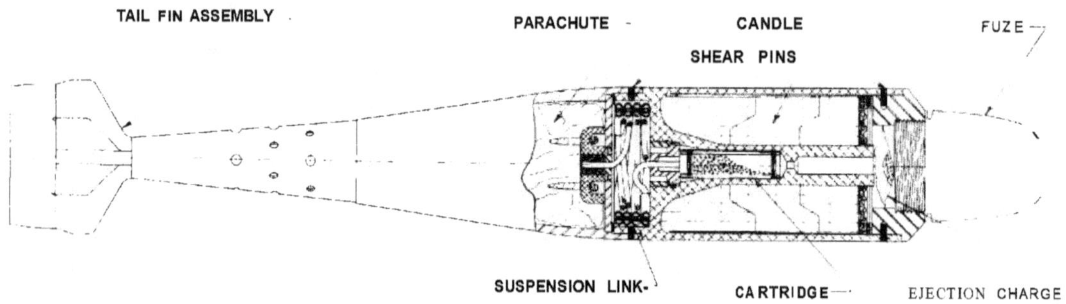

Figure 2-129. 80-mm T214 illuminating shell

In this shell the strength of the candle case presented a special problem. In addition to withstanding the setback load, the case must stand up under the pressures created by the pressing of the composition. It was stated by the development contractor, after initial test of candles molded with chopped-glass fiber, that the candle composition tended to crack the case upon curing. The experience of Picatinny Arsenal indicated that such cracking is most likely a result of improper curing of the molded case. Picatinny also suggested that phenolics other than glass-filled could be used for better moldability and better control of burning characteristics, because setback stresses are not as severe in the candle case as in the tail cone. Further work along these lines by the contractor produced a satisfactory case molded of an asbestos-filled phenolic.

COLORED MARKER SHELL

2-308. Introduction. The colored marker shell produces a colored cloud due to vaporization and condensation of the dye. The dye is dispersed by a burster charge consisting of a high-explosive core (baratol is the present standard) which is cast in the center of the pressed (present standard is 12,000 psi) or cast dye. The high-explosive core is separated from the dye by a double coating of acid-proof black paint. Unlike the base-ejection smoke shell, the cloud produced by the colored marker shell is not caused by the burning of a dye composition, but the energy of the burster charge both vaporizes the dye and fractures the shell wall.

2-309. Tactical Requirements. Colored marker shell are used in daylight as:

a. Target indicators for supporting air forces

b. Target indicators and general markers for ground forces

c. Target markers in artillery practice.

Figure 2-131 illustrates a representative type of colored marker shell. It is desired that the colored marker shell be designed to meet the following requirements:

a. Shell burst is to occur anywhere between ground level and 250 feet above ground.

b. Marker cloud is to be dense and compact, and is to persist in recognizable form from 15 seconds for 75-mm to 75 seconds for 155-mm shell.

c. Against any background, a color must be available which will be visible to air observers at from 3,000 to 8,000 feet, depending on the caliber.

d. Marker cloud is to form immediately upon shell burst without any build-up period.

e. Red, yellow, and green marking colors must be unmistakably recognizable in hue.

2-310. Terminal Effects' Limitations. Three characteristics determine the value of colored clouds:

a. Color

b. Duration

c. Size.

Color recognition depends on the individual observer and background, but instrumentation is being devised to eliminate human error. The colors which have been found most practical

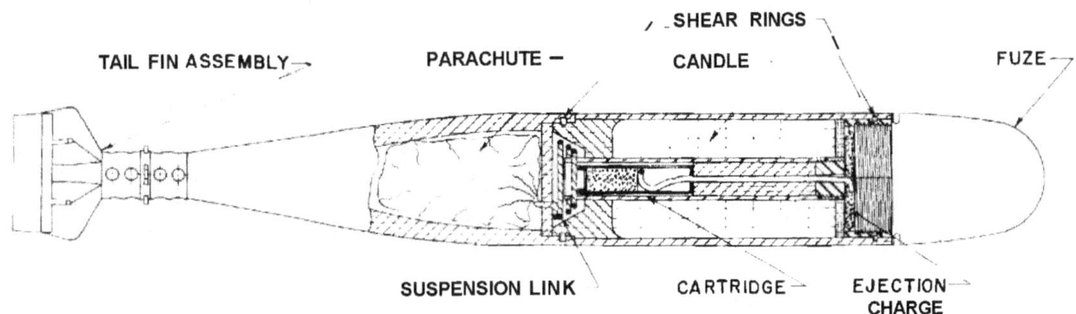

Figure 2-130, 60-mm T213 illuminating shell

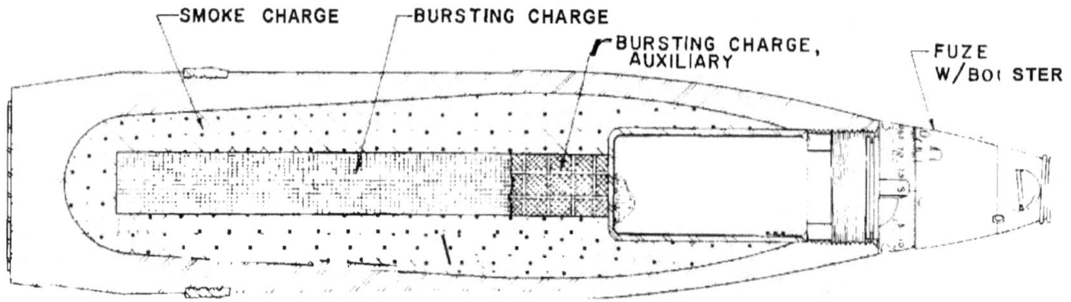

Figure 2-131. 105-mm M1 colored marker shell

to use, in terms of visibility and unmistakable identity, are red, yellow, and green. Violet is used, but is not recommended because of atmospheric selective absorption of the blue end of the spectrum. Haze, fog, and rain reduce visibility. For optimum visibility, it is desirable to have as dense a cloud as possible.

Against a sky background all colors become gray and indistinguishable at a distance sufficient to counterbalance the color intensity. With respect to the position of the observer relative to the sun, it may be stated that looking in a direction normal to, or obliquely into, the sun, the color of the cloud will appear of a lighter hue than if seen with the sun to the rear of the observer. Ground impact usually reduces the vividness of the color because of the mixture of dirt and debris in the cloud.

The greater the color saturation, the longer will a given color be recognizable as a specific color; that is, red will be seen as red rather than pink. The higher the wind velocity, the shorter the duration of the cloud. It has been observed that clouds experience a longer duration at altitudes above 1,500 feet.

The relative effectiveness of burning- versus bursting-type shell has not yet been investigated.

2-311. Shell Metal Parts Design. The body of marker shells is usually identical with the HE shell body. The loading assembly is designed to match the ballistics of standard HE shell within ±0.5 pounds. Both characteristics facilitate firing in standard weapons with the use of HE firing tables. The colored marker shell does not require a special canister to retain the dye composition, but uses the shell body for that purpose. Consequently, any dye chosen must be compatible with the metal or plastic of the shell.

2-312. Accessory Parts Design. The marker shell should be designed with a-deep cavity to accommodate either a mechanical time fuze, VT fuze, or point-detonating fuze. Fuzing should be set to function just above ground or in the air.

The explosive train should be initiated by supplementary charges for mechanical time fuzes and by standard boosters for VT fuzes. As a general rule, a booster should be used that provides just sufficient energy to transmit the initial shock wave to the initiating charge without interfering with the operation of the item under development.

2-313. Filler Design. The following factors are to be considered in the design of the filler for marker shell.

a. Kind of burster
b. Kinds of dyes
c. Ratio of dyes to diluent

d. Ratio of burster charge to smoke charge (static evaluation tests have been conducted to determine the best ratios).

Research has shown that the problem of colored smoke is too complex to be solvable by varying dye, diluent, and explosive ratios. Basic chemical and physical control of colored smoke cloud has not yet been reached. A research and development program to explore the mechanism of production of colored smoke clouds is being continued by **SFAL.** It is not known at present whether the colored cloud is a dust cloud or a vapor cloud, or a mixture of both. However, the dye composition is a mixture of an organic dye and a salt. It is believed that the salt is vaporized by the heat of the explosion to form nuclei on which the volatized dye may condense. As standardized, the colored clouds are nontoxic in ordinary field concentrations.

2-314. Burster. It is known that the shape and/or arrangement of the burster charge affects the cloud characteristics. However, no investigation along these lines has been conducted for the colored marker shell. This type of investigation has only been made for photoflash bombs. The burster diameter should be adjusted to propagate efficiently throughout the burster column, and at the same time should not be larger than is necessary to just burst the shell at high order. Paragraph 2-315 below describes a method of determining the weight of any tetryl burster required, since it is assumed that the heat energy of the explosive furnishes the energy to vaporize the dye. The required weight of any explosive other than tetryl may be established by the introduction of a conversion factor which establishes the ratio of caloric difference between tetryl and the explosive under trial. In all tests to which this principle has been applied, the results have proven satisfactory.

2-315. Determination of Weight of Burster Charge. Sufficient explosive must be included in the burster charge so that when it is detonated it will just burst the shell at high order. At the same time, the charge must not be so great that it causes excessive dispersion and/or burning of the filler. The following method was derived by assuming that the energy of the explosive charge is proportional to the strain energy required to burst the shell. In order to simplify calculations, conversion factors and constants are included in a dimensionless factor (K). The charge of tetryl burster required is given by

$$w_c = KW (Y + U) e \cdot K'$$

where

w_c = weight of tetryl required, in grams (including initiator)

K = a constant, 11.4×10^{-6} to 11.4×10^{-5}, depending on caliber. (The exact K can be found by empirical evaluation only.)

W = weight of steel components of shell (excluding fuze and base), lbs

Y = yield stress of shell steel, lb per sq in.

U = ultimate stress of shell steel, lb per sq in.

e = elongation factor expressed in percent

K' = ratio of caloric value of tetryl to caloric value of other explosive used.

For instance, if value of tetryl is 1,100 cal/gm and baratol is 900 cal/gm, the formula would be*

$$w_c = KW (Y + U) e(11/9)$$

2-316. Filler Materials. Various high explosives have been tried as burster materials, such as Composition B, amatol, cyclotol, 60-mm ignition powder, silas mason explosive, and others, but to date it has been found that baratol 67/33 produces the best clouds if initiation is sufficient and over the total diameter of the column.

The current method of producing coloredsmoke is to volatilize a mixture containing an organic dye. Nonorganic dyes are being investigated. To be suitable, dyes should have the following characteristics:

a. They should belong to one of the following groups:

Azine
Azo
Quinaline
Xanthene

b. The following groups may be present:

NH_2
RNH
R_2N
Aryl

---

*For the relative energy values of high explosives, see table 2-8.

Alkyl
Chloro
Bromo
Alkoxy
Hydroxy

c. The following groups must be absent:
Sulfonic
Hydrochloride
Nitro
Nitroso
Quaternary Ammonium
Oxonium

d. The molecular weight must not be more than 450.

e. The dye must not undergo auto-oxidation.

f. The dye should be volatile or should sublime readily.

g. The dye should have good thermal stability and a high flash point.

h. The dye should have a high color saturation.

Typical smoke compositions are given in table 2-36.

2-317. Discussion of Existing Designs. A method has been developed to evaluate color, size, and persistence of smoke clouds under test conditions. It has been found possible to match satisfactorily smoke clouds produced by static tests with ballistic tests for the same rounds.

2-318. Discussion of Future Designs. There have been suggestions of other approaches to the design of colored marker shell, including:

a. Homogeneous propellant dye mixes and burning-type smoke mixes

b. Use of base ejection shell metal parts in present and future filler designs

c. Application of conical, spherical, or contour-shaped bursters

d. Consideration of the application of various methods of mass production loading.

WP SMOKE SHELL

2-319. Introduction. WP smoke shell has a filler of white phosphorus that produces a white cloud when dispersed from the shell by a high explosive contained in a metal lburster tube in the center of the shell. The burster charge is designed to open the shell and disperse the WP, so as to provide an effective shield. A WP smoke shell is considered to have been designed satisfactorily if it disseminates roughly a 30-feet wide by 15-feet high WP cloud, lying low to the ground, which serves as a shield for about one minute.

Table 2-36

*Standard colored **smoke** compositions*

Ingredient	Percent		
	Red smoke MIL-STD-518	Yellow smoke MIL-STD-519	Green smoke MIL-STD-517
Red dye*	42.5	...	...
Yellow dye†	...	40.0	...
Green dye‡	...	...	41.0
Potassium chlorate	27.4	25.0	23.8
Sugar	...	20.0	...
Sulfur	10.6	...	9.2
Potassium bicarbonate	19.5	15.0	26.0

*Red dye: l-methylaminoanthraquinone, 90%; dextrin, 10%.
†Yellow dye: beta naphthalene azo dimethyl-aniline.
$Green dye: 1-4 di-p-toluidine anthraquinone, 70.7%; auramine hydrochloride, 29.3%.

2-320. Tactical Requirements. WP smoke shell are normally used for:

a. Shielding troop movements

b. Marking targets for supporting ground or air forces.

2-321. Terminal Effects' Limitations. The terminal effectiveness of this shell is limited by the resultant cloud density, which is affected by atmospheric conditions and wind velocity.

2-322. Shell Metal Parts Design. The WP smoke shell is designed to match the ballistics of the corresponding HE shell for the same gun. The shell body is made of the same steel as that of the HE shell and, if possible, the shell body is made of the same forgings. In appearance, the shell body is the same as its HE counterpart, except in the following two instances.

a. No base cover is used for the WP shell. Since the WP is not an explosive, the problem of leakage of propellant gases through the base of the shell to cause premature detonation does not exist.

b. A smooth, cylindrical, press-fit surface about 1/2-inch in length is provided to the rear of the fuze threads (see paragraphs 2-323 and 2-325). When a fuze with a smaller thread diameter than the fuze normally used in HE shell is used in a WP shell, a steel adapter is brazed to the inside of the shell body, just behind the fuze threads, to accommodate the smaller fuze base. When an adapter is used, the smooth, cylindrical sealing surface is incorporated in it.

2-323. Accessory Parts Design. The burster casing (see figures 2-132, 2-133, and 2-134) is made of steel or aluminum. At the present time, all standard WP shell are designed to incorporate a steel casing, which has a smooth, cylindrical, press-fit surface corresponding to a similar surface of the shell body. The diameter of the casing surface is slightly larger than that of the shell, allowing for a press fit. When using a steel burster casing, a minimum difference in diameter (called "minimum interference") of 0.003 inch provides the seal required, whereas use of an aluminum casing requires a 0.004-inch interference. After assembly, this press-fit seal prevents leakage of the WP from the shell. Current designfavors an extruded-aluminum burster casing that can be press-fitted directly into the shell body without the need for an adapter to accommodate the fuze. This represents a considerable saving in weight and metal parts and also facilitates weight matching of the WP shell tothe HE shell.

To eliminate whipping and bending of the burster casing during firing, a circularwellis machined in the base of the shell cavity. The casing is designed to extend the complete length of the cavity to seat in the well.

2-324. WP Filler Loading. To prevent oxidation of the white phosphorous filler, the shell is filled by water displacement, and the burster casing is pressed into position with a 1/8-inch layer of water over the WP filler. The interference fit of the burster casing prevents passage of air into, and leakage of filler out of, the shell.

2-325. Sealing of Chemical (WP) Shell. The seal of these shells is achieved by a press fit and a microsurface finish of the mating parts. A lubricant is used to facilitate the press-fit operation. White lead is used as the lubricant

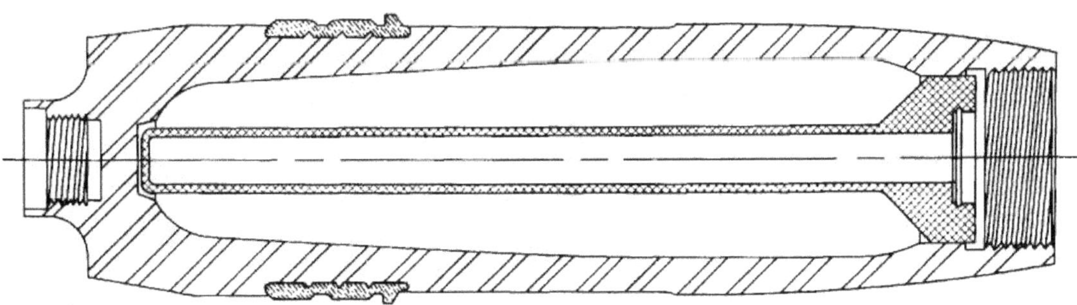

Figure 2-132. 76-mm T140E4 WP smoke shell using single-piece aluminum burster tube

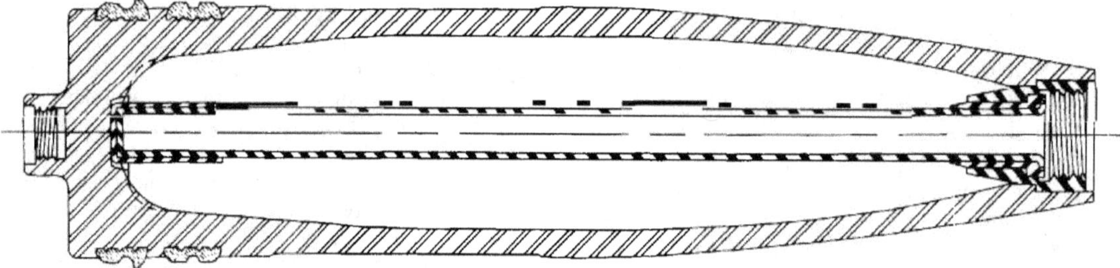

Figure 2-133. 120-mm T16E3 WP smoke shell using steel burster tube, requiring sleeve and adapter

for steel-to-steel, and molybdenum disulfide is used as the lubricant for aluminum-to-steel. Materials similar to "Molykote, type G" have been found satisfactory for aluminum-to-steel. The contact areas of both shell and casing are approximately 1/2 inch in width and receive a 32-microinch finish. The steel casing requires an adapter and sleeve, while the aluminum casing, which is still experimental, may be cast or extruded in one piece. (See paragraph 2-327.)

2-326. Comparison of Aluminum-to-Steel Closure Versus Steel-to-Steel. It is believed that for WP filling an aluminum-to-steel press-fit closure should be superior to steel-to-steel closure, provided that the aluminum burster casing can be assembled to final engagement without scoring or galling.[4] The basis for this belief is that the coefficient of expansion of aluminum is twice that of steel, and any increase in the temperature of the filled shell above the point at which the WP is molten and subject to leakage would tighten the seal by differential expansion of the aluminum. Although lowering of the temperature results in a looser fit between the casing and the shell, the WP is solidified at low temperatures and is less likely to leak. From a design standpoint, the prevention of scoring or galling appears to be primarily a matter of providing adequate radii on casing and shell at the entering edge of the press-fit surface, and suitable surface finish on the entire press-fit surfaces. Other factors that might have a bearing on prevention of scoring or galling are relative hardness of the mating surfaces and relative wall thickness of the press-fit portion of the casing and shell. All of these factors appear to have been satisfactorily met in the design of the 76-mm T140E4 shell (figure 2-132).

2-32?. Ejection Charge Design. The burster charge is a high explosive contained in a cylindrical metal tube which is inserted into the burster casing. Both 70/30 tetrytol and tetryl are in common use as burster charges in present shell. However, the former is preferred because it is less brisant and does not tend to cause pillaring of the WP cloud as does tetryl. Nevertheless, it is necessary to use tetryl in burster tubes of 1/2 inch or less in diameter because of loading and propagation difficulties. (See table 2-37.) Confinement should be kept in mind, since it plays an important part in the propagation of the

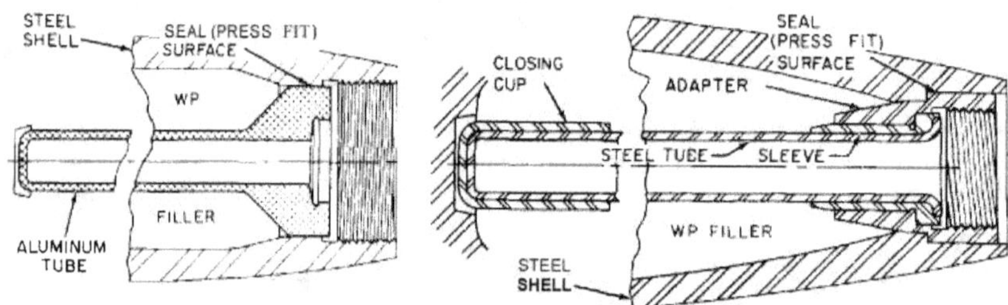

Figure 2-134. Cross sections of aluminum and steel burster tubes

detonation wave through a column of explosive.* (See table 2-8.) Table 2-37 gives the limits of propagation versus minimum column diameter known to date.

Table 2-37

Burster explosive	Limiting minimum diameter
Tetryl (pressed)	May be used in tubes 1/2-inch or less in diameter
70/30 Tetrytol (cast)	May be used in tubes larger than 1/2-inch diameter (is current standard of reference)
TNT (cast)	Not used in charges less than 1/2-inch in diameter
Baratol 67/33 (cast)	Not used in tubes less than 3/4 inch in diameter.*

*Colored marker shell use 1.1-inch column, unconfined.

2-328. Determination of Weight of Burster Charge. A formula similar to that used for the colored marker shell (paragraph 2-315) is used to approximate the weight of charge required to disseminate the WP cloud.3

$$w_c = KW\ (Y + U)e$$

where the symbols have the same meanings in both formulas except that:

K = 11×10^{-4} and yields w_c in grains

W = weight of steel components of shell, excluding base, fuze, burster casing, and tube, in lbs.

Burster charges designed by this method have functioned favorably. The quantity (Y + U)e is roughly equal to twice the strain energy absorbed by one cubic inch of steel. This equation is applicable when a tetryl charge is used, and may be applied to explosives other than tetryl by applying a correction factor to the calculated value of w_c. (See paragraph 2-315.)

*The charge used in WP and liquid-filled shells is confined in a burster tube, whereas the charge used in the colored marker shell is not so confined.

COLORED SMOKE SHELL

2-329. Introduction. Two types of colored smoke shell are used, the colored marker shell described previously, and the colored smoke shell described here. Although their terminal effects are the same, the mechanism of colored smoke generation differs for each type shell. The colored cloud produced by the former is due to vaporization and condensation of the dye, while the cloud generated by the latter is actually the result of burning. Both of these are known as signal smokes. Four colors are available: red, green, yellow, and violet. The last of these is of limited practical use.

2-330. Tactical Use. Signal smokes may be used to:

a. Identify friendly units
b. Identify targets
c. Coordinate fire
d. Control the laying and lifting of battery fire.

2-331. Terminal Effects' Limitations. (See paragraph 2-310.)

a. Setback may cause cracking of cast or pressed compositions, allowing flame to penetrate the composition and possibly cause detonation.

b. The signal visibility of burning type smoke shell has never been determined.

c. Haze is always detrimental to signal visibility.

d. The effect of background is important, but has not been quantitatively determined.

e. Qualitatively, the larger the smoke cloud the better. No density versus visibility data are available.

2-332. Shell Metal Parts Design. The colored smoke shell is designed to match the corresponding HE shell ballistically. In fact, initial design may start with a modified HE shell. In the interests of field interchangeability, the smoke shell may be identical to its corresponding illuminating shell, with the illuminant assembly merely replaced by smoke canisters (see figure 2-135). No coating or sheathing of the shell walls is needed with filler in place.

2-333. Accessory Parts Design. The smoke composition of colored smoke shell is contained in canisters within the shell body. The canisters are ejected from the base of the shell when the shell functions, and the burning dye composition is spread on the ground. Current smoke canister construction consists of a solid wall container vented at the top and bottom. The smoke mixture is ignited by quickmatch in a flash tube, which in turn is fired by a black powder initiator from the fuze. It has been found by the using services that the scattering of the canisters is undesirable. A steel canister is recommended, since steel does not react with the smoke composition. Magnesium is not recommended because it burns too fast. Currently, plastic canisters which burn at the same rate as the smoke compositions are being considered. To date all of this type shell use only a mechanical time fuze.

2-334. Filler Design. The main considerations of colored smoke compositions are treated in paragraphs **2-316** and **2-350.** However, it may be stated here that the increment boundary must either be so small that no stoppage of burning occurs, or must be provided with a material that transfers the heat of burning across the boundary from one increment to another.

2-335. Ejection Charge.

a. For smoke shell using base ejection, it is desirable to have ejection occur at a velocity equal and opposite to the forward velocity of the shell. (See paragraph **2-303.**)

b. It is difficult to determine the gas volumes produced. However, ejection gas pressure is in accordance with normal loading density versus pressure curve of black powder. (See paragraphs **4-21** through **4-51,** "Theoretical Methods of Interior Ballistics.")

2-336. Discussion of Current Designs. Figure **2-135** shows the body loading assembly of the 105-mm **M84** colored smoke shell. The using services are dissatisfied with the scattering of the canisters and the resultant dispersion of the smoke signal. It is also desired that a marker shell be capable of pinpointing a target for more accurate adjustment of HE fire. Another problem is that of sensitivity of the burning type smoke composition. The yellow smoke in particular has been borderline, and has given numerous prematures. With the low order detonation of the black-powder ejection charge and the smoke mix the prematured shell does not severely damage the howitzer. The components are not believed to separate until they emerge from the muzzle, and the shell body does not rupture.

PROPAGANDA SHELL

2-337. Introduction. The role of psychological warfare has become increasingly important. Strategic use is made of warning leafletsplaced in the target area in preparation for heavy artillery attack. Surrender leaflets are effectively distributed by the same means.

The purpose of the propaganda shell is to disseminate information-bearing leaflets over a

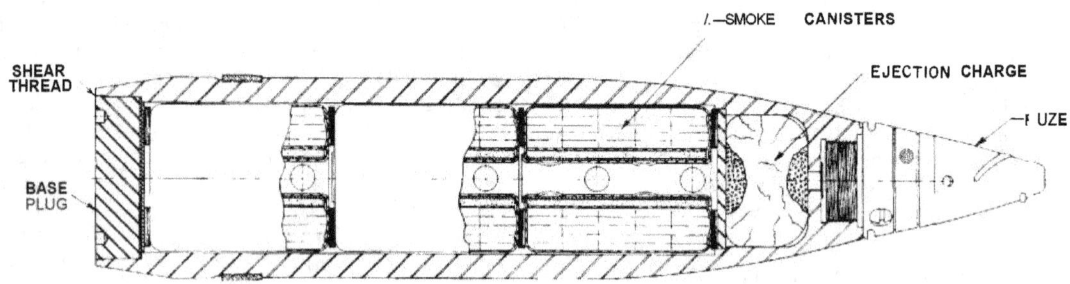

Figure 2-135. 105-mm M84 colored smoke shell

specified area. The area may range in size from an isolated position a few feet in diameter to something as large as a town or village.

2-338. Shell Metal Parts Design. The shell metal parts used for propaganda shell are the same as those used for the colored smoke or illuminating shell (figures 2-135 and 2-136). In the former case, smoke canisters are removed and replaced by propaganda leaflets, and in the latter case the illuminant assembly is replaced by the leaflets.

2-339. Accessory Parts Design. Up to the present time, this type of shell has been designed to function with a mechanical time fuze. It is expected that future design will consider the VT fuze.

2-340. Filler Design. To be carried in an artillery shell, the leaflets may be of any suitably practical shape. Rectangular 3 inch by 4 inch or 5 inch by 8 inch paper sheets have been found suitable to be rolled into tubes and inserted into shell. Circular sheets of a diameter equal to the I. D. of the shell may also prove suitable. This shell does not require special containers for the leaflets, but:

a. The positioning of the leaflets in the shell must provide resistance to setback and must eliminate wrinkling of the leaflet rolls. In this respect, a single roll of leaflets may be damaged by its own setback weight. For example, the setback weight of a 1.5-lb roll of paper at 10,000 g's is 1.5 x 10,000 = 15,000 lb.

b. A suggested method of reinforcing leaflet rolls is to insert a wedge in the core of the roll, which will automatically tighten the roll on setback.

2-341. Ejection Charge. The contents of the shell must be capable of shearing the baseplug. To deliver the load into the air, sufficient powder must be used to:

a. Break the shear pins or threads.

b. Provide plus 20 percent excess powder as a design safety factor. (See paragraph 2-303.)

2-342. Discussion of Existing Designs. The 105-mm M84 propaganda shell is simply the M84 colored smoke shell (figures 2-135 and 2-136) that is requisitioned in the field, usually by the Psychological Warfare Service, and converted into a propaganda shell. The base plug is unscrewed, the smoke canisters removed, and the leaflet rolls stuffed into the cavity.

Effective use has been made of the M84, but it has not proved very efficient, since approximately half of the leaflets remain crimped together instead of being dispersed over an area. The reason is not hard to find. The setback weight of the front roll is sufficient to severely crimp the rear roll. Even a single roll of leaflets is slightly damaged by its own setback weight. The damage appears as a granular network of cracks with the cracks increasing in number toward the rear of the shell.

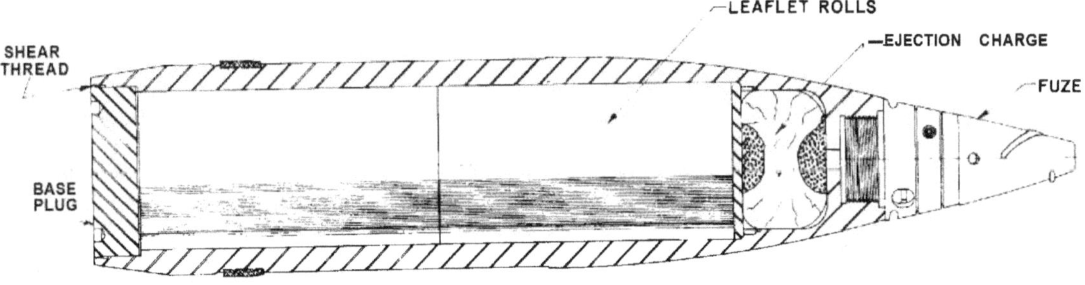

(FIELD **ASSEMBLY)**

Figure 2-136. 105-mm M84 propaganda shell

2-343. Direction of Future Design. The 105-mm T107 propaganda shell (figure 2-137) uses the same shell metal parts design as the T107 illuminating shell, but the illuminating canister is replaced by the propaganda leaflets and accessories. The illustration shows the use of split sleeves to support a steel disk which prevents the setback weight of the front roll of leaflets from damaging the rear roll of leaflets. Thus, the chief problem encountered in the M84 propaganda shell, previously described, has been eliminated. Obviously, when the weight of the paper is multiplied by the 10,000-g load, the paper is stressed beyond its column strength and will buckle in every possible direction until it can move no further. Initial success has been obtained with two wedge designs: one is a centercore filling of loose sand, the other is a wooden-core assembly consisting of a tapered dowel that is driven into a hollow split plug to give an initial tightening action to the roll during assembly.

LIQUID-FILLED SHELL

2-344. Tactical Requirements and Terminal Effects Limitations. Postwar development of liquid fillers required new chemical shell for various weapons. The ultimate objective is the design of a liquid-filled shell to be used for the optimum dispersion of persistent and non-persistent gases. The achievement of the desired terminal effects is limited by many factors, all of which have not yet been established. Those which have been recognized include the following.

a. Erratic flight due to variation between rotational velocities of the metal parts and of the filler.

b. Special fuzing (not yet available) is required to make the shell function fast enough to preclude cratering.

c. Because of alterations in the shell due to shell design and manufacturing difficulties, the shell fail to match the HE shell weightwise. Thus, the range of the heavier liquid-filled shell is less than the range of the lighter HE shell.

2-345. Shell Metal Parts Design. The shell is designed to match the ballistics of standard HE shell as closely as possible. The liquid-filled shell uses the shell metal parts of standard chemical or of standard HE shell with modifications to provide for fuzing and for sealing of the liquid filler. One example of a liquid-filled shell is the shell with the GB filler. The T77 (M121) shell, made from a modified 155-mm HE shell, was established as a prototype (see figure 2-138). The internal contour of the shell was changed from conventional hemispherical shape to a flat bottom with side walls conforming to the dimensions of the flat base of the burster casing, thus preventing whipping of the casing in flight. To prevent leakage of the filler, one-piece constructions without any brazed fittings were recommended, such as a body adapter integral with the shell body, and one-piece burster casing construction.

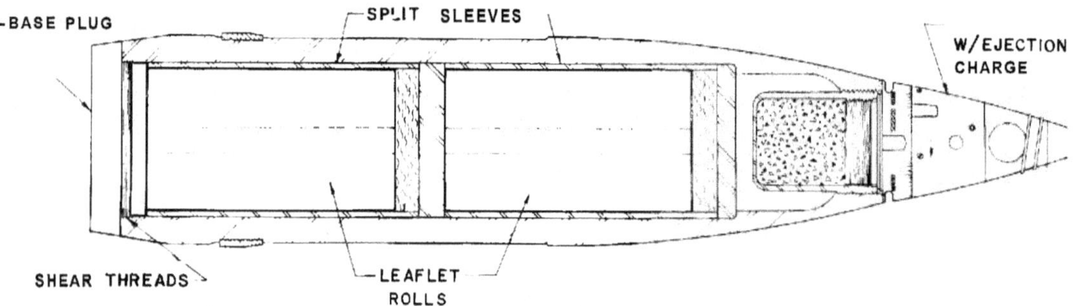

Figure 2-137. 105-mm T107 propaganda shell

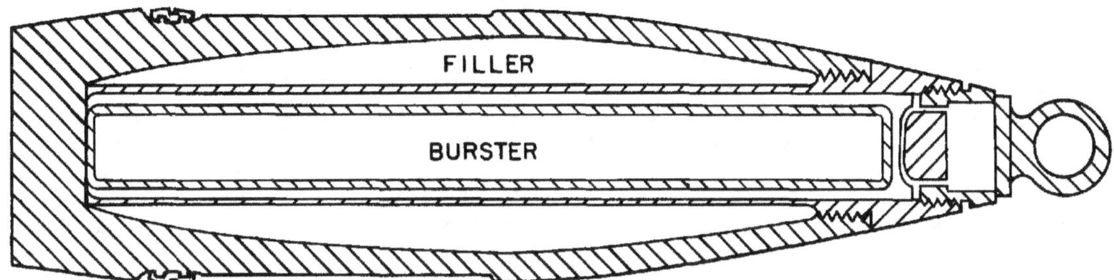

Figure 2-138. 155-mm M121 chemical shell

2-346. Accessory Parts Design. The design of the liquid-filled shell burster casing is similar to the burster casing used in the WP shell, except in size. To prevent the burster casing from whipping, in the liquid-filled shell the casing fits snugly between the shell walls at the base of the shell. The modification from HE shell consisted principally in providing a new nose adapter to fit the increased burster size required for proper dissemination of the filler, and of machining the interior base of the shell to satisfy the close tolerance required between the bottom of the burster casing and the wall of the shell.

The same considerations of construction of the burster casing apply as in the WP shell; however, the sealing requirements are more stringent. The interferences of all present or anticipated shell sizes are tabulated below. All contact surfaces are steel to steel, that is, steel casing to steel shell, and have a 32-microinch finish.

105-mm	155-mm	8-inch
0.003 to 0.005	0.003 to 0.005	0.010 to 0.012

2-347. Filler Design. GB agents are odorless, colorless, water-like liquids.[1]

2-348. Burster Charge. Sufficient charge is required to open the shell and to disseminate the liquid, which is either in the form of a persistent or a nonpersistent gas.

2-349. Direction of Future Design. In order to match HE and liquid-filled shell weightwise, components in the latter shell will tend toward lighter-than-steel metals. Improved closures are being developed to reduce any occurrences of leakage.

THE CHARACTERISTICS OF PYROTECHNICS COMPOSITIONS

2-350. Introduction. Previous portions of this section have described the design of pyrotechnic items from the engineering viewpoint. Without minimizing the important engineering aspects of pyrotechnics, it is obvious that the most carefully designed item cannot accomplish its purpose without an equally well designed pyrotechnic composition. Therefore, a discussion of the chemistry of pyrotechnic compositions is in order.

2-351. Constituents of Pyrotechnic Compositions. The constituents generally employed in pyrotechnic compositions are listed below; they are classified as (1) oxidizing agents, (2) fuels (or reducing agents), (3) color intensifiers, (4) retardants, (5) binding agents, (6) waterproofing agents, and (7) dyes for smokes.

a. Oxidizing Agents include nitrates, perchlorates, peroxides, oxides, chromates, and chlorates. These are all substances in which oxygen is available at the high temperature of the chemical reactions involved.

b. Fuels include metal powders, metal hydrides, red phosphorus, sulfur, charcoal, boron, silicon, and silicides. When these substances are finely divided, they readily undergo an exothermal oxidation with the formation of the corresponding oxides and the evolution of radiant energy.

c. Color Intensifiers are mainly highly chlorinated organic compounds, such as hexachloroethane (C_2Cl_6), hexachlorobenzene (C_6Cl_6), polyvinylchloride, Arochlors.

d. Retardants include inorganic salts, plastics, resins, waxes, oils. These are used to slow down the reactions between the oxidizing agent and the powdered metal to control the burning rate. Some behave merely as inert diluents while others participate in the reaction at a much slower rate than the main constituents.

e. Binding Agents include resins, waxes, plastics, oils. These are added to prevent segregation and to obtain more uniformly blended compositions. They also serve to make finely divided particles adhere to each other when compressed into pyrotechnic items, and help to obtain maximum density and efficiency in burning. Binders frequently desensitize mixtures which are otherwise sensitive to impact, friction, and static.

f. Waterproofing Agents include resins, waxes, plastics, oils, dichromating solutions. These are used as protective coatings on metals (such as magnesium) to reduce their reaction to atmospheric moisture.

g. Dyes for Smokes, such as azo and anthraquinone dyes.

Many of the above substances perform more than one function, thus simplifying the composition of some pyrotechnic mixtures.

2-352. Properties of Typical Pyrotechnic Compositions. Most pyrotechnic compositions can be defined as physical mixtures of finely-powdered compounds and elements, which upon ignition readily undergo chemical reactions in which a considerable amount of heat, light, smoke, and/or sound are produced in a relatively short period of time. The amount of heat evolved may vary from as little as 200 calories per gram for a delay fuze composition to 2,500 calories per gram for a photoflash composition. The reaction temperatures attained may vary from 200°C for smoke compositions to well over 3,500°C for photoflash mixtures and metal dust flashes. A comparison of some of the properties of typical pyrotechnic compositions with similar properties of a few of the better known explosives is given in table 2-38. Flares, signals, smokes, tracers, and illuminating shell are pyrotechnic compositions, which are generally pressed into candle cases and which burn in cigarette fashion from one end to the other at relatively slow burning rates. Thus, while the amount of energy at the upper limit is considerable and often extremely dangerous, it is not released in so destructive a fashion as is the energy of an explosive. Furthermore, the amount of gaseous products from the burning pyrotechnic compositions is appreciably less than that obtained from explosives. For photoflash munitions and spotting charges, however, loose pyrotechnic compositions are employed which may be extremely dangerous and may react with destructive violence but not with the force that explosives exhibit, as indicated by the brisance values. The pyrotechnic compositions do not have the sensitivity to heat that the explosives do, as shown by their higher ignition temperatures. As for the impact values, some of the pyrotechnic: compositions appear to be as sensitive as the explosives. Although radiant energy is emitted from most pyrotechnic compositions in the ultraviolet, visible, and infrared regions of the spectrum, until recently only radiation in the visible region (from 4,000 to 7,000 Angstrom Units) has been utilized in flares, signals, photoflashes, and tracers.

2-353. Required Characteristics of Pyrotechnic Compositions. The important characteristics required of pyrotechnic compositions are shown in table 2-39.

Luminous intensity (candlepower), burning rate, and color value are the usual. military requirements which must be met in pyrotechnic compositions used for illuminating and signalling purposes. Sensitivity to impact, static, and friction should be minimum for safety, while ignition temperature, ignitibility, stability, and hygroscopicity are important in determining the certainty of functioning. Standard tests have been developed for measuring these characteristics.

2-354. Factors Affecting: Light Characteristics and Stability. Some of the factors which affect the light characteristics and stability of pyrotechnic compositions are shown below.

a. Granulation of ingredients

1. Average particle diameter
2. Specific surface

Table 2-38

Comparison of some properties of pyrotechnic compositions with explosives

Composition		Heat of reaction cal/gram	Gas volume cc/gram	Brisance grams sand crushed	Ignition temperature °C*	Impact RM: cm:	PA: inch
Delay Barium chromate, % Boron, %	90 10	480	13	0	650	...	12
Delay Barium chromate, % Zirconium-nickel alloy, % Potassium perchlorate, %	60 26 14	497	12	0	485	56	23
Flare Sodium nitrate, % Magnesium, % Laminac, %	38 50 5	1456	74	8	640	60	19
Smoke Zinc, % Potassium perchlorate, % Hexachlorobenzene, %	69 19 12	616	62	8	475	23	15
Photoflash Barium nitrate, % Aluminum, % Potassium perchlorate, %	30 40 30	2147	15	7	700	100	26
Black powder		684	272	8	288	32	16
TNT		1060	1000	48	475	100	14
RDX		1240	600	60	260	13	5

* 5-second value

Table 2-39

Important characteristics of pyrotechic compositions

Military	Safety	Certainty of functioning
Luminous intensity	Sensitivity to impact	Ignitibility
Burning rate	Sensitivity to friction	Stability
Color value	Sensitivity to static	Hygroscopicity
(Smokes) Color visibility	Ignition temperature	Heat of reaction
		Efficiency

3. Particle shape
4. Particle size distribution

b. Burning surface area
c. Purity of ingredients
d. Flare case material and shape
e. Loading pressure
f. Presence of moisture
g. Degree of confinement during combustion

The average particle diameter, specific surface, shape, and distribution affect the burning rate and luminous intensity. The burning surface area will influence the total luminosity. The purity of ingredients and the presence of moisture are important to the shelf life of the stored composition. The type of flare case material and its shape will affect the burning efficiency. The loading pressure and degree of confinement will influence the burning rate.

In addition, the heat of reaction and the burning rate of composition are of fundamental importance, since sufficient heat must be evolved to make the composition burn propagatively and the rate of reaction must be rapid enough to more than compensate for heat losses.

No one of these factors can be said to be more important than any of the others, and all must be given careful consideration when formulating a pyrotechnic composition to meet specific requirements.

In any composition, the building block is a combination of finely-powdered oxidant and fuel; therefore, the chemistry of the mixture is primarily the chemistry of the reaction of these two types of ingredient. For an illuminant that will give a yellow flame, sodium nitrate and magnesium are used; for a green flame the sodium nitrate; and for a red flame composition, strontium nitrate is used.

2-355. Factors Affecting Luminous Intensity. Since the luminous intensity of these compositions seems to depend primarily on the amount of magnesium present, one would expect that the intensity values for all the compositions would be approximately the same. It is known that the values are not the same, but range from 119,000 candles to 780,000 candles for the optimum combinations of potassium nitrate and sodium nitrate, respectively, with magnesium. The difference in these luminosity values can be explained by the contribution that the metal of the oxidant makes. It should be noted that certain salts, when heated to excitation, give emissions in the visible region. Since sodium has the most intense lines in the visible region, it might be expected to contribute most to the luminous intensity. Potassium, which has practically no emission in the visible region, contributes little to the total luminosity. Qualitatively, this is borne out by the results obtained. The other oxidants are intermediate in their contribution.

2-356. Burning of Pressed Compositions. The manner in which these pressed compositions burn can best be shown by reference to figure 2-139. When a pressed composition is ignited, several things occur very rapidly. The composition is raised to its ignition temperature and, if conditions are favorable, it will continue

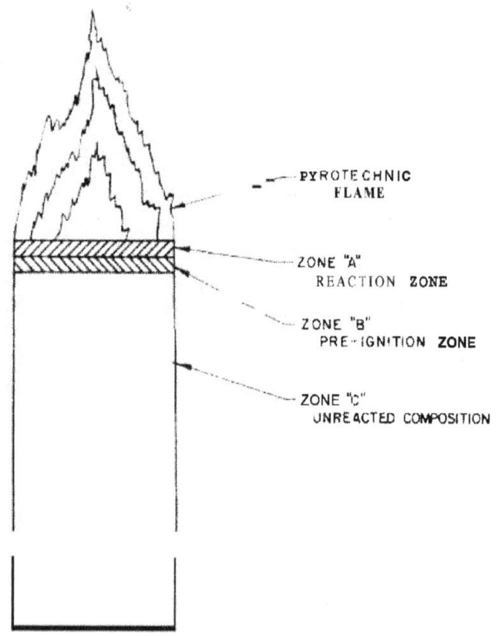

Figure 2-139. Pyrotechnic flame burning zones

to burn propagatively. From experimental evidence, three distinct zones have been established. In zone A the ingredients areundergoing exothermal chemical reactions, resulting in a volatilization of the excess fuel, which reacts with the oxygen in the air to form the oxide and to give a luminous flame. In the case of magnesium, the nitride will also be formed. At the same time, part of the radiant energy is preheating the composition directly beneath it, shown in zone B, which can be called the pre-ignition zone. Analysis of the radiant energy indicates that less than 10 percent is in the visible region and the remainder is in the infrared region. Directly beneath zone B is the remainder of the unreacted composition, or zone C. The rate at which the mixture will burn will depend on the heat evolved by the composition, the rate at which heat is evolved, the particle size of the ingredients, their thermal conductivity, the degree of consolidation, the type of container, and diameter of charge, and the state of the reaction products. Obviously, sufficient heat must. be produced by the mixture to heat the ingredients to a state of exothermal reaction, and the rate must be sufficiently rapid to more than compensate for heat losses, and to make the composition burn propagatively.

2-357. Properties of Aluminum and Magnesium-Aluminum Fuels. In the preceding discussion, the only fuel stressed was magnesium. Binary mixtures similar to those prepared with magnesium have been prepared with other fuels such as aluminum, and magnesium-aluminum alloys. They have not given as high a luminous intensity as those in which magnesium was used. Since aluminum has a higher heat of reaction with oxygen than magnesium, it might be expected to give a greater luminous intensity. A comparison of the physical properties and thermal parameters of these two metals, shown in table 2-40, will indicate why aluminum is not as efficient as magnesium.

Table 2-40

Some physical properties and thermal parameters of magnesium and aluminum

	Metal		Oxide		
Element	M.P. (°C)	B.P. (°C)	M.P. (°C)	B.P. (°C)	(Kcal)
Mg	651	1,120	2,102	3,077	143.8
Al	660	2,450	2,027	3,627	399.6

Element	AHE (Kcal)	ΔH (Cal/gm of metal)	AH (Cal/ml of metal)	Adiabatic temp. (°C)	Latent heat vaporization gm)
Mg	71.9	2,900	1,700	12,000	1,800
Al	66.6	3,700	1,370	13,000	3,800

Although the melting points of the two metals are quite similar, the boiling points are markedly different, that of aluminum being much higher. In addition, the amount of heat required to vaporize aluminum is twice that required for magnesium. Therefore, less aluminum is vaporized for the same amount of heat evolved by the composition, giving a smaller flame and consequently less luminous intensity. In addition, the boiling point of aluminum oxide is higher than that of magnesium oxide. The higher temperature in the flame produced by the condensation of the aluminum oxide vapors produces flames of higher temperature and a lower value.

2-358. Effect of Specific Surface of Reactants. The chemical law of mass action expresses reaction rate as a function of the concentration of the reactants. In the solid state chemistry of pyrotechnics, the specific surface of the ingredients may be considered roughly as their concentration. This is usually expressed as cm^2 per gram or M^2 per gram of material.

The equation for specific surface can be written as

$$S = \frac{6 \times 10^4}{Dd}$$

where

S = specific surface in cm^2/gram
D = density, gm/cm^3
d = diameter, microns.

This equation is based on the assumption that the particles are spheres. Any deviation from sphericity, as in a particle that has cracks or fissures through it, will result in a specific surface value greater than that for a comparable sphere.

In figure 2-140 are shown three varieties of finely divided magnesium which have exactly the same granulation, namely, 100 percent through a No. 50 U. S. Standard sieve and 100 percent on a No. 60 sieve. The top sample is milled, the one on the lower right is ground and the one on the lower left is atomized spherical magnesium.. The spherically shaped magnesium has the smallest specific surface, the highest purity, the highest apparent density, and the greatest resistance to atmospheric moisture. The effect of specific surface or average particle size and the shape of the particle on the performance of a typical composition is illustrated by the example in table 2-41.

Table 2-41 shows the burning characteristics of similar compositions prepared with ground and atomized magnesium.

It can be seen from this data that the ground magnesium which has the greater specific surface gives the greater luminous intensity and the faster burning rate. Generally, it is

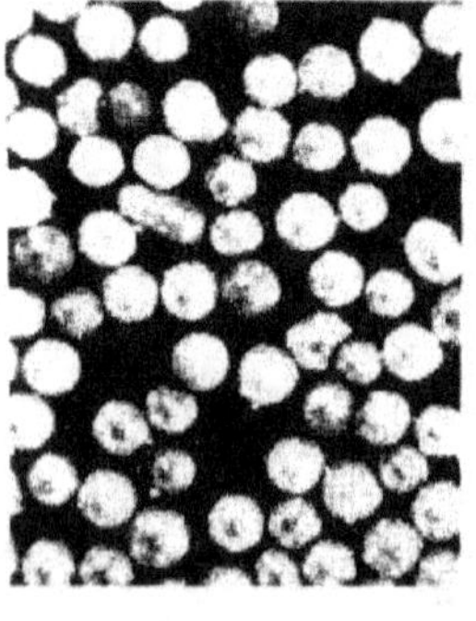

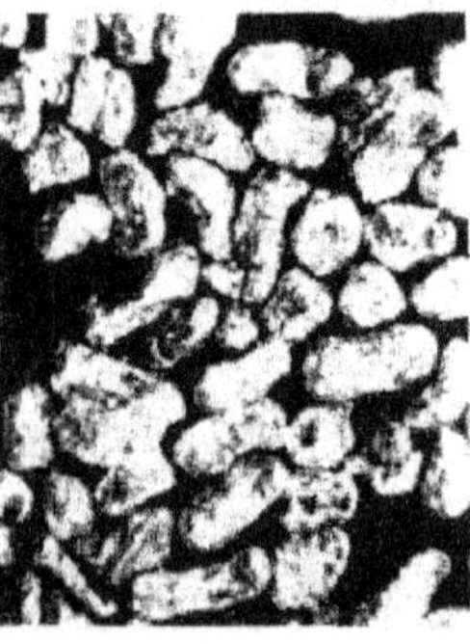

Figure 2-140. Magnesium particle sizes

Table 2-41

Effects of change in specific surface of magnesium

Composition (percent by weight)		
Ground magnesium	66.6	...
Atomized magnesium	...	66.6
Sodium nitrate	28.6	28.6
Resin	4.8	4.8
Characteristics of specified composition		
Luminous intensity (candles per sq in.)	200,000	178,000
Burning rate (in. per min)	9.4	5.7
Density	1.56	1.65
Efficiency (candle-seconds per gm)	50,000	65,200

true for all pyrotechnic compositions that, as the specific surface of the ingredients increase, the luminous intensity and the burning rate increase. It should be noted, however, that in this instance the efficiency of the composition containing the atomized magnesium is approximately 28 percent higher. The performance characteristics of any pyrotechnic composition are similarly affected by the specific surface of the oxidant as well as by that of the fuel.

2-359. Effect of Moisture on Shelf Life. A very important point to be considered when developing a composition is the shelf life or stability of the mixture. The most detrimental factor to the stability of a composition is atmospheric moisture. In the presence of moisture the oxidant will corrode the finely powdered metal and coat it with a layer of either the metal hydroxide or the metal oxide or with both. This layer reduces the ignitibility of the metal, with the result that it either fails to ignite or functions with reduced luminous intensity. For oxidants, the sensitivity to moisture can be determined by obtaining the critical relative humidity at room temperature. The higher the critical relative humidity, the less sensitive the substance is to moisture. (Roughly, the higher the critical relative humidity the less soluble the oxidant.) By subjecting samples of the salt to atmospheres of known relative humidity and determining the weight gain, the critical relative humidity of the salt can be obtained. Small traces of impurities in the oxidant will lower its critical relative humidity.*

The effect of moisture on a finely powdered metal is determined by placing a sample in distilled water and maintaining the system at a specified constant temperature. By collecting the gas evolved at constant pressure and noting the rate of evolution, the rate of corrosion of the metal can be established. For atomized magnesium it was found that the rate of corrosion in water was an exponential function during the initial stages.

$$t = Ae^{N/K}$$

where

t = time
A = constant
N = amount of H_2 formed
K = constant.

*Until recently, there was a reluctance to use sodium nitrate in pyrotechnic compositions because of its low critical relative humidity. In fact, specification grade sodium nitrate, which is a commercial grade, has a critical relative humidity of less than 50 percent at normal temperature. Therefore, compositions employing this grade were very hygroscopic. This was partially overcome by the use of commercially available U. S. P. double-refined sodium nitrate, which has a critical relative humidity of 75 percent at normal temperature or the same as the purest nitrate.

In other words, the corrosion proceeds at an increasingly faster rate with time. However, over the temperature range of 30-65°C there was only a small increase in the rate of corrosion. Atmospheres of nitrogen and oxygen had slight influence; hydrogen had a retarding affect; and carbon dioxide had an accelerating effect. The pH of the solution and the specific surface of the metal dust were found to affect the results materially. The smaller the particle or the greater the specific surface the faster the rate of corrosion.

2-360. Protection Against Effects of Moisture. The deleterious effect of moisture on finely divided magnesium can be reduced by coating the metal with a chromate film. If the powdered magnesium is treated in a hot acid bath of potassium chromate for a short period of time, and then washed thoroughly to remove all traces of acid, a thin layer of chromate remains, acting as an effective protective film against moisture.

To further reduce the effects of moisture, only the purest ingredients are used. These are processed in air-conditioned rooms and, when pressed into the candle cases, are covered with a nonhygroscopic first fire that further protects the composition. In addition to the use of pure materials, self-hardening resins are employed in place of linseed oil, which was previously used as a binder and a coating agent for the magnesium. Because of the slow continuous oxidation of linseed oil during storage of the pyrotechnic item, the compositions were found to harden and change in their burning characteristics. Polyester resins containing some styrene, such as Laminac, which are self-hardening at room temperature when a catalyst like Lupersol is added, are now used in place of the linseed oil. They polymerize at room temperature within a relatively short period of time, depending on the amount of catalyst; hence, very little change occurs in the burning characteristics of the composition upon storage.

2-361. Heat Sensitivitv of Pvrotechnic Compositions. An understanding of the effect of heat on the behavior of a pyrotechnic composition is obtained when a small sample of the mixture is heated and the time to ignition measured. It has been found that the time to reaction decreases with increasing temperatures. If the logarithm of the time to reaction is plotted against the reciprocal of the absolute temperature, a straight line is usually obtained. This indicates that the reaction rate of the composition is an Arhennius function. This type of plot enables a comparison of various compositions with respect to their heat sensitivity. By means of this technique (which was first applied to explosives and pure compounds and only recently to pyrotechnic compositions) a better understanding of the reaction mechanism of pyrotechnic mixtures is being arrived at.

2-362. Tracer and Igniter Compositions. What has just been said about flare compositions applies equally well to mixtures used in tracers. A standard tracer composition, **R-45,** consists of 56 percent strontium nitrate, 37 percent magnesium-aluminum alloy, and 7 percent polyvinyl chloride. Although strontium nitrate has been the principal oxidant for tracer compositions, sodium nitrate is now being used experimentally when greater luminous intensities are required.

Another type of pyrotechnic mixture that is important is the igniter composition, which includes first fires. As the name implies, the function of this composition is to ignite the main charge. The igniter is formulated to produce heat and slag rather than to give a high luminosity. In fact, the igniter may have relatively little flame. Zirconium and titanium or their hydrides and silicon are used in place of magnesium or aluminum. The former metals are more resistant to moisture and in addition are very ignitable, with high heats of reaction. The igniter, or first fire, must be capable of igniting those compositions with which it is in contact. Since this mixture is placed between the exposed end of the main charge and the atmosphere, it is advantageous to employ an igniter which is nonhygroscopic. Such a composition, when pressed over the exposed surface of the main charge, will act as a protective covering for the more moisture sensitive main charge. A first fire having many of these desirable properties has been in use for a number of years and has been found to be very satisfactory. This composition, which is commonly called FF-101L, contains barium nitrate, zirconium hydride, silicon, and TNC (tetranitrocarbazole) is added to increase the ignitability of the composition.

2-363. Radiation Effectiveness of Pyrotechnic Compositions.[7] To be suitable for military signalling and illuminating, pyrotechnic compositions are designed to meet the following considerations (table 2-42).

Table 2-42

Parameters	Units
Luminous intensity	Candlepower
Burning rate	Inches per minute
Color value (color saturation)	Ratio of apparent luminosity through an appropriate colored glass filter to the total luminous intensity
Efficiency	Candleseconds per gram (or per milliliter) of composition (can also be expressed as candles per square inch of burning surface)

The general rule is to design for visibility under the worst possible condition, that is, maximum sky brightness. The following nomogram (figure 2-141) is used to obtain a rough approximation of candlepower requirements to observe signals at various distances. To use the nomogram, the curve is selected that represents maximum sky brightness (10^3 foot-lamberts) and through an abscissa equivalent to the specified target distance, say 15,000 yards, a line is drawn from an ordinate corresponding to the minimum visibility, say 15,000 yards, on the meteorological range axis, and prolonged to the intensity axis. The meteorological range is assumed to be at least equal to the target distance. In the example cited, the required luminous intensity would be about 2.5×10^6 candles.

2-364. Summary. Space has not permitted a more detailed discussion of the chemistry of pyrotechnics. To summarize, it can be said that when a new item is developed or an old one improved no one factor can be considered the most important. All the factors discussed must be taken into account. The ingredients must be of high purity and processed under carefully controlled conditions. The particle sizes and proportions of the ingredients must be adjusted and balanced to give the desired burning characteristics. The type and size of container must be selected to obtain the optimum results with the compositions. Finally, provisions must be made to seal the composition adequately so that moisture will be excluded. All of these considerations require the scientific approach and the full exploitation of chemical, physical, and engineering principles.

The results of field tests5 to determine the visibility threshholds for signal color lights over a two-mile range are summarized in table 2-43. Since the colored filters used in these tests did not duplicate the color emission of burning flares, the results may only be correct in order of magnitude, assuming wavelength attenuation over a two-mile distance to be similar to that experienced at a great distance (about 5 to 10 miles).

Table 2-43

Minimum source candlepower visible from two-mile distance

		Filter		
	White	Red	Yellow	Green
Candlepower	10,200	10,500	13,200	23,600
Ratio to white	1.00	1.03	1.29	2.32

PYROTECHNIC PARACHUTE DESIGN

2-365. Introduction. Parachutes are airbrakes used to reduce the free-flight velocity of a body to a desired low value. Parachute-type airbrakes are aerodynamic shapes, fabricated usually from textiles into umbrella-like canopies that are inflated and maintained in shape by the pressure of entrapped air. This section will discuss generally the external and internal parameters necessary for design. More detailed information on the more complex parameters will be found in references 2, 6, 9, 10, and 11.

2-366. Concept of Optimum Height in the Design of Parachute-Supported Flares.[6] The reference covers the derivation of optimum height of burning flare and the design of a parachute-supported flare, making use of the concept of optimum height.

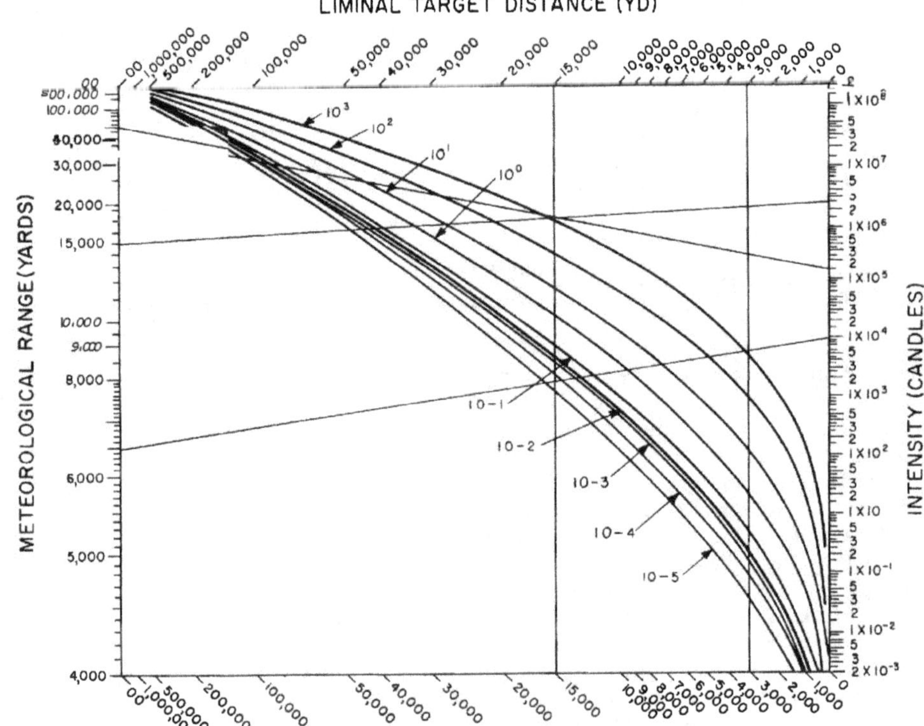

NOMOGRAM FOR THE VISUAL RANGE OF LIGHTS BASED ON THE TIFFANY DATA (PROBABILITY OF DETECTION 95 PERCENT). THE NINE CURVES REFER TO DIFFERENT VALUES OF BACKGROUND LUMINANCE IN FOOT-LAMBERTS, FROM 10^3 TO 10^{-5}

Figure 2-141. Meteorological range versus light intensity

This reference concludes that to obtain the maximum illumination of the periphery of a circle of ground radius R_o (figure 2-142), with the center of the circle directly under the flare, the flare should burn as closely to the optimum height h, which is 0.7 R_o, as possible. This reference also states that the minimum candlepower required for this radius will be at the height of 0.7 R_o and will be equal to 2.58 $F_c R_o^2$, where F_c is the number of foot-candles of illumination desired at R_o.

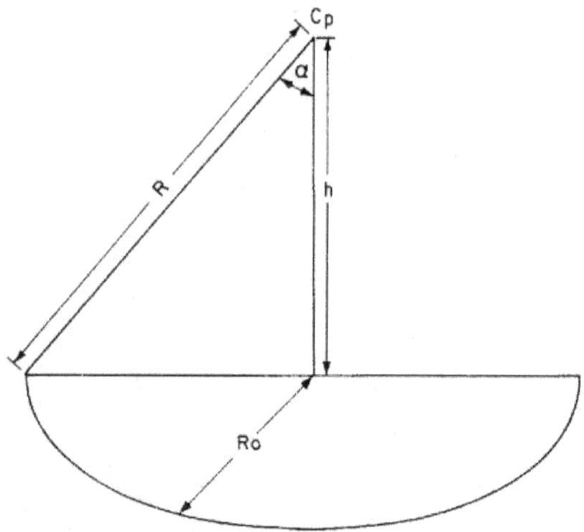

Figure 2-142. Optimum height for area illumination

2-367. Types of Parachutes. The major types of parachutes, their construction features, and their characteristics are briefly outlined as follows.

a. A Standard Flat Parachute is constructed from a number of wedge-shaped gores, each with an apex angle equal to 360° divided by the number of gores; thus, it forms a flat circular canopy when it is uninflated. Usually it contains a small vent in the crown. Inflated, it assumes a near-parabolic shape, with a diameter of about 3/4 its constructed diameter. Aerodynamics: usually the quickest opening type of parachute, with a very good coefficient of drag, but less stable than most shaped parachutes. Examples are standard paratroop and lifesaving parachutes; it is also used on aircraft flares and most illuminating shell.

b. Shaped Parachutes are constructed from a number of shaped gores that depart slightly from the wedge shape to form an uninflated geometric shape with a curved skirt. Inflated, a shaped parachute forms a shape varying from elliptical to spherical, depending upon the particular design. Aerodynamics: takes advantage of shape €or emphasis on stability, shock load, or other parameters of design. Examples of shaped parachutes are the baseball type, where the fabric is shaped into a half sphere to relieve stress at the crown; British R type, where geometry is designed to equalize stresses at all points as if the fabric were inextensible; ribbon-type, where concentric rings of ribbon (tape or webbing) are bound together in a canopy, for ruggedness, to allow a sort of geometric porosity to function where fabrics such as canvas would preclude any advantage of porosity; airfoil type, designed for maximum aerodynamic efficiency.

c. Parasheet is constructed from a single piece of material, or a patchwork of as few pieces as its size will allow, in order to avoid the complex and costly operations of sewing many gores together. It is usually constructed to lie flat, but is sometimes shaped by gathering the hem. Its inflation is similar to that of parachutes. Aerodynamics: similar to parachutes, but usually weaker than an equivalent parachute because of a lack of symmetry of weave in its inflated shape. Examples: octagonal, hexagonal, square, and triangular.

2-368. Factors Affecting Parachute Design. Parachute design is affected by many factors, some of which are briefly outlined as follows.

a. Weight. The combined weights of all parts of the free-flight system are equal to the drag force during steady-state suspension. The weight of a flare is reduced during burning. The equation of motion for vertical descent is

$$W - (1/2) \rho C_D A V^2 = (W/g)\ (dv/dt)$$

b. Velocity. See figure 2-144 for velocity curve and definitions. The velocity squared is proportional to the drag force. V_1 is controlled by design of the packaged item. V_2 is affected by the opening of the package. V_3 is reduced by drag during squidding. $V_4 = V_c$ when squidding occurs first. $V_5 = Z$, a special symbol for terminal velocily (steady-state suspension).

c. Drag Coefficient. See figure 2-143 for effects. In order of magnitude of C_D, types of parachutes are: flat parachute, shaped parachute, ungathered parasheet, and gathered parasheet.

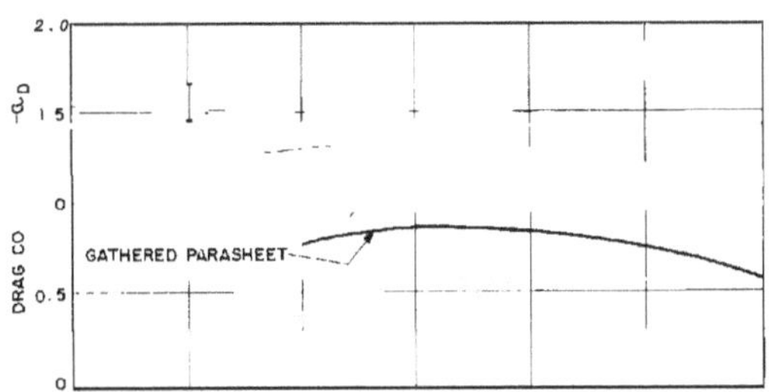

d. Air Density. See subparagraph 2-370c for effects.

e. Diameter. See figure 2-144 for illustration; dimensional changes at various stages during functioning.

f. Length of Lines. Due to angularity of lines, a force component tends to collapse the skirt of the parachute; increased length reduces tendency to collapse; minimum L/D = 1, which value is commonly used to avoid excess bulk of lines.

g. Number of Lines. Affects parachute opening characteristics; more lines give more control of opening.

h. Deployment Methods. "Canopy first" imposes greater shock loads than "lines first"; "canopy first" allows the parachute to start opening before lines are fully extended, which allows the possibility of coincidence of F_S and F_O. (See figure 2-144.)

i. Reefing. Used for control of sequence of deployment and also to control shape of canopy before and/or after inflation. Reefing devices include: elastic line loops, "chain stitch" shortening of lines, short taschengurts and pocket vents, skirt bands, deployment bags, and center cords.

j. Porosity. Porosity refers to the amount of air that can pass through a fabric at a given pressure. It is a quality of canopy textiles that affects squidding, opening, and stability characteristics; for extensible fabrics (nylon has about 30 percent elongation before breaking), porosity varies with variation of drag force.

2-369. Stages in the Opening of a Parachute. Figure 2-144 shows a typical parachute suspension system. The performance sequence is briefly described by stages as follows.

a. Stage 1 represents the packaged parachute at the instant of release from its container. It represents the last known velocity before drag force of the airstream takes over.

b. Stage 2 represents the first step of deployment, where the lines are taut and absorb the inertia force of the bulk parachute before opening.

c. Stage 3 represents the squidded shape when the velocity is above V_c. Drag force is considered to be one-tenth the equivalent for the fully open parachute. Diameter is considered to be one-fourth ($D3 = D_5/4$). Stability of this shape is due to balance of air pressure flow outward through the pores against the external air pressure impact.

d. Stage 4 represents the critical opening velocity of the parachute where the squid balance is destroyed and the canopy "jibes."

e. Stage 5 represents the steady-state suspension where suspended weight (of the system) balances the drag force of the canopy. Diameter is from six-tenths to nine-tenths the constructed diameter, depending on shape of canopy.

f. Stage 6 represents the constructed size and shape of the parachute (uninflated).

2-370. Forces Acting When Parachute Opens. The drag formula states

$$F = \frac{1}{2}\rho C_D A V^2$$

where

- F = drag force on the parachute at a given instant
- ρ = density of air at the altitude of the parachute
- C_D = drag coefficient established by wind tunnel experiment
- A = projected area of parachute as inflated at a given instant
- V = velocity of the parachute at a given instant.

In two ways the above formula can be redefined to furnish approximations of the diameter required to support a given load at a given rate of

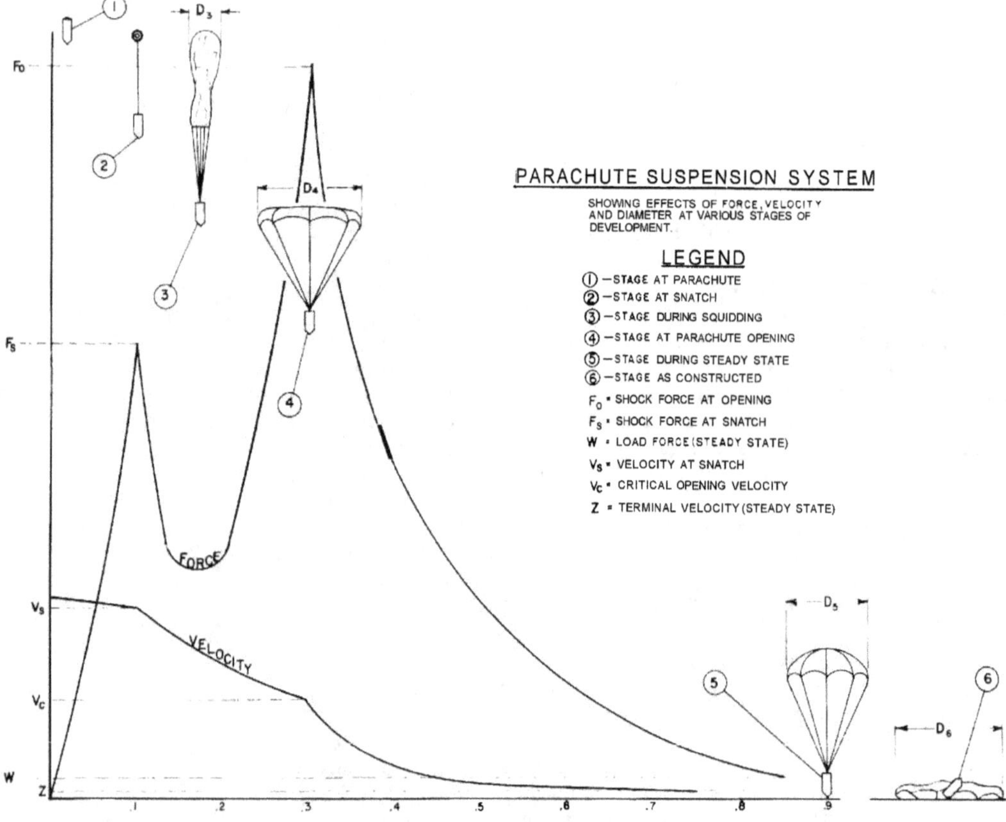

Figure 2-144. Parachute shock force uersus time

descent, and the maximum shock force upon the system during opening. The conditions are set by figure 2-144, stages 5 and 4, respectively. Reduced and simplified, the approximations are as follows..

Diameter	Shock force
$D_5 = \sqrt{W/0.03Z}$	$F_o = 7.85 \times 10^{-4} D_4{}^2 V_c{}^2$

Examples

D_5	W	Z	F_o	D_4	V_c
0.75	0.2	20	5	1.0	260
2.33	2.0	20	6,420	2.6	1,100
7.50	20.0	20	530	10.0	260
23.33	200.0	20	35,875	26.0	260

Additional empirical formulas are furnished by the British (as a rough guide and admittedly resulting in overdesign) as follows:

a. Tensile Strength of Fabric.

$$T_f = \frac{3}{192} \frac{\rho C_D}{k_s} D V_c{}^2$$

or, at sea level,

$$T_f = \frac{3}{8} \frac{D}{k_s} \left(\frac{V_c}{100}\right)^2$$

where k_s is a shock load factor.

Type of Parachute	Value of k_s
Shaped parachute	0.9
Flat parachute	0.5
Ungathered parasheet	0.2
Gathered parasheet	0.35

b. Tensile Strength of Shroud Line.

$$T_l = \frac{9}{32} \frac{\pi \rho C_D}{n} D^2 V_c{}^2$$

or, at'sea level,

$$T_l = 21 \left(\frac{D^2}{n}\right)\left(\frac{V_c}{100}\right)^2$$

2-371. Method of Calculating Air Density at Any Altitude. Density of air at sea level is directly proportional to the pressure and inversely proportional to the absolute temperature. In the standard atmosphere

$$\rho_o = 0.002378 \text{ slugs per cu ft}$$

so that, at a pressure p mm of mercury and temperature T_c (degrees Centigrade), the density is

$$\rho = \frac{0.002378 \times 288.6p}{760(273 + T_c)}$$

Relative density is defined as

$$\frac{\rho}{\rho_o} = \frac{288.6p}{760(273 + T_c)}$$

Therefore, at any altitude, $\rho = \rho_o$ x relative density. Accordingly, the relative densities for certain altitudes are found to be as follows.

Height (ft)	Relative density	Height (ft)	Relative density
0	1.000	30,000	0.374
5,000	0.862	35,000	0.310
10,000	0.738	40,000	0.246
15,000	0.629	45,000	0.193
20,000	0.532	50,000	0.152
25,000	0.448		

REFERENCES AND BIBLIOGRAPHY

1. Military Chemistry and Chemical Agents, TM3-215, August 1952 (Restricted).

2. Parachute Handbook, U. S. Air Force, Wright Air Development Center.

3. Personal Notes of J. Dubin, Artillery Ammunition Section, Picatinny Arsenal.

4. Bell, J. B., "Notes on Development of Aluminum Burster Casings in Closing and Sealing WP Filled Shell," 16 March 1955, Edgewood Arsenal, Md.

5. Ordark Research Project, Final Summary Report for the period 16 October 1952 to 15 October 1953.

6. Cohen, H. N., Parachute Design Notes, Parachute File, Pyrotechnics Section, Picatinny Arsenal, June 1953. Technical Reports No. 2043, 2080, 2081, and 2082.

7. Middleton, W. E. K., "Vision Through the Atmosphere," University of Toronto Press, 1952, p. 138 ff.

8. Field Behavior of Chemical Agents, TM3-240, May 1951, pp. 45-76.

9. Technical Information Report 6-9-8A2, Office of Chief of Ordnance, November 1954 (CONFIDENTIAL).

10. Technical Information Report 6-14-8A2, Office of Chief of Ordnance, January 1955 (CONFIDENTIAL).

11. Technical Information Report 6-14-8A4, Office of Chief of Ordnance, February 1955 (CONFIDENTIAL).

12. Brown, W. D., "Parachutes," Pitman, London, 1951, p. 99.

U. S. GOVERNMENT PRINTING OFFICE : 1957 O - 431165

www.ingramcontent.com/pod-product-compliance
Lightning Source LLC
Chambersburg PA
CBHW062103220526
45471CB00010B/3579